S0-BXL-527

# The Solution of the Inverse Problem in Geophysical Interpretation

## ETTORE MAJORANA INTERNATIONAL SCIENCE SERIES

Series Editor:

**Antonino Zichichi**
European Physical Society
Geneva, Switzerland

---

**(PHYSICAL SCIENCES)**

*Recent volumes in the series:*

**Volume 5** **PROBING HADRONS WITH LEPTONS**
Edited by Giuliano Preparata and Jean-Jacques Aubert

**Volume 6** **ENERGY FOR THE YEAR 2000**
Edited by Richard Wilson

**Volume 7** **UNIFICATION OF THE FUNDAMENTAL PARTICLE INTERACTIONS**
Edited by Sergio Ferrara, John Ellis, and Peter van Nieuwenhuizen

**Volume 8** **CURRENT ISSUES IN QUANTUM LOGIC**
Edited by Enrico G. Beltrametti and Bes C. van Fraassen

**Volume 9** **ENERGY DEMAND AND EFFICIENT USE**
Edited by Fernando Amman and Richard Wilson

**Volume 10** **INTERACTING BOSE–FERMI SYSTEMS IN NUCLEI**
Edited by F. Iachello

**Volume 11** **THE SOLUTION OF THE INVERSE PROBLEM IN GEOPHYSICAL INTERPRETATION**
Edited by R. Cassinis

# The Solution of the Inverse Problem in Geophysical Interpretation

Edited by

R. Cassinis
University of Milan
Milan, Italy

Plenum Press · New York and London

Library of Congress Cataloging in Publication Data

Main entry under title:

The Solution of the inverse problem in geophysical interpretation.
(Ettore Majorana international science series. Physical sciences; v. 11)
"Proceedings of the third course of the International School of Applied Geophysics . . . held March 27-April 4, 1980, in Erice, Sicily" — Verso of t. p.
Bibliography: p.
Includes index.
Contents: Introduction/R. Cassinis — Linearized inversion of (teleseismic) data/G. Nolet — The resolving power of seismic surface waves with respect to crust and upper mantle structural models/G. F. Panza — [etc.]
1. Geophysics — Congresses. 2. Prospecting — Geophysical methods — Congresses. I. Cassinis, R. (Roberto) II. International School of Applied Geophysics. III. Series.
QE500.S56 622'.15 81-4067
ISBN 0-306-40735-3 AACR2

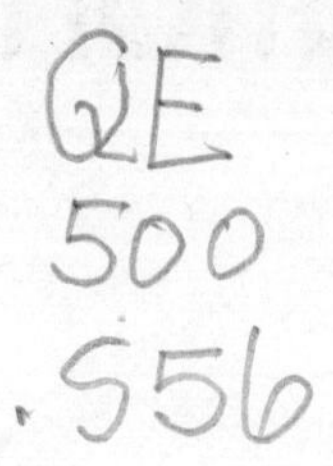


Proceedings of the Third Course of the International School of Applied Geophysics, on the Solution of the Inverse Problem in Geophysical Interpretation, held March 27—April 4, 1980, in Erice, Sicily


A Division of Plenum Publishing Corporation
233 Spring Street, New York, N.Y. 10013



Printed in the United States of America

## PREFACE

As is apparent from the table of contents, the lectures at the Third Course of the International School of Applied Geophysics, Erice, March 27-April 4, 1980 (the first part of this volume) dealt with several applications of inversion to different geophysical methods. For every field, the more general lectures come first, followed by those aimed at more specialized objectives. Not all topics are covered and the coverage is not uniform. The seismological section (especially the seismic reflection methods) is the most developed, and this is only partly due to the actual state of the art. Unfortunately, only abstracts are available for two of the lectures.

The second part of the volume contains some short notes and contributions presented either by the lecturers themselves or by other participants. They do not necessarily deal with the process of inversion itself but with the preparation and meaning of the data to be inverted or with some original treatments of problems that were discussed in the afternoon sessions.

The discussion sessions and the round table that followed the lectures were essential to the success of the Course and to an understanding of the different perspectives of the various specialists.

I hope that the group of very brilliant and willing geophysicists that made the meeting so interesting will stay in touch, grow closer, and meet again. Close scientific cooperation among them could contribute much to the "unification" of geophysical science.

Roberto Cassinis

CONTENTS

## GRAVITY AND MAGNETIC FIELDS



## GEOELECTRIC METHODS



## SHORT NOTES

# INTRODUCTION

## INTRODUCTION

Roberto Cassinis

Università di Milano

via G.B. Viotti 5
20133 Milano

Why a Course on the "inversion" of geophysical data? To my knowledge, this is the first time that this approach has been discussed among specialists in different geophysical methods. Nowadays, geophysical exploration covers a very broad field, and sometimes the work done by the specialists of one technique is almost unknown to those dealing with other methods. This is particularly true if we consider the objectives of our science (e.g., prospecting for oil or for water, mining, engineering).

For this reason I considered this fascinating though very difficult topic for the third Course of the School. It has been a unique opportunity to learn to what extent the interpretation of the highly sophisticated data gathered in the various fields of geophysics can benefit from equally sophisticated interpretation procedures in order to reach objectives of increasing complexity.

In geophysics, the inverse problem is that of gaining knowledge of the physical features of a concealed disturbing body by making observations of its effects (generally on the surface of the earth), i.e., finding the model from the observed data. This procedure is in contrast to the direct, forward, or normal problem, of calculating the observable effects from a given model (Sheriff, 1973). The forward approach has a unique solution while the inverse process, being carried out on the basis of hypotheses, is always characterized by a lack of uniqueness.

Inversion is not a new process in geophysical methods. During the pioneering years the inversion process was commonly used, though for an approximate and qualitative analysis. Recall, for example, the interpretation of negative gravity anomalies in North Germany or along the Gulf Coast, or the time residuals in fan shooting. This early use was occasioned by the fact that the objective (detection of the source) very clearly emerged from the background, or, in other words, the model was very simple.

Even the first reflection records obtained by Karcher in Oklahoma are an example of this "direct inversion." Because of the crudeness of the objective, noise was actually nonexistent and the migration immediate, only one constant velocity in the overburden being considered. As the objectives became more complicated, the need for a direct process (model matching) arose.

During the last two decades, the struggle to find oil under more and more difficult conditions has led to very sophisticated acquisition, processing, and interpretation procedures to filter the unwanted data and to "invert" the signal in order to reconstruct the underground lithology in great detail. Reflection seismology has been the leader in developing these techniques, the other methods lagging behind, in part because of lesser support from the oil industry.

In discussions of this problem, there is still some confusion about the words "direct" and "inverse." It is frequently said that the model is obtained by a <u>direct</u> computation from observed data, since this is actually the case. Perhaps it would be better to use the term "forward process" instead of "direct." Also, inversion is often taken to mean several transformation procedures carried out in order to prepare the data to be inverted rather than the inverse process itself.

Inversion, or the identification of the source, is a problem common to all geophysical methods, but the approach is quite different, depending on:

- the type of <u>field</u> measured (potential, transient)
- the type of <u>source of energy</u> (natural or artificial - <u>passive</u> or <u>active</u> methods)
- the type of <u>model</u>, which, in turn, must be defined according to physical (geological) constraints

The type of <u>source</u> seems to be a major factor determining the degree of ambiguity. To take an obvious example, let's compare the seismic exploration method (an "active" technique) with the seismological method (a "passive" technique). The former involves the use of a controlled source and of a predetermined and optimized pattern of receivers. Therefore the spatial coordinates of both the source and the receivers as well as the origin time are known;

this is not true in the case of seismology. In other words, the latter needs a much greater quantity of observations to reduce the ambiguity.

Time is an essential parameter in reducing the uncertainty that is absent when measurements are done of "static" (or very slowly changing) fields, such as the gravitational or even the magnetic field. In some passive methods (magnetotellurics or geomagnetics--the analysis of longer periods of magnetic field) time is also considered, but its use and significance are here quite different.

The type of model and its physical features are another essential bias determining the ambiguity. The two most popular "active" methods, seismic exploration and vertical electrical soundings, result in the lowest ambiguity because, generally, lateral homogeneity and subhorizontal boundaries are assumed, the main objective of the exploration being the depth and the trend of subhorizontal discontinuities. When lateral inhomogeneities are looked for, the picture becomes more and more complicated.

In passive techniques, where the measurement of "static" potential fields is involved, the ambiguity is very high, for the reason that, in general, it is impossible to disregard a mixed influence of both lateral and horizontal inhomogeneities (or of deep and shallow sources).

A further element that complicates the inverse process is the amplitude and the type of noise. The presence of noise generates instability in the inverse process. A common definition is that noise is any feature not needed by the interpreter or which does not comply with the hypothesis of the model. Again, in active techniques the definition of noise is generally easier mainly because the field of hypotheses is narrower.

It must be said, however, that it is very difficult to speak of "pure inversion"; the process of interpretation is in any case iterative, requiring a certain amount of "trial and error" procedures.

The problems and difficulties outlined above explain why the answers given by the lecturers are far from complete. However, they gave the community involved in geophysical exploration the chance to start on a new course, and that is an excellent achievement.

# PASSIVE AND ACTIVE SEISMOLOGY

# LINEARIZED INVERSION OF (TELESEISMIC) DATA

Guust Nolet

Vening Meinesz Laboratorium
P.O. Box 80.021
3508 TA Utrecht/The Netherlands

*Abstractness is not popular among physicists. Unfortunately, it lies hidden behind all their methods and, sometimes, it suddenly bites them.*

*P.C. Sabatier*

SUMMARY

This paper consists of two parts. First, I show how the inverse problem for teleseismic waves can be linearized. The second part deals with linear inversion methods, and is applicable to geophysical data in general.

The two principles of stationarity in seismology, Fermat's Principle and Rayleigh's Principle, can be used to establish a linearized relationship between perturbations in model parameters on the one hand, and the resulting perturbations in travel times or dispersion data on the other hand. Tables of formulae are presented for the calculations of the direct problem for Love and Rayleigh dispersion and for travel times, with flat and spherical Earth models and different interpolation rules between model points. Integral kernels for inversion of these data are also catalogued.

Backus-Gilbert theory for the inversion of linearized relationships between model and data perturbations in general is briefly reviewed and extended to the case of discretized models. Interpolation functions may be used to span a subspace of the Hilbert space of all possible Earth models. This leads to a discrete set of equations of much smaller size than the original Backus-Gilbert formulation, thus allowing for the simultaneous inversion of much larger sets of data. Orthonormalization of the basis of interpolation functions leads to simplifications in the theory, a more logical approach to discrete inversion, and allows the resolution calculations of Backus and Gilbert to be carried over from the domain of piecewise continuous models to be discretized case. The method presented here has considerable computational advantages over earlier methods.

## 1. INTRODUCTION

In this paper I shall use the term teleseismic data for measurements resulting from seismic sources of such energy that the waves are able to bridge large distances (say, > 200 km). On two ends of the energy scale, this definition comprises local seismic refraction experiments as well as earthquakes, of which the energy may travel the full circumference of the globe. These are not the usual type of sources one associates with applied geophysics, and the question arises: do teleseismic data have any relevance for geophysical exploration or other branches of applied geophysics?

The answer is a complicated "yes". If this positive answer is not unequivocal, it is only because teleseismic data do not have the standing of a basic technique to locate minerals or hydrocarbons, heavily refined by extensive and widespread commercial application. This is a position that teleseismic data shall never attain, because they are only sensitive to heterogeneities with wavelengths exceeding 1000 m or more. Nevertheless the analysis of teleseismic data does have significance for applied geophysics in several respects.

a. We now understand that not only the occurrence of earthquakes but also the presence of oil, gas and mineral deposits in the Earth is intimately connected with the tectonic history of the rocks. Our knowledge of the conditions favourable for the genesis of metal ores and hydrocarbons is growing fast (e.g. Strong, 1974; Fischer and Judson, 1975). We recognize plate boundaries as "fertile" regions. Using the principle of actualism we may expect to find similar deposits in the buried sutures of past episodes of continental break-up. Teleseismic data have already been used to map the tectonic division of the Earth to large depth, and to find the extent of sedimentary basins at shallow depth. They continue to do so with an increasingly good resolution.

b. Powerful techniques have been developed by Earth scientists for the interpretation of teleseismic waves. The applicability of mathematical methods, such as Backus-Gilbert inversion, is by no means restricted to the field of seismic data and have often been used elsewhere. It is conceivable that a growing interaction between seismologists and exploration geophysicists will lead to new and improved techniques in exploration (and vice-versa). Many of the advanced theories used by seismologists can be transferred directly to the high-frequency domain of exploration geophysics. The growing interest in the application of S waves for exploration purposes is a case in point. For instance, the interested reader will find in the present paper all that is necessary to construct a workable method to find static corrections for S waves from the groundroll in explosion seismics.

c. Seismic risk has become an important factor in the building of many structures such as dams, nuclear power plants, bridges, and pipelines for the transport of oil and gas. Improved local velocity models are necessary for a better determination of source location, fault plane orientation, and estimates of the stress drop.

The interpretation of teleseismic data is a complicated job, especially if many data of different type have to be processed simultaneously. Trial and error techniques are then too cumbersome to be of much use. Analytical inversion of the direct problem is only possible in a few special cases, and even then meets with many difficulties. Sometimes the problem is not stable against the influence of data errors, very often the range of models that satisfy the data is difficult to establish.

There is one branch of mathematics in which problems of uniqueness and stability have been studied extensively: linear algebra. The inversion of matrices - even singular ones - is well understood and stable algorithms exist that solve systems of linear equations as routine matter on digital computers. Moreover, if more than one solution exists to the system, it is possible to delineate the entire subspace of solutions. Evidently, we would make our task of interpretation much easier if we could find a way to cast our theory in the framework of linear algebra. In special cases, like the inversion of travel times in a medium with a monotonic increasing velocity function this is possible without any approximations (Garmany, 1979). In general, however, we must "linearize" the problem. This means that we must try to find an approximate linear relationship between a small *perturbation* of a model, and its *(small) effects on the measurable properties* calculated for that model.

Teleseismic data are a very good candidate for such an attempt, since the behaviour of teleseismic waves is governed by two principles of stationarity: Fermat's Principle and Rayleigh's Principle. We shall see in the next section that these principles enable us to find linear relationships that are fairly simple and usually quite accurate. In §3 we give a brief outline of the most fundamental aspects of Backus-Gilbert theory. This task has already been accomplished by many authors (e.g. Parker, 1977a), and the only reason to do it once more is to keep this paper as self-contained as possible. In §4 we deal with the discretization of the Backus-Gilbert method for efficient implementation on computers, and compare this method with the discrete methods proposed by Wiggins (1972) and Jackson (1972, 1976, 1979).

## 2. THE PRINCIPLES OF STATIONARITY IN SEISMOLOGY

The propagation of seismic waves in an elastic continuum in the absence of body forces is governed by the elastodynamic equations; using several different notations, these equations can take any of the following forms:

$$\rho \frac{\partial^2 s_i}{\partial t^2} = \partial_j \sigma_{ij} = \partial_j (c_{ijkl} \partial_k s_l) = (Hs)_i \qquad (1)$$

where $\rho$ is the density, $s_i$ the (small) displacement from equilibrium, t the time, $\partial_j$ the partial derivative with respect to a cartesian coordinate $x_j$, $\sigma_{ij}$ the stress, $c_{ijkl}$ the elastic tensor satisfying the symmetry relationships $c_{ijkl} = c_{jikl} = c_{ijlk} = c_{klij}$ and $H_i$ a vectorial differential operator.
If the medium is isotropic the elastic tensor satisfies:

$$c_{ijkl} = \lambda \delta_{ij} \delta_{kl} + \mu (\delta_{ik} \delta_{jl} + \delta_{il} \delta_{jk}) \qquad (2)$$

We shall assume isotropy of the medium in the following. If the continuum is bounded with $\underset{\sim}{n}$ a unit vector normal to its surface, the boundary conditions read:

$$\sigma_{ij} n_j = 0 \qquad (3)$$

If the continuum is unbounded, as in a halfspace, we have to assume radiation conditions instead of (3) in order to retain finite energy in the wavefield. In spite of their apparent simplicity, eqs. (1-3) are difficult to solve for realistic Earth models. Approximations to the theory, the model, or both, are necessary for practical reasons.

In classical seismology, as in most of present-day exploration seismics, one usually assumes that the energy of the waves with high frequency travels as "rays" according to the laws of geometrical optics. This is a rather drastic approximation to the theory since the effects of diffraction, the continuous reflections from velocity gradients and the continuous conversion of P to S waves are all neglected. But it has a great advantage because it allows one to handle inhomogeneous models of very great complexity. We shall assume that the reader is familiar with ray theory (see Cervený and Ravindra, 1971, for derivations), and with its important stationarity principle:
Fermat's Principle: *The seismic signal travelling between two points follows a path that renders the travel time stationary.*
Originally this principle was formulated as a minimum principle, which is not always true. The locus $\underset{\sim}{r}$ of the ray can be written as a function of the path length s and satisfies the system of differential equations:

$$\frac{d}{ds}\left(\frac{1}{\alpha} \frac{d\underset{\sim}{r}}{ds}\right) = \nabla\left(\frac{1}{\alpha}\right) \qquad (4)$$

where $\alpha = [(\lambda + 2\mu)/\rho]^{\frac{1}{2}}$ is the velocity of P-waves. Of course a similar system can be formulated for S-waves, replacing $\alpha$ by $\beta = (\mu/\rho)^{\frac{1}{2}}$. For models that are inhomogeneous only with respect to the depth coordinate (4) takes the familiar form of Snell's law, and the direct problem can be solved analytically for many analytical velocity functions (Appendix A). For 2- or 3 dimensional models numerical methods for integration of (4) are given by Cervený and Ravindra (1971) or Julian and Gubbins (1977). Fermat's Principle has important simplifying consequences for the linearization of travel time data. The travel time T between two points $P_1$ and $P_2$ at depths $z_1$ is given by

$$T = \int_{P_1}^{P_2} \frac{ds}{\alpha} = \int_{z_1}^{z_2} \frac{1}{\alpha} \left(\frac{ds}{dz}\right) dz \tag{5}$$

where ds/dz is determined by $\underset{\sim}{r}$ in (4).
If we perturb the Earth model and give it a velocity $\alpha' = \alpha + \delta\alpha$, then not only T will change to $T + \delta T$, but the path followed will be different from the original path. However, since T is stationary with respect to small perturbations in the path, we may use the original trajectory, and thus the same ds/dz in (5) to calculate the new travel time:

$$T + \delta T = \int_{z_1}^{z_2} \frac{ds/dz}{\alpha + \delta\alpha} \, dz$$

from which:

$$\delta T = - \int_{z_1}^{z_2} \frac{\delta\alpha}{\alpha^2} \left(\frac{ds}{dz}\right) dz \tag{6}$$

Perturbation formulae for several interpolation functions between model points are given in Appendix B.

This procedure was first followed by Julian and Anderson (1968). Johnson and Gilbert (1972) have developed similar formulae for the perturbation of the intercept time $\tau$, which is related to the horizontal slowness p and the epicentral distance $\Delta$ by $\tau = T - p\Delta$. Although a great advantage of $\tau(p)$ is that it is a monotonic function of p, this datum does not constrain the velocity model as strictly as $T(\Delta)$ does. It is tempting to use $\tau$, since there are some difficulties with the inversion of T. T may be a multivalued function of $\Delta$ if two rays arrive at the same location . Worse even, a ray may exist at a distance $\Delta$ for model $\alpha(r)$, but disappear in the perturbed model $\alpha(r) + \delta\alpha(r)$, especially if $\delta\alpha$ introduces a negative velocity gradient. It is clear that the concept of linearization degenerates in this case. However, it is my experience that these problems are easily solved ad hoc in practical applications, whereas I have some negative experiences with models satisfying the $\tau(p)$ curve but disagreeing with the original data $T(\Delta)$. I therefore strongly recommend to invert the travel time directly.

Aki et al. (1977) have developed a method in which (6) is used to obtain 3-dimensional velocity models from the travel times of P-waves from teleseismic events measured with an array of seismometers. Besides applications for the structure beneath the large seismic arrays LASA and NORSAR, the method has been used successfully to map velocity inhomogeneities in lithospheric blocks of about 100x100 km in California, Yellowstone and Hawaii (Aki, 1977).

Garmany (1979) uses a geometrical transformation to show that there is an exact linear relationship between the travel time functions and the depth as a function of velocity, provided the velocity increases with depth. The latter requirement is impractical for near-surface studies, and the method is difficult to synthesize with other data for simultaneous inversion. For these reasons I shall, in the present paper, not tread more deeply into the subject of exact linear inversion of travel times.

Many modern seismological methods do not make use of the ray-theoretical approximations, but try to solve (1) exactly for a simplified model instead. The direct problem is easy to solve using horizontal layers for a "flat Earth" or radial symmetry for a spherical Earth, and harmonic time dependence $\exp(i\omega t)$. This leads to dispersive waves, of which the surface waves are the more well-known and probably most widely applied type of waves. With $\underset{\sim}{y}$ a vector of displacements and stresses, (1) reduces to a system of first order differential equations of the form:

$$\underset{\sim}{y}' = A\underset{\sim}{y} \qquad (7)$$

where differentiation is with respect to the depth coordinate z or the radius r. A is a 2x2 matrix for horizontal (SH) displacements or Love waves, and a (4x4) matrix for the coupled P-SV or Rayleigh type waves (see Takeuchi and Saito, 1972, or Vlaar and Nolet, 1978). The system (7) is easily integrated on a digital computer using standard Runge-Kutta or predictor-corrector algorithms, although some precautions against numerical instabilities may be necessary at high frequencies (Neigauz and Shkadiskaya, 1972). The expressions for $\underset{\sim}{y}$ and A are compiled in Appendix C, together with the boundary conditions at the free surface. Jeffreys (1961), following earlier work of E. Meissner in the 1920's, pointed attention to the fact that Rayleigh's Principle can be used as a basis for a perturbation theory of seismic surface waves. This principle appears in many forms. It is intimately connected with the well-known property of harmonic free oscillations, in which kinetic and potential energy, averaged over a cycle, are equal. We may express this as $\omega^2 E_k = E_p$, where $E_k$ and $E_p$ are functionals of the displacement field. If we perturb the displacement field, $E_k$ and $E_p$ will also change in value, and we would

expect the frequency $\omega^2 = E_p/E_k$ to take on a different value. Rayleigh, however, found that this is not the case. He formulated his principle as follows:
Rayleigh's Principle: *The period of a conservative system, vibrating in a constrained type about a position of stable equilibrium is stationary in value when the type is normal.*
A very complete proof of Rayleigh's Principle is given by Woodhouse and Dahlen (1978). However, their derivation is of a very great algebraic complexity and of a more general nature than shall be needed here. To the best of my knowledge, no simple proof tailored to the case of seismic wave propagation in a continuous, elastic solid is available in the open literature. Since the principle has been misunderstood, and has even been wrongly applied on several occasions, it seems important that a correct proof for this case is formulated. This I shall do in the following.

We introduce a convenient notation by defining an inner product of two displacement vectorial functions $\underset{\sim}{u}(r)$ and $\underset{\sim}{v}(r)$:

$$(\underset{\sim}{u},\underset{\sim}{v}) \equiv \int_V \underset{\sim}{u}\cdot\underset{\sim}{v}^* dV = \int_V u_i(\underset{\sim}{r})v_i(\underset{\sim}{r})^* dV \qquad (8)$$

Here * denotes complex conjugation. The volume V is bounded by the stress free surface S. Using (8), we shall first prove an important Lemma: The vectorial differential operator H in (1) is Hermitian, i.e. $(H\underset{\sim}{u},\underset{\sim}{v}) = (\underset{\sim}{u},H\underset{\sim}{v})$, if V is bounded by stress free surface.
Proof:

$$\begin{aligned}(H\underset{\sim}{u},\underset{\sim}{v}) &= \int_V (H\underset{\sim}{u})_i v_i^* dV \\ &= \int_V \partial_j(c_{ijkl}\partial_l u_k)v_i^* dV \\ &= \int_S c_{ijkl}(\partial_l u_k)v_i^* n_j dS - \int_V c_{ijkl}(\partial_l u_k)(\partial_j v_i^*)dV \\ &= -\int_V c_{ijkl}(\partial_l u_k)(\partial_j v_i^*)dV\end{aligned}$$

where we have applied Gauss' theorem. The surface integral is zero because of vanishing stress. In the same way, using the symmetry properties of $c_{ijkl}$, we can show that

$$(\underset{\sim}{u},H\underset{\sim}{v}) = -\int_V c_{ijkl}(\partial_l u_k)(\partial_j v_i^*)dV$$

From which the lemma follows:

$$(H\underset{\sim}{u},\underset{\sim}{v}) = (\underset{\sim}{u},H\underset{\sim}{v}) \qquad (9)$$

If the medium is unbounded, as in a halfspace, we can make use of the periodic nature of the motion to show that the integral over S is zero if V is a volume bounded by surfaces one wavelength apart. If gravity influences the dispersion, the lemma remains true as well (e.g. Vlaar, 1976). We give a proof for Rayleigh's Principle that will also provide us with the basic formulae for use in linearized inversion. If we assume time harmonic motion in (1), we have

$$-\rho\omega^2 \underset{\sim}{s} = H\underset{\sim}{s} \qquad (10)$$

We shall now study the disturbance caused in $\underset{\sim}{s}$ by perturbations in the model (thus in $\rho$ and H). The perturbed equation than becomes formally:

$$-(\rho+\delta\rho)(\omega^2+\delta\omega^2)(\underset{\sim}{s}+\delta\underset{\sim}{s}) = (H+\delta H)(\underset{\sim}{s}+\delta\underset{\sim}{s})$$

Taking the inner product with s now gives:

$$-\omega^2(\underset{\sim}{s},\rho\delta\underset{\sim}{s}) - \delta\omega^2(\rho\underset{\sim}{s},\underset{\sim}{s}) - \omega^2(\delta\rho\underset{\sim}{s},\underset{\sim}{s}) = (\delta H\underset{\sim}{s},\underset{\sim}{s}) + (H\delta\underset{\sim}{s},\underset{\sim}{s}) + O(\delta^2)$$

Because of the lemma $(H\delta\underset{\sim}{s},\underset{\sim}{s}) = (\delta\underset{\sim}{s},H\underset{\sim}{s})$, thus

$$-(\delta\underset{\sim}{s},\omega^2\rho\underset{\sim}{s}+H\underset{\sim}{s}) - \delta\omega^2(\rho\underset{\sim}{s},\underset{\sim}{s}) = \omega^2(\delta\rho\underset{\sim}{s},\underset{\sim}{s}) + (\delta H\underset{\sim}{s},\underset{\sim}{s})$$

and, because of (10):

$$\delta\omega^2 = 2\omega\delta\omega = -\frac{(\delta H\underset{\sim}{s},\underset{\sim}{s}) + \omega^2(\delta\rho\underset{\sim}{s},\underset{\sim}{s})}{(\rho\underset{\sim}{s},\underset{\sim}{s})} \qquad (11)$$

Since $\delta\underset{\sim}{s}$ is absent in (11), $\omega^2$ is stationary if only $\underset{\sim}{s}$ is perturbed, and Rayleigh's Principle is valid in this case. Equation (11) also tells us how the eigenfrequency varies with small perturbations to the model, either in $\rho$, or the elastic constants or both. Since $\delta\underset{\sim}{s}$ is absent, the relationship (11) between $\delta\omega$ and $\delta\rho$, $\delta\mu$ or $\delta\lambda$ is linear. It is important to realize that the two integrals with $\delta\underset{\sim}{s}$ cancel in first order; in some early papers one may find the erroneous assumption that integrals containing $\delta\underset{\sim}{s}$ are of second order themselves. Another error, which has persisted until corrected recently by Woodhouse (1976), is that (11) can be used to calculate the effect of perturbations in depth to seismic discontinuities such as the Moho. A correct first-order theory has been formulated by Woodhouse (1976) and Woodhouse and Dahlen (1978) for this case. Cloetingh et al. (1980) demonstrated that first-order theory is only of limited validity when perturbations in the depth to discontinuities in the upper crust are effectuated. In this case the inversion may require other methods than simple linearized inversion (see Panza, this volume).

Equation (11) can be used to derive more specific formulae for perturbation or surface wave dispersion. As an example we shall derive the first order perturbations in the dispersive properties of the Love wave when the model is perturbed to density $\rho+\delta\rho$ and shear modulus

$\mu+\delta\mu$. From Appendix C, we can easily construct the second order differential equation in the displacement $y_1$ and find the expression for H:

$$H = \frac{d}{dz}\left(\mu \frac{d}{dz}\right) - k^2\mu$$

$$\delta H = \frac{d}{dz}\left(\delta\mu \frac{d}{dz}\right) - k^2\delta\mu$$

Substitution in (11) gives

$$\delta\omega(k) = -\frac{\int_V \{\frac{d}{dz}(\delta\mu \frac{dy_1}{dz})y_1 - k^2\delta\mu\}dV + \omega^2\int_V \delta\rho y_1^2 dV}{2\omega\int_V \rho y_1^2 dV}$$

Integration over the volume V can be replaced by integration over the depth coordinate z only, since the integration over r and $\phi$ gives a common factor to all integrals. Partial integration of the first integral and application of the boundary conditions then gives:

$$\delta\omega(k) = \frac{\int_0^\infty \{(\frac{y_2}{\mu})^2 + k^2y_1^2\}\delta\mu\, dz - \omega^2\int_0^\infty y_1^2\delta\rho\, dz}{2\omega\int_0^\infty \rho y_1^2 dz}$$

or, in a more concise notation:

$$\delta\omega(k) = \int_0^\infty \left(\frac{\partial\omega}{\partial\mu}\right)\delta\mu\, dz + \int_0^\infty \left(\frac{\partial\omega}{\partial\rho}\right)\delta\rho\, dz$$

Introducing the phase velocity $c = \omega/k$, we find for the differences between the original and the perturbed dispersion curves:

$$\delta c(\omega) = \frac{c^2}{\omega U}\,\delta\omega(k) \tag{12}$$

where $U = d\omega/dk$ is the group velocity. From $\lambda = \rho(\alpha^2 - 2\beta^2)$ and $\mu = \rho\beta^2$ we have:

$$\frac{\partial c}{\partial\alpha} = 2\alpha\rho\frac{\partial c}{\partial\lambda}\left(= \frac{2\alpha\rho c^2}{\omega U}\frac{\partial\omega}{\partial\lambda}\right)$$

$$\frac{\partial c}{\partial\beta} = 2\beta\rho\frac{\partial c}{\partial\mu} - 4\beta\rho\frac{\partial c}{\partial\lambda} \tag{13}$$

$$\frac{\partial c}{\partial\rho} = \beta^2\frac{\partial c}{\partial\mu} + (\alpha^2 - 2\beta^2)\frac{\partial c}{\partial\lambda} + \frac{\partial c}{\partial\rho}$$

This allows us to formulate the perturbation theory in terms of the more common datum $c(\omega)$ and the model velocities $\alpha$ and $\beta$ that are used in travel time studies as well. This approach enables us to perform simultaneous inversion of both types of data, and therefore gives us stronger constraints on the final model. For Love waves:

$$\frac{\partial c}{\partial \beta} = \frac{\beta \rho c^2}{\omega^2 UE}\{(\frac{y_2}{\mu})^2 + k^2 y_1^2\} \quad , \quad E = \int_0^\infty \rho y_1^2 dz$$

etcetera. A catalogue of partial derivatives is given in Appendix D.

Can one obtain the density from dispersive waves? A very simple mathematical exercise gives us insight into this problem. Suppose one multiplies equation (1) on both sides with a constant factor. Physically this means that we change the density and the elastic constants proportionally, but that we keep the velocities, i.e. the quotient of the two, fixed. It is evident that the solution to the equations remains unchanged. This shows an important redundancy in the information given by surface waves. The implication of our exercise is that the dispersion is invariant to constant relative perturbations $\delta\rho/\rho$ independent of the velocities. Hence, from surface waves only we can only obtain the *shape* of the density curve, not its absolute height. This conclusion is not as surprising as it may seem at first sight. It is a well known fact that the period of a simple pendulum is independent of the weight attached to it, since the restoring force is proportional to the weight. But if we vary $\rho$ in the Earth, and keep $\alpha$ and $\beta$ constant, we again keep the restoring forces $(\mu,\lambda)$ proportional to the weights of the volume elements.

Group velocity partial derivatives are most easily calculated using numerical differentiation of the phase velocity derivatives (Rodi et al., 1975):

$$\delta U(\omega) = \frac{U(\omega)}{c(\omega)}\{2 - \frac{U(\omega)}{c(\omega)}\}\,\delta c(\omega) + \frac{U^2(\omega)}{c^2(\omega)}\,\frac{\delta c(\omega+\Delta\omega) - \delta c(\omega-\Delta\omega)}{2\Delta\omega} \tag{14}$$

Both Fermat's and Rayleigh's Principle pertain to the behaviour of the wave in *time*. The *amplitude* behaviour is conspiciously absent, and so far no satisfactory theory for amplitude perturbation has been developed for body waves. It is even doubtful if one will ever succeed to do so, since these amplitudes depend on the second derivative of the time-distance curve and are therefore very sensitive to even small perturbations of the model. Surface wave amplitudes depend in a similar way on the second derivative of the phase velocity, and a perturbation theory with respect to the model seems equally difficult. For a given model, but an unknown source, the amplitudes may be used to find the source characteristics, however. The relationship between the surface wave amplitudes and the moment-tensor is fully linear, so that no perturbation theory is required (Mendiguren, 1977; Vlaar and Nolet, 1978).

## 3. BACKUS-GILBERT THEORY: DETERMINATION OF DATA RESOLUTION

In the previous section we have seen that linearization of the functionals $T(\Delta)$ and $c(\omega)$ results in relationships between the perturbations $g_i$ in the M data (T or c) and a perturbation $p(r)$ in one of the model parameters of the form:

$$g_i = \int_0^R G_i(r)p(r)dr \tag{15}$$

In a flat Earth the integration interval is $0 \leq z < \infty$, of course. Since we can always scale or transform the data, we can assume that the data $g_i$ are independent and univariant, i.e. they have the unit matrix I as their covariance matrix.

We have only a finite amount of data at our disposal, that are contaminated with errors. Thus the hope that there exists only one model perturbation $p(r)$ satisfying (15) is vain. In fact, Backus and Gilbert (1967) proved that the collection of models satisfying a finite amount of data (15) is either empty or infinite dimensional. It is important to realize that the uncertainty in the final model results from two sources: the insufficient (finite) number of data and the errors in the data. We arrive here at the fundamental problems of inverse studies: which model should we choose from the infinite dimensional space of models that are derived from the data, and how can one determine the degree of uncertainty in this model from the data and their error statistics? If no other constraints on $p(r)$ except (15) are present, it is usually the best policy to choose the model which is the "smallest" in a least-squares sense, that is we minimize the Euclidean norm of $p(r)$:

$$||p(r)|| = \int_0^R p(r)^2 dr \tag{16}$$

The smallest model perturbation gives us a final model which is "closest" to the starting model, and yet satisfies the data. The advantage of this approach is that it enables us to incorporate all sorts of common sense notions about the Earth in the starting model, with the provision that these notions will be maintained in the final model as far as these are allowed by the data.

With (16) we have reduced the inversion problem to the task of finding a minimum for $||p||$, subject to the constraints (15). From the calculus of variations (e.g. Matthews and Walker, 1973, Chapter 12) we know that this is equivalent to setting

$$||p|| - \sum_{j=1}^{M} 2\nu_j \int G_j(r)p(r)dr = \text{minimum} \tag{17}$$

where $\nu_j$ is a Lagrange multiplier and the factor 2 is only for future convenience.

We find:

$$\int_0^R \{2p(r) - 2 \sum_{j=1}^{M} \nu_j G_j(r)\}\delta p(r)dr = 0$$

or

$$p(r) = \sum_{j=1}^{M} \nu_j G_j(r) \qquad (18)$$

(18) is a very interesting expression. We may look at p(r) as a function in a linear 'vector' space of functions which is infinite dimensional. Such a space of functions is usually called a Hilbert space if the functions satisfy several criteria of good behaviour, none of which is violated in our geophysical applications. (18) tells us that kernels $G_i(r)$ as a basis span a finite dimensional subspace S of the Hilbert space, and that p(r) will be smallest in a least square sense if we require that $p(r) \in S$. We can determine the coefficients through substitution of (18) in (15):

$$\sum_{j=1}^{M} T_{ij}\nu_j = g_i \quad \text{with} \quad T_{ij} = \int_0^R G_i(r)G_j(r)dr \qquad (19)$$

We notice at once a difficulty: the integral kernels for the travel times (table 1) are not square integrable. The problem can be solved by partial integration of (15), making dp/dr the model instead of p(r) (Johnson and Gilbert, 1972). A less roundabout solution to this problem will be given in the next section. There we shall also briefly deal with the problems posed by matrices that are singular or ill-conditioned (which means almost singular), which is one of the reasons that (19) often has an unstable solution.

Once the model has been calculated it remains to determine its uncertainty: how far can one deviate from this model, and still fit the data within their errors? It will be intuitively clear that we cannot simply assign an uncertainty $\varepsilon(r_0)$ to the parameter p at $r_0$. The introduction of one very thin, very high velocity layer will always be possible at $r_0$ with negligible effect on the data. We say that very thin layers are not "resolved". Backus and Gilbert (1970) show that a managable way to define both uncertainty and resolution is to calculate the uncertainty of a local average of p(r), and show that there is a trade-off between the uncertainty (variance) and the width of the depth range over which the local average is calculated. We shall briefly describe the essential part of their theory. A local average of the model at $r_0$ can be written as:

$$\langle p(r_0)\rangle = \int_0^R A(r,r_0)p(r)dr \qquad (20)$$

with

$$\int_0^R A(r,r_0)dr = 1 \tag{21}$$

When $A(r,r_0)$ equals a delta function centered at $r_0$, the local average equals the parameter value of the true Earth $p(r_0)$. An example of a kernel is shown in figure 1. We ignore at this point that there may be independent information on $p(r)$, such as physical limitations, and assume that $p(r)$ must be constructed from the data, and from nothing else. Backus and Gilbert (1968) proved that a linear average of $p(r)$ must also depend in a linear way on the data, hence:

$$\langle p(r_0)\rangle = \sum_{i=1}^{M} a_i(r_0)g_i = \int_0^R \sum_{i=1}^{M} a_i(r_0)G_i(r)p(r)dr$$

from which

$$A(r,r_0) = \sum_{i=1}^{M} a_i(r_0)G_i(r) \tag{22}$$

We shall wish to minimize the width or spread of $A(r,r_0)$ and therefore introduce the following measure of this spread:

$$S(r_0) = 12 \int_0^R A(r,r_0)^2(r-r_0)^2dr \tag{23}$$

where the factor 12 was introduced to make S numerically equivalent to the width of $A(r,r_0)$ when this is box-shaped and centered at $r_0$. Since the data have errors, these will propagate in the calculated average. We find for the variance $\varepsilon(r_0)^2$ of the average:

$$\varepsilon(r_0)^2 = \mathrm{var}\{\langle p(r_0)\rangle\} = \sum_{i=1}^{M} a_i(r_0)^2 \tag{24}$$

since we assumed the data to be univariant and independent. The

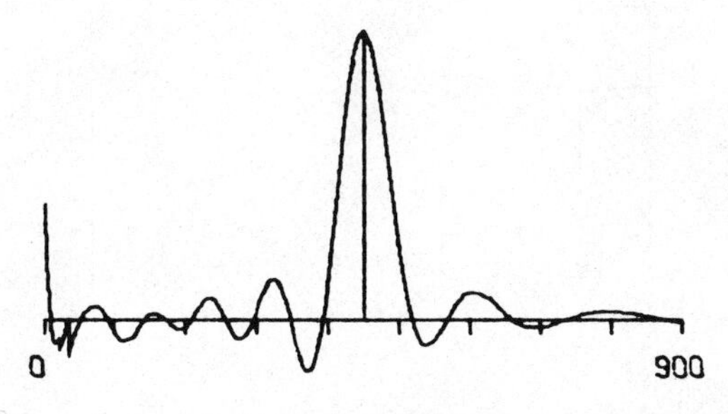


Figure 1. An example of $A(r,r_0)$. Horizontal axis gives the depth.

factors $a_i(r_0)$ can now be determined from:

$$S(r_0)\cos\theta + \varepsilon(r_0)\sin\theta = \text{minimum}$$

with (21) as a constraint. $\theta$ is a weighting factor, that allows us to emphasize either the minimization of the resolving length or the minimization of the variance of $\langle p(r_0)\rangle$. In this way, $\theta$ parametrizes a trade-off between the two. A typical trade-off curve is given in figure 2. (24) can of course be minimized using the technique of Lagrange multipliers. Gilbert (1971) shows that this can be done very efficiently if the kernels $G_j(r)$ are first transformed to an orthonormal base in Hilbert space. In the following, we shall use similar techniques, starting, however, from a model that is discretized a priori. The discretized approach is considerably more efficient in use of computer memory, and makes simultaneous inversion of truly large data sets easily possible, even on computers of moderate size.

## 4. DISCRETIZATION: CHOOSING A BASIS IN HILBERT SPACE

Even before we have a knowledge of our data and the associated kernels $G_j(r)$ we may reject the notion that our models should in principle be able to lie anywhere in the infinite dimensional function space of models. In particular, heavily oscillating functions may be so useless for interpretation that we prefer to discard them right away. In terms of Backus-Gilbert theory, we would say that we are not interested in averaging kernels of very small spread, since the variance of the average is much too large to be of any use.

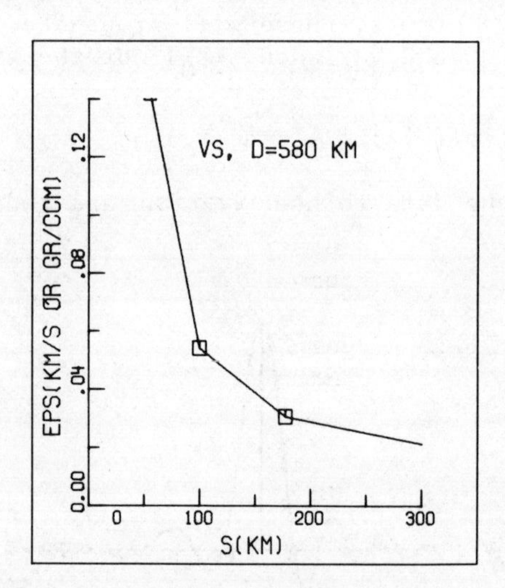


Figure 2. Example of a trade-off curve for the S velocity at a depth of 580 km (from Nolet, 1978).

By discarding heavily oscillating models we ignore that part of the trade-off curve that has very large $\varepsilon(r_0)^2$, but leave the remaining part of the curve intact. Mathematically, we may describe this process by introducing a basis of piecewise continuous and relatively smooth functions $h_1^*(r)$, $h_2^*(r)$,....,$h_N^*(r)$, that spans a subspace of the Hilbert space. From the viewpoint of inversion theory, it would lead to convenient simplifications if the basis were orthonormal. In practical applications, the basis $\{h_i^*(r)\}$ if often intimately connected with interpolation rules in an interval and possibly with continuity conditions at the boundary between intervals. In that case the basis $\{h_i^*(i)\}$ is in general not orthonormal, but it can of course be orthonormalized. Define

$$H_{kl} = \int_0^R h_k^*(r)h_l^*(r) \tag{25}$$

Without really binding our hands, we can require that H is a positive definite matrix. H is symmetric, so that we can always find a transformation U such that:

$$U\ H\ U^T = I \tag{26}$$

Define a new basis:

$$h_i(r) = U_{ik}h_k^*(r) \tag{27}$$

then

$$\int_0^R h_i(r)h_j(r)dr = U_{ik}H_{kl}U_{il} = \delta_{ij} \tag{28}$$

so that $\{h_i(r)\}$ is an orthonormal basis. With the basis we may project all possible models on the subspace. This will automatically smooth out unwanted oscillations:

$$p(r) = \sum_{k=1}^{N} p_k h_k(r) \tag{29}$$

An example of a set of functions that performs linear interpolation is shown in figure 3a. Their orthonormalized equivalent is shown in figure 3b.

We have discretized the function p(r) to a vector p. The linearized integral equation (15) reduces now to a system of linear algebraic equations:

$$\underset{\sim}{g} = G\underset{\sim}{p} \tag{30}$$

where

$$G_{ik} = \int_0^R G_i(r)h_k(r)dr = \sum_{j=1}^{N} U_{kj}G_{ij}^* \tag{31}$$

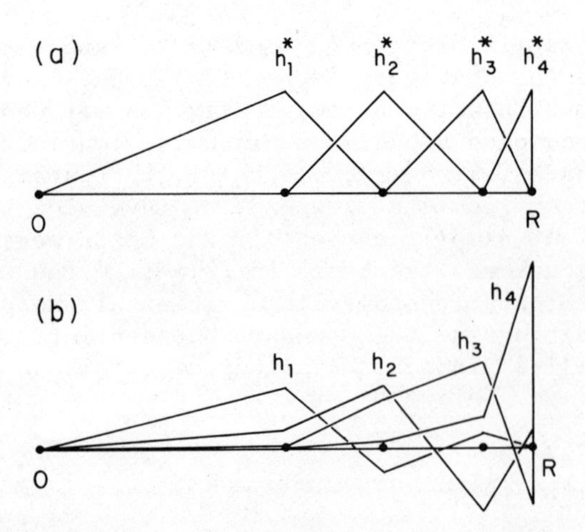


Figure 3. The basis of interpolation functions (a) and its orthonormalized transform (b).

with

$$G^*_{ij} = \int_0^R G_i(r)h^*_j(r)dr$$

In many applications, the behaviour of the kernel $G_i(r)$ is such that a rather narrow spaced model grid with many points is necessary to obtain sufficiently accurate quadrature. In Backus-Gilbert theory, the matrix $T_{ij}$ in (19) can only be calculated after calculation and subsequent storage of the kernels $G_i(r)$. This procedure claims an unnecessarily large part of computer memory. Large data sets will be impossible to handle without time-consuming writing to mass storage, even in very big machines.

The number N of basis functions $h_i(r)$ can be much smaller than the number of grid points necessary for an accurate description of the kernels and the model, if the behaviour of the $h_i(r)$ between grid points is described analytically, as with common interpolation schemes (linear, quadratic, splines, etc.). This means that we can immediately calculate one row of the matrix G in (30) once we have calculated $G_i(r)$ and accomplish a considerable reduction in memory requirement. This enables us to handle large data sets for simultaneous inversion even on computers of moderate sizes.
If we have a very rough discretization, and the number of data (M) is large so that M > N, we may solve (30) by minimizing $|\underset{\sim}{g} - G\underset{\sim}{p}|^2$, which gives us Gauss' normal equations or the "least squares" solution:

$$G^T G \underset{\sim}{p} = G^T \underset{\sim}{g} \tag{32}$$

or

$$\underset{\sim}{p} = (G^T G)^{-1} G^T \underset{\sim}{g} \tag{33}$$

In this way we select a very smooth model, i.e. we choose a point on the trade-off curve with large spread s, small error $\varepsilon$, and $\theta \simeq 0$. By limiting the subspace to a very small dimension N we have discarded the option of a significant trade-off. For a good insight into the true resolution of the system this is not desirable. Moreover, the model may be too smooth to be a good representation of the real Earth. We do not know a priori how careful we must be in our choice of N, and we may encounter unpleasant surprises when we find that $G^T G$ is an ill-conditioned matrix, i.e. when its determinant is so small that small errors in $\underset{\sim}{g}$ will blow up to gigantic errors in the parameter vector $\underset{\sim}{p}$. Ordinary least squares forces one to stay on the conservative side and choose a very small number of parameters. We would like to be more generous in the parametrization of the model, yet keep the errors under control.

This is a well known situation in geophysical inverse problems, and it can be handled by using the generalized inverse of $G^T G$, an approach first introduced into the geophysical literature by Wiggins (1972) and Jackson (1972), and very similar to the method of winnowing proposed by Gilbert (1971). This approach allows us also to take $N > M$. Here we shall follow Nolet (1980) in extending the idea of trade-off to the case of models with an arbitrary parametrization. We inspect the case that $\det(G^T G) = 0$. Some of the eigenvalues of $G^T G$ will then be equal to 0 (the others are positive). The matrix is symmetric and therefore can be diagonalized with an orthogonal matrix V:

$$G^T G = \hat{V}\hat{\Lambda}\hat{V}^T \tag{34}$$

If the rank of $G^T G$ is K, the diagonal matrix $\hat{\Lambda}$ has K nonzero elements which we rank in order of decreasing magnitude: $\lambda_1 \geq \lambda_2 \geq \ldots \geq \lambda_k > 0$. Now (32) becomes:

$$\hat{V}\hat{\Lambda}\hat{V}^T \underset{\sim}{p} = G^T \underset{\sim}{g} \tag{35}$$

Because the last (N-K) elements of $\hat{\Lambda}$ are 0, this system is equivalent to the truncated system

$$V \Lambda V^T \underset{\sim}{p} = G^T \underset{\sim}{g} \tag{36}$$

where V is the NxK matrix whose columns form the first K eigenvectors of $G^T G$, and $\Lambda$ is a KxK diagonal matrix of nonzero eigenvalues. In practice we also skip very small eigenvalues that are contaminated by machine precision errors, to avoid any problems with numerical stability. We notice that $V^T V = I$, since the eigenvectors are orthonormal, but that $VV^T \neq I$. Setting

$$\underset{\sim}{p}^* = V^T\underset{\sim}{p} \tag{37}$$

we may invert

$$\underset{\sim}{p}^* = \Lambda^{-1}V^TG^T\underset{\sim}{g} \tag{38}$$

The K-dimensional vector $\underset{\sim}{p}^*$ is determined uniquely by the data. The covariance matrix C of $\underset{\sim}{p}^*$ is easily calculated

$$C^* = \Lambda^{-1}V^TG^TI(\Lambda^{-1}V^TG^T)^T = \Lambda^{-1}$$

Thus

$$\mathrm{var}\{p_i^*\} = \lambda_i^{-1} \tag{39}$$

The components of $\underset{\sim}{p}^*$ may be looked upon as transformed data because of (38). Alternatively, $\underset{\sim}{p}^*$ is the projection of all fitting models on a subspace, spanned by the orthogonalized set of data kernels formed by the columns of V. There are, however, infinitely many models $\underset{\sim}{p}$ that give the same projection $\underset{\sim}{p}^*$. We may again choose the model that is closest to the starting model. The Euclidean norm of p(r) has a very simple expression in terms of the vector components:

$$||p(r)|| = \int_0^R p(r)^2 dr = \sum_{k=1}^N p_k^2 = |\underset{\sim}{p}|^2$$

Wiggins (1972) and Jackson (1972) show that the smallest p in a least squares sense is given by the vector

$$<\underset{\sim}{p}> = V\underset{\sim}{p}^* = VV^T\underset{\sim}{p} \tag{40}$$

with

$$\mathrm{Var}\{<p>_i\} = \sum_{j=1}^K V_{ij}^2\lambda_j^{-1} \tag{41}$$

(40) bears a resemblance to the definition of a local average in Backus-Gilbert theory (see equation (20)). $VV^T$ is therefore often called the resolution matrix. However, it is erroneous to interprete $VV^T$ as a set of discretized averaging kernels, a habit which is unfortunately gaining widespread popularity. What is missing in the Wiggins-Jackson approach is the unimodularity condition (21). In the "most squares" method developed by Jackson (1976) one may define unimodular kernels, but their unimodularity will be destroyed if $K < N$. To remove this shortcoming, Jackson (1979) must add a priori guesses for all parameter values with their uncertainties to the data, so that $K = N$ (see also the discussion at the end of this section).
Another difficulty in the interpretation of $VV^T$ as a set of averaging kernels is that the components of $\underset{\sim}{p}$ may consist of parameters that have nothing to do with the actual model values. This is the case when $\underset{\sim}{p}$ consists of coefficients of some polynomial, for instance.

A direct adaption of Backus-Gilbert theory to the discrete case was attempted by Kennett and Nolet (1978). This was, however, only derived for piecewise constant functions $h_i^*(r)$, and in the efficient method proposed by Nolet (1978) equal layer thickness was required as well. Nolet (1980) develops a resolution theory for the discrete case which is valid for arbitrary discretization and singular $G^TG$. The method is as follows: A local average for the model in $r_k$ is

$$\langle p(r_k)\rangle = \int_0^R A(r,r_k)p(r)dr \tag{20'}$$

$$\int_0^R A(r,r_k)dr = 1 \tag{21'}$$

Of course these expressions must be discretized as well.

$$A(r,r_k) = \sum_{l=1}^{N} a_l(r_k)h_l(r) \tag{42}$$

$$\langle p(r_k)\rangle = \sum_{l=1}^{N}\sum_{j=1}^{N} a_l(r_k)p_j \int_0^R h_l(r)h_j(r)dr$$

$$= \sum_{l=1}^{N} a_l(r_k)p_l \tag{43}$$

We may express a in terms of orthonormal basis V:

$$\underset{\sim}{a} = V\underset{\sim}{b} \tag{44}$$

thus:

$$\langle p(r_k)\rangle = \sum_{l=1}^{N}\sum_{j=1}^{K} V_{lj}b_j(r_k)p_l$$

$$= \sum_{j=1}^{K} b_j(r_k)p_j^* \tag{45}$$

and because of (39):

$$\varepsilon_k^2 = \mathrm{Var}\{\langle p(r_k)\rangle\} = \sum_{j=1}^{K} b_j^2(r_k)\lambda_j^{-1} \tag{46}$$

The spread of $A(r,r_k)$ in the discretized version can be defined analogous to (23), and the resulting minimization problem can then be solved for $\underset{\sim}{a}$. There is a less elegant but much more efficient method. We want $A(r,r_k)$ to resemble $\delta(r-r_k)$. The closest to this delta function that we can get in the subspace of interpolation functions is:

$$D(r,r_k) = \sum_{i=1}^{N} h_i(r_k)h_i(r) \tag{47}$$

for which $||D(r,r_k) - \delta(r-r_k)||$ is a minimum.
Instead of the spread $S(r_k)$ we now use the difference between $A(r,r_k)$ and $D(r,r_k)$:

$$\Delta_k = \int_0^R \{A(r,r_k) - D(r,r_k)\}^2 dr$$

$$= \sum_{i=1}^{K} b_i^2(r_k) - 2\sum_{i=1}^{K} u_i(r_k)b_i(r_k) + \sum_{i=1}^{N} h_i(r_k)^2 \qquad (48)$$

where

$$u_i(r_k) = \sum_{l=1}^{N} V_{li}h_l(r_k) \qquad (49)$$

and minimize:

$$\Delta_k \cos\theta + \varepsilon_k^2 \sin\theta$$

subject to the constraint:

$$\sum_{l=1}^{K} b_l(r_k)t_l = 1 \qquad (50)$$

where

$$t_l = \sum_{j=1}^{N} V_{jl} \int_0^R h_j(r)dr \qquad (51)$$

Using a Lagrange multiplier $\eta$ we find $b_m$ through

$$\frac{\partial}{\partial b_m(r_k)} \{\Delta_k\cos\theta + w\varepsilon_k^2 \sin\theta + \eta\sum_l b_l(r_k)t_l\} = 0 \qquad (52)$$

A weight $w$ is introduced to ensure that a regular distribution of $\theta$ in $[0,\pi/2]$ gives a properly distributed set of points on the trade-off curves. I suggest to try $w = \lambda_q$, with $q \simeq K/2$, which gives good results in inversion of teleseismic data. The solution is:

$$b_m(r_k) = \frac{2u_m(r_k)\cos\theta - \eta t_m}{2\cos\theta + 2\lambda_m^{-1} w\sin\theta} \qquad (53)$$

where

$$\eta = \frac{\sum_{m=1}^{K} \frac{2u_m(r_k)\cos\theta\, t_m}{\cos\theta + \lambda_m^{-1} w\sin\theta} - 2}{\sum_{m=1}^{K} \frac{t_m^2}{\cos\theta + \lambda_m^{-1} w\sin\theta}} \qquad (54)$$

$A(r,r_k)$ can now be constructed and plotted. Although visual inspection of the resolving kernels gives the most satisfactory impression of the true resolution, a more compact display in terms of trade-off kernels may be preferable. Since $\Delta_k$ does not have dimensions of length, it cannot be used as a measure of the spread. We may calculate $S_k$ directly from (23), but must keep in mind that we have not minimized $S_k$ but $\Delta_k$, so that a strictly monotonic behaviour of $S_k$ with $\varepsilon_k^2$ is not guaranteed. A very simple alternative is to estimate

$$S_k \simeq A(r_k, r_k)^{-1} \tag{55}$$

which would be the exact averaging length of a box-shaped kernel centered at $r_k$. If $A(r,r_k)$ deviates strongly from a box-shape (55) may not be exact enough. $S_k$ may even become negative if there is little or no resolution at $r_k$ so that $A(r,r_k)$ is peaked at a radius that differs very much from $r_k$. But whenever there is a reasonable resolution of the data (55) has proven to be an adequate criterion of resolving power.

This inversion method is by no means restricted to the case of teleseismic data. Nor is its application limited to the case of one-dimensional functions of one variable. Extension of the method to n-tuples such as $p(r) = (\alpha(x,y,z), \beta(x,y,z), \rho(x,y,z))$ is treated in Nolet (1980).

Parker (1977b) and Jackson (1979) have advocated the use in the resolution calculations of a priori knowledge on the Euclidean norm of the model and the model parameter values, respectively. Of the two methods, Jackson's appears to be more efficient from the point of view of computational effort, and it has the advantage that it can be easily incorporated into the present method if one wants to do so. Briefly, this can be done as follows. Suppose we have independent knowledge that the model at some point, say at $r_s$, cannot exceed a lower bound $c_1$ and an upper bound $c_2$. For the resolving power calculations we may handle this knowledge in an approximate way by assuming that $p(r_s)$ is an unknown random variable with a Gaussian probability centered at $(c_1 + c_2)/2$, with a variance $(c_2 - c_1)^2/6$, which would be the variance of a uniform probability density function between $c_1$ and $c_2$. We then have:

$$p(r_s) = \sum_{k=1}^{N} p_k h_k(r_s) = \frac{c_1 + c_2}{2}$$

which is just another linear equation in the parameters $p_k$ that can be added (after scaling to unit variance) to the system (30). Thus, a priori notions about the model are introduced in the inversion as extra data. It is very much a matter of personal taste whether one prefers to deal with a priori knowledge before or after the actual inversion. The reader who wants to make up his mind is advised to study the paper by Jackson (1979), and the poetic comments on it made by Sabatier (1979).

The method presented here is an attempt to revive the correct use of the concept of "resolving length", as defined originally by Backus and Gilbert, through removal of some of the calculational disadvantages. By extending the use of unimodular averaging kernels to the discretized case, the method is much more efficient in the use of computer memory then the original Backus-Gilbert approach. The calculation of unimodular kernels does involve very little extra computational effort once the matrix $G^TG$ has been diagonalized for a stable inversion. The orthogonalization of the basis-functions in Hilbert space has made it possible to formulate both the inverse problem and the resolving power calculations for models of arbitrary parametrization.

## REFERENCES


Aki, K., 1977, Three dimensional seismic velocity anomalies in the lithosphere. Method and summary of results, J. Geophysics, 43:235.

Aki, K., A. Christoffersson, and Husebye, E.S., 1977, Determination of three-dimensional seismic structure of the lithosphere, J. Geophys. Res., 82:277.

Backus, G., and Gilbert, F., 1967, Numerical application of a formalism for geophysical inverse problems, Geophys. J.R. astr. Soc., 13:247.

Backus, G., and Gilbert, F., 1968. The resolving power of gross Earth data, Geophys. J.R. astr. Soc., 16:169.

Backus, G., and Gilbert, F., 1970, Uniqueness in the inversion of inaccurate gross Earth data, Phil. Trans. Roy. Soc. Lond., A266:123.

Cervený, V., and Ravindra, R., 1971, Theory of seismic head waves, Univ. of Toronto Press.

Cloetingh, S.A.P.L., Nolet, G., and Wortel, M.J.R., 1980, Standard graphs and tables for the interpretation of Rayleigh wave group velocities in crustal structures, Proc. Roy. Neth. Ac. Sci., B83(1):101.

Fischer, A.G., and Judson, S. (eds.), 1975, Petroleum and Global Tectonics, Princeton University Press.

Garmany, J., 1979, On the inversion of travel times, Geophys. Res. Lett., 6:277.

Gilbert, F., 1971, Ranking and winnowing gross Earth data for inversion and resolution, Geophys. J.R. astr. Soc., 23:125.

Jackson, D.D., 1972, Interpretation of inaccurate, insufficient and inconsistent data, Geophys. J.R. astr. Soc., 28:97.

Jackson, D., 1976, Most squares inversion, J. Geophys. Res., 81:1027.

Jackson, D.D., 1979, The use of a priori data to resolve non-uniqueness in linear inversion, Geophys. J.R. astr. Soc., 57: 137.

Jeffreys, H., 1961, Small corrections in the theory of surface waves, Geophys. J.R. astr. Soc., 6:115.

Johnson, R.E., and Gilbert, F., 1972, Inversion and inference for teleseismic ray data, Meth. Comput. Phys., 12:231.

Julian, B.R., and Anderson, D.L., 1968, Travel times, apparent velocities and amplitudes of body waves, Bull. Seism. Soc. Am., 58:339.

Julian, B.R., and Gubbins, D., 1977, Three dimensional seismic ray tracing, J. Geophysics, 43:95.

Kennett, B.L.N., and Nolet, G., 1978, Resolution analysis for discrete systems, Geophys. J.R. astr. Soc., 53:413.

Matthews, J., and Walker, R.L., 1973, Mathematical methods of physics, 2nd. ed., W.A. Benjamin Inc., Menlo Park, CA.

Mendiguren, J.A., 1977, Inversion of surface wave data in source mechanism studies, J. Geophys. Res., 82:889.

Neigauz, M.G., and Shkadinskaya, G.V., 1972, Method for calculating surface Rayleigh waves in a vertically inhomogeneous half-space, in: Computational Seismology, ed. V.I. Keilis-Borok, Consultants Bureau, N.Y..

Nolet, G., 1978, Simultaneous inversion of seismic data, Geophys. J. R. astr. Soc., 55:679.

Nolet, G., 1980, Backus-Gilbert theory for models with arbitrary parametrization, submitted for publication.

Parker, R.L., 1977a, Understanding inverse theory, Ann. Rev. Earth Plan. Sci., 5:35.

Parker, R.L., 1977b, Linear inference and underparametrized models, Rev. Geophys. Space Phys., 15:446.

Rodi, W.L., Glover, P., Li, T.M.C., and Alexander, S.S., 1975, A fast, accurate method for computing group-velocity partial derivatives for Rayleigh and Love modes, Bull. Seism. Soc. Am., 65:1105.

Sabatier, P.C., 1979, Comment on "The use of a priori data to resolve non-uniqueness in linear inversion" by D.D. Jackson, Geophys. J.R. astr. Soc., 58:523.

Strong, D.F. (ed.), 1974, Metallogeny and Plate Tectonics, The Geological Association of Canada, Special paper 14.

Takeuchi, H., and Saito, M., 1972. Seismic surface waves, Meth. Comp. Phys., 11:217.

Vlaar, N.J., 1976, On the excitation of the Earth's seismic normal modes, Pure Appl. Geophys., 114:863.

Vlaar, N.J., and Nolet, G., 1978, Seismic surface waves, in: Modern problems in elastic wave propagation, ed. J. Miklowitz and J.D. Achenbach, J. Wiley and Sons.

Wiggins, R.A., 1972, The general linear inverse problems: implication of surface waves and free oscillations for Earth structure, Rev. Geophys. Space Phys., 10:251.

Woodhouse, J.H., 1976, On Rayleigh's principle, Geophys. J.R. astr. Soc., 46:11.

Woodhouse, J.H., and Dahlen, F.A., 1978, The effect of a general aspherical perturbation on the free oscillations of the Earth, Geophys. J.R. astr. Soc., 53:335.

APPENDIX A

## A1. General formulae for rays in a layered medium with velocity v

(P = ray parameter, $v(z(p)) = p^{-1}$, i = angle with vertical)

*flat earth* (depth $0 < z < \infty$)

Snell's law $$p = \frac{\sin i(z)}{v(z)} \quad \text{(sec/km)}$$

2-way travel time $$T = 2 \int_0^{z(p)} \frac{v^{-2}\,dz}{(v^{-2}-p^2)^{\frac{1}{2}}}$$

Epicentral distance $$X = 2 \int_0^{z(p)} \frac{p\,dz}{(v^{-2}-p^2)^{\frac{1}{2}}}$$

Perturbation formula $$\delta T = -2 \int_0^{z(p)} \frac{v^{-3}\delta v}{(v^{-2}-p^2)^{\frac{1}{2}}}\,dz$$

*spherical earth* (radius $0 < r < R$)

$$p = \frac{r \sin i(r)}{v(r)} \quad \text{(sec/rad)} \qquad T = 2 \int_{r(p)}^{R} \frac{dr}{v(1-p^2v^2/r^2)^{\frac{1}{2}}}$$

$$\Delta = 2 \int_{r(p)}^{R} \frac{pv\,dr}{r^2(1-p^2v^2/r^2)^{\frac{1}{2}}} \qquad \delta T = -2 \int_{r(p)}^{R} \frac{v^{-2}\delta v}{(1-p^2v^2/r^2)^{\frac{1}{2}}}\,dr$$

## A2. T and X or Δ in one layer, $z_1 < z < z_2$ or $r_1 < r < r_2$ *)

*flat earth*

(a) v = constant

$$T = \int_{z_1}^{z_2} \frac{z}{v(1-v^2p^2)^{\frac{1}{2}}} \qquad X = \int_{z_1}^{z_2} \frac{z}{(v^{-2}p^{-2} - 1)^{\frac{1}{2}}}$$

(b) linear v between $v_1 = v(z_1)$ and $v_2 = v(z_2)$

$$T = \frac{z_2-z_1}{v_2-v_1} \ln \left\{\frac{v_2}{v_1} \cdot \frac{1 + (1-p^2v_1^2)^{\frac{1}{2}}}{1 + (1-p^2v_2^2)^{\frac{1}{2}}}\right\}$$

$$X = \frac{z_2-z_1}{v_2-v_1} \{(p^{-2}-v_1^2)^{\frac{1}{2}} - (p^{-2}-v_2^2)^{\frac{1}{2}}\}$$

*) Replace $z_2$ or $r_1$ by the turning point if this is within the layer.

*spherical earth*

(a) v = constant

$$T = \int_{r_1}^{r_2} \left(\frac{r^2}{v^2} - p^2\right)^{\frac{1}{2}} \qquad X = -\int_{r_1}^{r_2} \sin^{-1}(pv/r)$$

(b) linear v = b + ar

$$a = (v_2 - v_1)/(r_2 - r_1) \qquad c = 1 - |a^2p^2|^{\frac{1}{2}}$$

$$b = (v_1r_2 - v_2r_1)/(r_2 - r_1) \qquad i = \sin^{-1}(pv/r)$$

where

$$T = \int_{r_1}^{r_2} \frac{\ln tg(i) - J(i)}{a} \qquad X = -\int_{r_1}^{r_2} \{i + apJ(i)\}$$

where

$$J(i) = \frac{2}{c}\tan^{-1}\left\{\frac{-ap\tan(i/2) + 1}{c}\right\} \qquad a^2p^2 > 1$$

$$= \frac{1}{c}\ln\left\{\frac{-ap\tan(i/2) + 1 - c}{-ap\tan(i/2) + 1 + c}\right\} \qquad a^2p^2 < 1$$

$$= ap\cos i/(\sin i - ap) \qquad |ap| = 1$$

(c) See Julian and Anderson (1968) for expressions with $v = b + ar^2$ or $v = ar^b$.

APPENDIX B

Perturbation formulae for seismic rays (symbols as in Appendix A).

*flat earth*

(a) $v, \delta v$ = constant: $$\frac{\partial T}{\partial v} = -\int_{z_1}^{z_2} \frac{z}{v^2(1 - v^2p^2)^{\frac{1}{2}}}$$

(b) $v = b + az$ : $$\frac{\partial T}{\partial a} = -\int_{z_1}^{z_2} \frac{1}{a^2}\{\ln\tan(i/2) + bp\cot i\}$$

$$\frac{\partial T}{\partial b} = \int_{z_1}^{z_2} \frac{p}{a}\cot i$$

*spherical earth*

(a) v, δv = constant: $$\frac{\partial T}{\partial v} = -\int_{r_1}^{r_2} \frac{r}{v^2}(1 - p^2v^2/r^2)^{\frac{1}{2}}$$

(b) v = b + ar : $$\frac{\partial T}{\partial a} = \frac{1}{a}\int_{r_1}^{r_2} \{p \cot i - \frac{1}{a} \ln tg(i/2) + \frac{1}{a} J(i)\}$$

$$\frac{\partial T}{\partial b} = -\frac{p}{b}\int_{r_1}^{r_2} \cot i$$

See also Nolet (1978, appendix).

APPENDIX C

Dispersion equations

*flat earth*

Cylindrical vector harmonics:

$$\underset{\sim}{R}_k^m(r,\phi) = Y_k^m \hat{e}_z$$

$$\underset{\sim}{S}_k^m(r,\phi) = k^{-1}\partial_r Y_k^m \hat{e}_r + (kr)^{-1}\partial_\phi Y_k^m \hat{e}_\phi$$

$$\underset{\sim}{T}_k^m(r,\phi) = (kr)^{-1}\partial_\phi Y_k^m \hat{e}_r - k^{-1}\partial_r Y_k^m \hat{e}_\phi$$

where

$$Y_k^m(r,\phi) = J_m(kr)\exp(im\phi)$$

and

$J_m$ = Besselfunction, m = 0, ± 1, ± 2,....

Love waves:

$$\underset{\sim}{s} = y_1(z)\underset{\sim}{T}_k^m(r,\phi) \;,\; \underset{\sim}{\sigma}_z = y_2(z)\underset{\sim}{T}_k^m(r,\phi)$$

Boundary condition $y_2(0) = 0$

Matrix A (eq. 7) $$\begin{pmatrix} 0 & \frac{1}{\mu} \\ k^2\mu-\omega^2\rho & 0 \end{pmatrix}$$

Rayleigh waves:

$$\underset{\sim}{s} = y_1(z)\underset{\sim}{R}_k^m(r,\phi) + y_3(z)\underset{\sim}{S}_k^m(r,\phi)$$

$$\underset{\sim}{\sigma}_z = y_2(z)\underset{\sim}{R}_k^m(r,\phi) + y_4(z)\underset{\sim}{S}_k^m(r,\phi)$$

Boundary condition $y_2(0) = y_4(0) = 0$

Matrix A
$$\begin{pmatrix} 0 & \frac{1}{\sigma} & k\lambda & 0 \\ -\omega^2\rho & 0 & 0 & k \\ -k & 0 & 0 & \frac{1}{\mu} \\ 0 & \frac{k\lambda}{\sigma} & k^2(\sigma-\frac{\lambda^2}{\sigma})-\omega^2\rho & 0 \end{pmatrix}$$

*spherical earth*

Spherical vector harmonics:

$$\underset{\sim}{R}_l^m(\theta,\phi) = Y_l^m\hat{e}_r$$

$$\underset{\sim}{S}_l^m(\theta,\phi) = \partial_\theta Y_l^m\hat{e}_\theta + (\sin)^{-1}\partial_\theta Y_l^m\hat{e}_\phi$$

$$\underset{\sim}{T}_l^m(\theta,\phi) = (\sin\theta)^{-1}\partial_\phi Y_l^m\hat{e}_\theta - \partial_\theta Y_l^m\hat{e}_\theta$$

where

$$Y_l^m(\theta,\phi) = P_l^{|m|}(\cos\theta)\exp(im\theta)$$

and

$$P_l^m = \text{Legendre function, } m = 0, \pm 1, \ldots, \pm l$$

Love waves

$$\underset{\sim}{s} = y_1(r)\underset{\sim}{T}_l^m(\theta,\phi) \quad , \quad \underset{\sim}{\sigma}_z = y_2(r)\underset{\sim}{T}_l^m(\theta,\phi)$$

Boundary condition $y_2(R) = 0$

Matrix A
$$\begin{pmatrix} \frac{1}{r} & \frac{1}{\mu} \\ (L^2-2)\frac{\mu}{r^2} - \omega^2\rho & -\frac{3}{r} \end{pmatrix}$$

Rayleigh waves

$$\underset{\sim}{s} = y_1(r)\underset{\sim}{R}_1^m(\theta,\phi) + y_3(r)\underset{\sim}{S}_1^m(\theta,\phi)$$

$$\underset{\sim}{\sigma}_z = y_2(r)\underset{\sim}{R}_1^m(\theta,\phi) + y_4(r)\underset{\sim}{S}_1^m(\theta,\phi)$$

Boundary condition $y_2(R) = y_4(R) = 0$

$$\text{Matrix A} \begin{pmatrix} -\frac{2\lambda}{\sigma}\frac{1}{r} & \frac{1}{\sigma} & \frac{\lambda}{\sigma}\frac{L^2}{r} & 0 \\ -\rho(\omega^2+\frac{4g}{r})+\frac{4}{r^2} & -\frac{4\mu}{\sigma r} & (-\frac{2\xi}{r^2}+\frac{\rho g}{r})L^2 & \frac{L^2}{r} \\ -\frac{1}{r} & 0 & \frac{1}{r} & \frac{1}{\mu} \\ \frac{\rho g}{r}-\frac{2\xi}{r^2} & -\frac{\lambda}{\sigma r} & -\omega^2\rho+\frac{4L^2\mu(\lambda+\mu)}{\sigma r^2}-\frac{2\mu}{r^2} & -\frac{3}{r} \end{pmatrix}$$

where

$$\sigma = \lambda + 2\mu \qquad \xi = \frac{(3\lambda + 2\mu)\mu}{\lambda + 2\mu} \qquad L^2 = 1(1+1)$$

$g$ = gravity accelleration

APPENDIX D

Variational parameters for surface wave phase velocities.

*flat earth*

Love waves:

$$q = \frac{c/U}{\omega^2 \int_0^\infty \rho y_1^2 dz}$$

$$\frac{\rho}{c}\left(\frac{\partial c}{\partial \rho}\right) = \frac{q}{2}\{-\omega^2\rho y_1^2 + k^2\mu y_1^2 + \frac{1}{\mu}y_2^2\}$$

$$\frac{\beta}{c}\left(\frac{\partial c}{\partial \beta}\right) = q\{k^2\mu y_1^2 + \frac{1}{\mu}y_2^2\}$$

Rayleigh waves:

$$q = \frac{c/U}{\omega^2 \int_0^\infty \rho(y_1^2 + y_3^2)dz}$$

$$\frac{\rho}{c}\left(\frac{\partial c}{\partial \rho}\right) = \frac{q}{2}\,\{-\omega^2\rho(y_1^2 + y_3^2) + \frac{1}{\sigma}\,y_2^2 + \frac{1}{\mu}\,y_4^2 + k^2(\sigma - \frac{\lambda^2}{\sigma})y_3^2\}$$

$$\frac{\alpha}{c}\left(\frac{\partial c}{\partial \alpha}\right) = q\,\{\sigma\left(\frac{dy_1}{dz}\right)^2 + k\sigma y_3(ky^3 - \frac{dy_1}{dz})\}$$

$$\frac{\beta}{c}\left(\frac{\partial c}{\partial \beta}\right) = q\,\{4k\mu y_3\,\frac{dy_1}{dz} + \frac{1}{\mu}\,y_4^2\}$$

*spherical earth*

Love waves:

$$q \doteq \frac{c/U}{\omega^2 \int_0^\infty \rho y_1^2 r^2 dr}$$

$$\frac{\rho}{c}\left(\frac{\partial c}{\partial \rho}\right) = \frac{q}{2}\,\{-\omega^2\rho r^2 y_1^2 + \frac{r^2}{\mu}\,y_2^2 + (L^2-2)\mu y_1^2\}$$

$$\frac{\beta}{c}\left(\frac{\partial c}{\partial \beta}\right) = q\,\{(L^2-2)\mu y_1^2 + \frac{r^2}{\mu}\,y_2^2\}$$

Rayleigh waves:

$$q = \frac{c/U}{\omega^2 \int_0^\infty \rho(y_1^2+L^2y_3^2)r^2 dr}$$

$$\frac{\rho}{c}\left(\frac{\partial c}{\partial \rho}\right) = \frac{q}{2}\,\{-\omega^2\rho r^2(y_1^2 + L^2y_3^2) + \frac{r^2}{\sigma}\,y_2^2 + \frac{L^2}{\mu}\,r^2y_4^2 + \xi(2y_1 - L^2y_3)^2$$

$$+\, 1(1^2 - 1)(1 + 2)\mu y_3^2) \qquad \text{(gravity effect neglected)}$$

$$\frac{\alpha}{c}\left(\frac{\partial c}{\partial \alpha}\right) = q\,\{\sigma r^2\left(\frac{dy_1}{dr}\right)^2 + \sigma(2y_1 - L^2y_3)(2r\,\frac{dy_1}{dr} + 2y_1 - L^2y_3)\}$$

$$\frac{\beta}{c}\left(\frac{\partial c}{\partial \beta}\right) = q\,\{2\mu L^2y_3(y_1 - y_3) - 2\mu(2r\,\frac{dy_1}{dr}+y_1)(2y_1-L^2y_3)+ \frac{L^2r^2}{\mu}\,y_4^2\}$$

# THE RESOLVING POWER OF SEISMIC SURFACE WAVES WITH RESPECT TO CRUST AND UPPER MANTLE STRUCTURAL MODELS

Giuliano F. Panza

Istituto di Geodesia e Geofisica
Università di Bari
70100 Bari, ITALY

## 1. INTRODUCTION

The recording and, to a large extent, the processing of the data to obtain phase and group velocities is a relatively straightforward procedure. To draw inferences from the velocities about the physical properties of the Earth is, however, a much more equivocal process involving serious questions of non uniqueness. A rather complete review of surface wave measurements is given by Kovach (1978) on a world wide scale and by Panza et al.(1978) for the European area. The primary goal of this paper is to provide a brief summary on the inversion techniques relevant to seismic surface wave studies and to show an example of their applicability to other indirect geoexploration methods.

Two different approaches to inversion are used mainly at present. The first is the so called "linearized inversion" due to Backus and Gilbert (1968,1970) and their numerous successors. The method is essentially based on the iterative procedure of refering the Earth model by small disturbances of an initial a priori known Earth structure. The second approache combines different trial-and-error techniques, searching for fitting the observational data among a predetermined set of possible Earth models. The well-known Monte Carlo random search (Keilis-Borok and Yanovskaya,1967; Press 1968) or the "Hedgehog" random-deterministic search (Valyus, 1972; Valyus et al.,1969; Knopoff, 1972) fall into this category.

To define a set of possible Earth models it is necessary to describe them by a limited number of parameters. Such a parametrization of the Earth is to be based on a priori knowledge of some features of velocity and density distribution in the Earth's interior. Results of both approaches to inversion are thus essentially dependent on a priori information. The situation is very different for body waves since the inversion technique of travel-times is free from this essential disadvantage. The velocity cross section of the Earth comes out in this case as a straightforward analytic solution of Abel's integral equation, and to get it no a priori knowledge of the structure is necessary. Quite recently Brodskii and Levshin (1979) have proposed a new straightforward approach to free oscillation data inversion similar to the $\tau(p)$ method for body waves (Bessonova et al., 1974,1976). The approach is essentially an asymptotic one and for this reason it does not make possible the retrieval of all information concerning the Earth's structure from the observational data. However the method is very interesting since it provides an efficient way to generate an initial model for more refined inversion and the significant part of the observed Earth's free oscillation spectra actually possesses asymptotic properties.

## 2. GENERAL SCHEME FOR THE INVERSION

In general two schemes of inversion of geophysical data can be identified: 1) the trial-and-error method, 2) the method of direct inversion. The trial-and-error method is basically the following. The unknown cross-section is replaced by a set of parameters and the determination of the cross-section is reduced to the determination of numerical values of the parameters. The possible limits of these parameters, i.e. the region where the real cross-section exists, are indicated. Different cross-sections are chosen in a consecutive order in this region. For each cross-section, theoretical values are computed for comparison with real data and the discrepancy between the computed data and the observed ones is calculated. The set of cross sections for which this discrepancy is sufficiently small is the solution of the problem. So the problem is to find the zone of minimum of a multidimensional function (the discrepancy between the computed and the observed data) in the space of unknown parameters of the cross-section. The method of direct inversion has similar logics. However the curves (travel times, dispersion relations, free oscillations, etc.) and not the structures are represented through a finite set of parameters. The possible limits for each parameter, i.e. the region where the real curve lies, are indicated. Different sets of parameters are chosen inside this region. For each set of parameters the curves are computed. If the discrepancy between computed curves and observations is sufficently small, the curve is inverted into structure; a set of all such structures is a solution of the inversion problem. As has already been mentioned, the possibilities of this method are limited to travel-time curves and to free oscillation in the asymptotic approach. For this reason it may be useful to invert some of the data by the direct inversion method, and then to test the obtained structure by the trial-and-error technique using the rest of the data.

Each operation of the methods described is connected with

specific problems which are now discussed.

## 2.1 Treatment of observations

The raw observations should be used; in other words the clouds of measured points have to be used without drawing the above mentioned functions characterizing the wave-propagation. For example the arrival times should be given as a cloud of dots in the ( $t, \Delta$ ) plane for each depth h, or better, for each range of depths, without drawing the curve $t(\Delta)$, even less its loops; also surface wave velocities c (phase velocity) or U(group velocity) should be given as a cloud of dots in the (c,T) or (U.T) planes, without draving the dispersione curve. The error should be estimated for each dot and the dots should represent measurements which do not predetermine each other. The identification of body wave type, or of mode number, is not necessary and not even desirable as it is sometimes a powerful source of error. The waves should only be divided into body or surface waves, free oscillations into spheroidal or torsional modes.

## 2.2 Parametrization and flow of cross-sections

This step involves the representation of the cross-section through a finite number of numerical parameters and the indication of a priori limits for them. This requires the division of the structure into layers and the approximation of each physical function in each layer. The parameters of the structure are thus the parameters of the approximating functions in each layer as well as the depth of the layer boundaries. Usually the physical function in each layer is approximated by a constant, a straight-line, a parabola or some higher polynomial in h, where h is the depth.

The choice of parameters may be decisive for the success of inversion, since too simple or too complicated an approximation of the structure can produce absolutely meaningless results. It is important to keep the number of parameters N as small as possible, for an easy understanding of the inversion results, but if the number of parameters is too small, the result is too rough an approximation of the structure, and possibly some of its interesting elements are missed. The choice of parameters must be correlated with the physical task of inversion. If, for example, a search is made to find out whether some elements of the structure exist ( a wave-guide, a discontinuity of velocity within some depth interval, etc.), then the assumed system of parameters has to allow the structure both with these elements and without them. The probability of meeting the considered element by random choice should be about 1/2. Finally the information contained in the given observations is the decisive factor in parametrization.

As a result of parametrization the cross-section is represented as a point in the space of its unknown parameters, indicating their limits, thus determining the region in which the point lies. Limits can also be imposed on the values of parameters and on any function of them. The problem of inversion is to narrow this region as far as the observations allow. A set of combinations of the unknown parameters has to be found for which the discrepancy between observations and the computed properties is sufficiently small. This is a widely developed problem of computational mathematics, though usually it is not the region of the minimum but only the minimum itself which is sought.

The lack of a general theory of solution of this problem makes the following empirical method the simplest and the only one absolutely reliable. The investigated multidimensional region is divided by a net, in each knot of the net the function of parameters is computed and finally the points are chosen in which the function is sufficiently small. The simplest method of construction of this net is to place the knots along co-ordinate axes at equal intervals. The step of the net must have a magnitude of the order of the error that is allowed for the determination of the boundaries of the minimum volume. If the number of parameters N is large (more than four or five) the number of knots, given by K to the N-th power, where K is the number of different values which can be assumed by each parameter, becomes so large that the described method can not be applied in practice. An alternative way of constructing a net is to use the Monte Carlo method, i.e. taking random points as knots. The number of knots of this net can be much less than in the previous case. However, it must also increase with increasing N, otherwise the knots would be distributed unevenly and the distance between them would be too large, so that a part of the investigated volume might be missed. A great disadvantage of the Monte Carlo method is that the results of the trials already made are not used in the next trial. Thus when N is large the guided methods for the search of the minimum are used. Among these methods a relevant position is occupied by the " Hedgehog method " developed by V.Valyus. The method, accordingly with Keilis-Borok and Yanovskaya (1967) and Keilis-Borok (1971), can be briefly described as follows.

A single point, X, of the minimum region

$$X\ (P_1, P_2 \ldots\ldots P_N)$$

where $P_i$ are the parameters of the cross-section, is found by the Monte Carlo or some other technique. Then the neighbouring points X', are tried

$$X'\ (P_i + \alpha_i dP_i),\ i = 1, \ldots\ldots N \qquad (1)$$

where $\alpha_i$=0 or 1, and certain combinations of $\alpha_i$ are picked in turn. Points which fall within this minimum region are selected. The same procedure is applied to every selected point until the whole region is covered. Upon this, return is made to the Monte Carlo technique, (omitting, of course, the previously found region from further investigation) another minimum region is reached and so on.

## 3. THE RESOLVING POWER OF A GIVEN DATA SET, WITH RESPECT TO STRUCTURAL PARAMETERS

In this section the results of the following experiments are summarized. First the phase (c) and group (U) velocities of fundamental Love and Rayleigh modes have been computed for a known structure, random phases have been added to the results and then an inversion has been performed. Second, the phase velocities for the first six Rayleigh modes have been computed for a known structure, random phases have been added to the results and then an inversion has been performed both for each mode separately and for all the modes simultaneously. The results of these experiments have a general validity and are a very useful guide in the solution of the problem of the choice of parametrization to be used in the inversion of real data.

Table 1a. Crust-upper mantle structure for continent

| Depth(km) | Thickness(km) | $\beta$(km/sec) | $\alpha$(km/sec) | $\varrho$(g/cm$^3$) |
|---|---|---|---|---|
| 0 | 10 | 3.49 | 6.05 | 2.75 |
| 10 | 20 | 3.67 | 6.35 | 2.85 |
| 30 | 20 | 3.85 | 7.05 | 3.08 |
| 50 | 65 | 4.65 | 8.17 | 3.45 |
| 115 | 250 | 4.30 | 8.35 | 3.54 |
| 365 | 85 | 4.75 | 8.80 | 3.65 |
| 450 | 200 | 5.30 | 9.80 | 3.98 |
| 650 | 400 | 6.20 | 11.15 | 4.43 |
| 1050 | 240 | 6.48 | 11.78 | 4.63 |
| 1290 | $\infty$ | 6.62 | 12.02 | 4.71 |

Table 1b. Crust-upper mantle structure for ocean

| Depth(km) | Thickness(km) | $\beta$(km/sec) | $\alpha$(km/sec) | $\varrho$(g/cm$^3$) |
|---|---|---|---|---|
| 0 | 4 | 0.00 | 1.52 | 1.03 |
| 4 | 1 | 1.00 | 2.10 | 2.10 |
| 5 | 5 | 3.70 | 6.41 | 3.07 |
| 10 | 50 | 4.65 | 8.10 | 3.40 |
| 60 | 150 | 4.15 | 7.60 | 3.40 |
| 210 | 240 | 4.75 | 8.80 | 3.65 |
| 450 | 200 | 5.30 | 9.80 | 3.98 |
| 650 | 400 | 6.20 | 11.15 | 4.43 |
| 1050 | 240 | 6.48 | 11.78 | 4.63 |
| 1290 | $\infty$ | 6.62 | 12.02 | 4.71 |

### 3.1 Experiment n.1 (Knopoff and Chang,1977)

Let us consider a known geophysical structural cross-section for which the dispersion relations U(T) and c(T) are computed with ease by standard methods, $T = 2\pi / \omega$ is the period. To simulate standard inversion procedure, the dispersion values are digitized into discrete phase and group velocity samples at selected periods, $T_i$. The velocity samples $U(T_i)$ and $c(T_i)$ are assumed to have standard deviations $\sigma_U(T_i)$ and $\sigma_c(T_i)$ respectively. The errors in the "data" are assumed random and uncorrelated at each of the sample periods. These "noisy data" are then inverted by a linear inversion procedure using as a starting model the "exact" solution, that is the models given in Table 1. The calculations have been performed for both continental and oceanic models. The phase and group velocities derived for both Rayleigh and Love waves in the fundamental mode are shown in Fig.1. All four of these curves have been digitized at the 17 periods 20(5) 40(10) 100(25) 250 sec. The values in parentheses represent the period interval between adjacent "digitized" values. The inversion has been restricted to six parameters in the crust and upper mantle for the continental structure, and to five parameters for the oceanic structure. The six parameters are the lid, channel and subchannel S-wave velocities $\beta_{LID}$, $\beta_{CH}$, $\beta_{SUB}$ and the crust, lid and channel thicknesses $h_{CR}$, $h_{LID}$, $h_{CH}$. For $h_{CR}$ it has been assumed that the three infrastructural layers of the crust are always found in the ratio of the thicknesses as they appear in the original model 1:2:2. For the Oceanic structure the five parameters are $\beta_{LID}$, $\beta_{CH}$, $\beta_{SUB}$, $h_{LID}$, and $h_{CH}$.

A description of model variances corresponding to the data variances in six-dimensional parameter space requires the specification of a large number of numbers which is difficult for the casual reader to assess. To simplify this task it is possible to choose to list the diagonal elements of the model error matrix:

$$\left[\frac{I}{N} \sum_{i=1}^{N} \left(\frac{\partial c(T_i)}{\partial P_j}\right)^2 \sigma(T_i)^{-2}\right]^{-\frac{1}{2}} \qquad N=17 \qquad (2)$$

which are the intercepts of the solution ellipsoid with the parameters axes $P_j$. Evidently, if parameter $P_j$ is allowed to vary by an amount $\partial P_j$ from its starting value, while the others are held fixed at the starting value, then the rms difference between the exact result and the model result is:

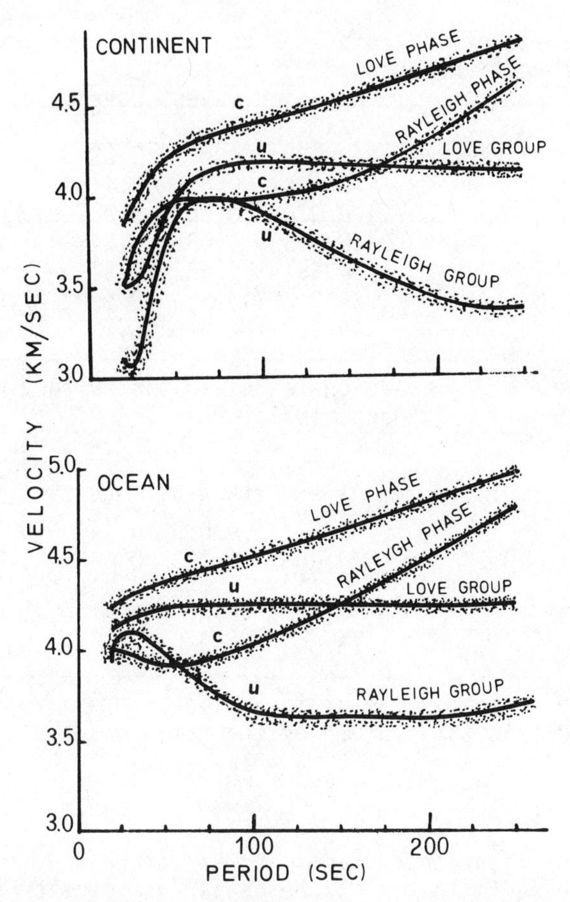


Fig. 1 Rayleigh and Love wave phase and group velocities for continental and oceanic structures, listed in Table 1, after Knopoff and Chang(1977).

$$\frac{1}{N} \sum_{i=1}^{N} \left( \frac{\partial c(T_i)}{\partial P_j} \right)^{\frac{1}{2}} \delta P_j \qquad N=17 \qquad (3)$$

which can be set equal to the preassigned value $\sigma_c$. The symbol U can be substituted for c as appropriate, without any change in the reasoning. The quantities given by (2) are the standard deviations in the parameters for the cases in which all the other parameters are kept fixed at their starting values. Thus the tabulation of the items (2) does give some rough information regarding resolution of the parameters $P_j$ by the data set, and the quantities (2) can be called the resolution, despite the fact that this definition

Table 2a. Results of inversion for continental structure (Knopoff and Chang, 1977)

| | Rayleigh waves | | Love waves | |
|---|---|---|---|---|
| | c | U | c | U |
| rms error in data(km/s) | 0.03 | 0.03" | 0.03 | 0.03" |
| $\delta\beta_{LID}$(km/sec) | 0.144 | 0.086 | 0.116 | 0.101 |
| $\delta\beta_{CH}$(km/sec) | 0.070 | 0.062 | 0.067 | 0.077 |
| $\delta\beta_{SUB}$(km/sec) | 0.373 | 0.435 | 0.541 | 1.34 |
| $\delta h_{CR}$(km) | 4.55 | 2.46 | 4.90 | 3.47 |
| $\delta h_{LID}$(km) | 30.9 | 16.5 | 39.9 | 27.8 |
| $\delta h_{CH}$(km) | 49.2 | 52.6 | 75.2 | 142.2 |

Table 2b. Results of inversion for oceanic structure (Knopoff and Chang, 1977)

| | Rayleigh waves | | Love waves | |
|---|---|---|---|---|
| | c | U | c | U |
| rms error in data(km/s) | 0.03 | 0.03" | 0.03 | 0.03" |
| $\delta\beta_{LID}$(km/sec) | 0.127 | 0.063 | 0.144 | 0.130 |
| $\delta\beta_{CH}$(km/sec) | 0.068 | 0.055 | 0.046 | 0.035 |
| $\delta\beta_{SUB}$(km/sec) | 0.113 | 0.092 | 0.116 | 0.194 |
| $\delta h_{LID}$(km) | 11.9 | 6.4 | 37.0 | 24.6 |
| $\delta h_{CH}$(km) | 25.0 | 17.2 | 21.3 | 23.1 |

"For comparison with results of inversion of phase velocity data, multiply values in this column by the ratio $\sigma_U/\sigma_c$.

is inconsistent with other usages in the literature.

From here it follows that the proper use of the results given by (2) is their application as guiding criteria in the structure parametrization to be used with the hedgehog inversion procedure (see section 5). In fact, in such a way, a complete inversion can be obtained with the advantage of minimizing the necessary computer time. In order to preserve the linearity of the inverse process the computation (2) has been performed with $\sigma_c = \sigma_U = 0.03$ km/sec, even if $\sigma_U$ is expected to be larger than $\sigma_c$ (Knopoff and Chang, 1977; Knopoff, 1977), at least at periods far from group velocity extrema. The case for larger $\sigma_U$ can be assessed by a simple multiplication on the results for $\sigma_U = 0.03$ km/sec.

The results of the inversion are given in Table 2. The values listed provide some valuable insights into the relative resolution of different kinds of dispersion data with the same period range. First it is possible to see that at the same level of error in the data, the inversion of Rayleigh wave group velocities leads to a better resolution of the continental crustal thickness and lid thickness and S-wave velocity, as compared with phase velocities.

For the oceanic model the group velocity data uniformly provide greater resolution for upper mantle structure than the phase velocity data. In the case of Love waves, the phase and group velocity data are about equivalent in resolving the structural parameters of the oceanic structure, while in the case of the continental structure phase velocities seem to have a generally higher resolving power. Considering the fact that the intrinsic level of error may in general be higher for group velocities than for phase velocities, we may conclude that in practice phase velocities are superior in resolving upper mantle parameters. A further result indicated in Table 2 is that, for the continental structure, Love and Rayleigh wave phase velocity data give about the same resolution for crustal thickness, channel velocity and lid thickness. Love wave phase velocity data give a poorer resolution of sub-channel S.-wave velocity than do Rayleigh wave phase velocities but both values of $\delta\beta_{SUB}$ are so large that it can be concluded that data set considered gives very little information regarding sub-channel velocities. Love wave phase velocity data give slightly better resolution regarding $\beta_{LID}$ and Rayleigh wave phase velocity data give better resolution of channel thickness. In the case of the oceanic structure the resolving power of the two data sets are roughly comparable, except for the greater resolving power of the Rayleigh wave data for the lid thickness. The oceanic data provide reasonable resolving power for all five parameters considered.

3.2 Experiment n.2 (Knopoff and Panza, 1977)

Limitations of band width of the World-wide Standard Seismographic Network (WWSSN) have made it difficult to record routinely surface wave signals with periods greater than 150 sec; this feature, as has been shown, limits the depth to which exploration of the S-wave structure of the upper mantle can be made to about 200 to 250 km, using fundamental mode Rayleigh waves. The recording of 4th or 5th order higher modes to periods as great as 50 sec permits sampling of the mantle to much greater depths, since the eigenfunctions of these modes at these periods are significantly nonzero down to depths as great as 1000 km. Thus the instruments of the WWSSN and similar instruments can be used in a regional exploration of possible inhomogeneities to much greater depths than hitherto. However the analysis of recordings of higher mode surface waves in the presence of a strong fundamental, requires the use of arrays of long-period seismographs, a subject which has received attention in recent years (Nolet, 1975, 1977; Nolet and Panza, 1976; Mitchel, 1977; Panza and Scalera, 1978). Here the attention is focussed on the improvement in the resolution of structural parameters to be obtained by adding phase velocity data in the higher modes. The starting continental model is given in Table 1. The same procedure described in the previous section has been applied to the phase velocities for the first six Rayleigh

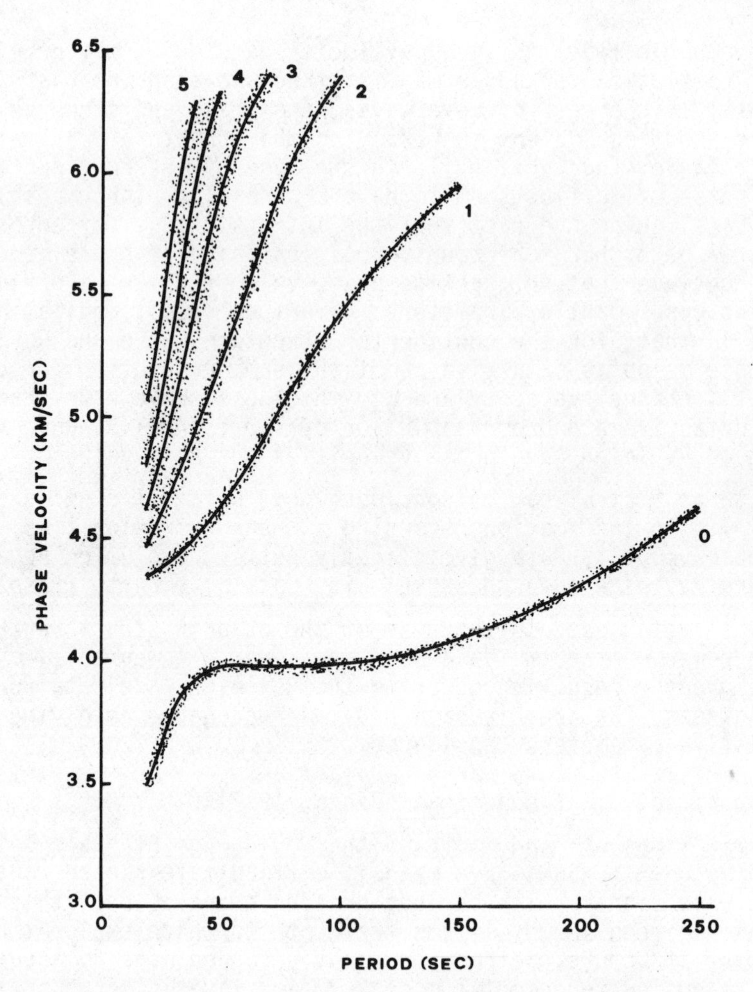


Fig. 2 First 6 Rayleigh mode phase velocities for the continental structure, listed in Table 1, after Knopoff and Panza(1977)

modes (Fig.2). Mode 0 is the fundamental mode, mode 1 is the first higher mode etc. Each of these curves has been assumed to been digitized at the periods indicated in Table 3. These period ranges are not inconsistent with those obtained in the recent applications of stacking procedures to isolate the higher modes of Rayleigh waves cited above. The values in parentheses represent the period interval between adjacent "digitized values". The number of digitized values in each mode is also indicated. To embark on the inversion, it can be assumed that the value of $\sigma$ is uniformly 0.03 km/sec for all values. As a measure of resolution the following extension of (2) to the case of different modes can be

Table 3. Digitization scheme for the data of Fig.2. N is the number of samples for each mode (Knopoff and Panza, 1977)

| MODE | PERIOD RANGE (sec) | N |
|---|---|---|
| 0 | 20 (5) 40 (10) 100 (25)250 | 17 |
| 1 | 20 (5) 40 (10) 100 (25)150 | 13 |
| 2 | 20 (5) 40 (10) 100 | 11 |
| 3 | 20 (5) 40 (10) 60 | 7 |
| 4 | 20 (5) 40 (10) 50 | 6 |
| 5 | 20 (5) 40 | 5 |

used

$$\left\{ \frac{1}{N} \sum_{i,n} \left( \frac{\partial c(T_{i,n})}{\partial P_j} \right)^2 \left[ \sigma (T_{i,n}) \right]^{-2} \right\}^{-\frac{1}{2}} \qquad n= \text{mode number} \quad (4)$$

In this experiment the ability of the data set, or parts of the data set, to resolve thirteen different structural parameters is considered. The parameters are the thickness, the S-wave velocity and the density of lid, channel, subchannel and "spinel" layers. The possibility of studying the density resolution of higher modes has been introduced since the higher modes are reputed to be more sensitive to density influences than is the fundamental mode. The results of the analysis for the resolution parameter (4) are listed in Tables 4-6. In the upper parts of each table are listed the results of the calculation for each of the modal phase velocity curves taken separately. In the lower parts of the

Table 4 Layer Thickness Resolution (km)(criterion (4)) (Knopoff and Panza,1977)

| EARTH \ MODE | 0 | 1 | 2 | 3 | 4 | 5 |
|---|---|---|---|---|---|---|
| LID | 30.9 | 138. | 59.6 | 38.2 | 40.3 | 52.0 |
| CHANNEL | 49.2 | 113. | 56.9 | 44.5 | 48.1 | 60.5 |
| SUBCHANNEL | 48.8 | 39.2 | 30.6 | 29.3 | 28.4 | 28.1 |
| "SPINEL"LAYER | 66.4 | 45.8 | 39.3 | 33.0 | 30.3 | 26.6 |
| **EARTH \ MODES** | **01** | **012** | **0123** | **01234** | **012345** | |
| LID | 38.3 | 40.8 | 40.1 | 40.1 | 40.7 | |
| CHANNEL | 54.0 | 52.5 | 50.9 | 50.4 | 50.8 | |
| SUBCHANNEL | 43.1 | 37.8 | 35.7 | 34.4 | 33.4 | |
| "SPINEL"LAYER | 54.7 | 48.9 | 44.7 | 42.1 | 39.7 | |

Table 5. Shear Wave Velocity Resolution(km/sec) (criterion(4))(Knopoff and Panza,1977)

| EARTH \ MODE | 0 | 1 | 2 | 3 | 4 | 5 |
|---|---|---|---|---|---|---|
| LID | 0.145 | 0.526 | 0.227 | 0.168 | 0.195 | 0.271 |
| CHANNEL | 0.070 | 0.037 | 0.045 | 0.055 | 0.054 | 0.057 |
| SUBCHANNEL | 0.371 | 0.211 | 0.154 | 0.136 | 0.145 | 0.158 |
| "SPINEL"LAYER | 0.329 | 0.147 | 0.118 | 0.099 | 0.085 | 0.077 |

| EARTH \ MODES | 01 | 012 | 0123 | 01234 | 012345 |
|---|---|---|---|---|---|
| LID | 0.187 | 0.195 | 0.190 | 0.190 | 0.195 |
| CHANNEL | 0.048 | 0.047 | 0.048 | 0.049 | 0.049 |
| SUBCHANNEL | 0.269 | 0.216 | 0.193 | 0.186 | 0.183 |
| "SPINEL"LAYER | 0.199 | 0.163 | 0.144 | 0.131 | 0.122 |

tables are indicated the effects of increasing the data sets by including more and more data, starting with the fundamental mode and adding successively the higher modes in order. It can be seen that the fundamental mode data do the best job in resolving the thickness and S-wave velocity of the lid, that the third higher mode is best for determining channel thickness and that the third higher mode is best for determining channel thickness and that the fifth higher mode is best for determining the subchannel and "spinel layer" thickness and the S-wave velocity in the "spinel layer". In fact, additional data in even higher modes might be more appropriate to resolve these layer thickness and velocity but the calculations have not been carried out that far, in view of present capabilities of data analysis. The S-wave velocity in the channel is best resolved by the first higher mode data and the subchannel velocity is best given by the third higher mode. In no case does the data set described gives good resolution

Table 6. Density Resolution ($g/cm^3$)(criterion(4)) (Knopoff and Panza, 1977)

| EARTH \ MODE | 0 | 1 | 2 | 3 | 4 | 5 |
|---|---|---|---|---|---|---|
| LID | 0.361 | 1.62 | 0.815 | 0.665 | 0.745 | 1.52 |
| CHANNEL | 0.399 | 0.497 | 0.798 | 0.909 | 1.03 | 1.24 |
| SUBCHANNEL | 0.885 | 0.668 | 0.648 | 0.733 | 0.736 | 1.02 |
| "SPINEL"LAYER | 0.706 | 0.636 | 0.595 | 0.777 | 1.15 | 1.59 |

| EARTH \ MODES | 01 | 012 | 0123 | 01234 | 012345 |
|---|---|---|---|---|---|
| LID | 0.471 | 0.519 | 0.537 | 0.551 | 0.573 |
| CHANNEL | 0.434 | 0.482 | 0.513 | 0.535 | 0.554 |
| SUBCHANNEL | 0.768 | 0.730 | 0.730 | 0.731 | 0.746 |
| "SPINEL"LAYER | 0.673 | 0.649 | 0.665 | 0.691 | 0.716 |

Table 7. Layer Thickness Resolution (km)(criterion(5)) (Knopoff and Panza, 1977)

| EARTH \ MODE | 0 | 1 | 2 | 3 | 4 | 5 |
|---|---|---|---|---|---|---|
| CRUST | 3.3 | | | | | |
| LID | 14.(40) | 37.(60) | 24.(25) | 19.(20) | 22.(40) | 21.(20) |
| CHANNEL | 36.(125) | 33.(20) | 31.(20) | 31.(30) | 29.(25) | 26.(20) |
| SUBCHANNEL | 26.(225) | 25.(60) | 28.(80) | 26.(50) | 24.(35) | 24.(25) |
| "SPINEL"LAYER | 24.(250) | 23.(90) | 23.(90) | 22.(100) | 25.(40) | 20.(25) |

of density in the upper mantle. Although the ability to resolve layer thickness to within 30 or 40 km may not seem useful to practitioners experienced in other areas of seismic data analysis, nevertheless, such results present an accurate, if unflatteringly realistic picture of what can be done with surface wave studies.

When higher modes are added to the data set, the ability to resolve those parameters that are well resolved with fundamental mode data under criterion (4) decreases. This is not unexpected because of the rms nature of the criterion. In most typical uses of least square criteria, the addition of irrelevant data to a data set weakens the quality of a parametric fit. Since most modern linear inversions are dependent on one form or another of a least squares fit to the data, this result suggests that perhaps a more careful selection of the data would lead to a better resolution of model parameters than simply an indiscriminate rush to acquire more and more data in the hope that this would improve resolution. Under a least squares fit this criterion is not assured.

Consider now the problem of searching for the most suitable data set for the determination of any given model parameter. If the premise is correct, namely that the inclusion of too much data is deleterious to the program of optimizing resolution, then the problem of finding the proper subset of data can be approached as follows. Starting with a large data set, which from Tables 4 to 6, seemingly does not do as good a job as a smaller data set, it is possible to reject those data which make the largest

Table 8. Shear Wave Velocity Resolution (km/sec) (criterion (5)) (Knopoff and Panza,1977)

| EARTH \ MODE | 0 | 1 | 2 | 3 | 4 | 5 |
|---|---|---|---|---|---|---|
| LID | 0.08(40) | 0.35(60) | 0.20(30) | 0.11(20) | 0.14(40) | 0.22(25) |
| CHANNEL | 0.05(150) | 0.03(20) | 0.04(20) | 0.04(30) | 0.04(25) | 0.04(20) |
| SUBCHANNEL | 0.17(250) | 0.12(70) | 0.11(40) | 0.10(60) | 0.11(25) | 0.10(20) |
| "SPINEL"LAYER | 0.12(250) | 0.08(100) | 0.07(60) | 0.07(40) | 0.60(35) | 0.06(25) |

contributions to the estimate of (4). For example, from Table 4, if the intention is to provide the data set that optimizes the resolution of the thickness of the lid in the upper mantle structure of Table 1, all the higher mode data should be discarded and only the fundamental mode data, should be retained, since the value of 30.9 km is less than 38.3, etc. for <u>all</u> the cases of higher modes combined with the fundamental. <u>But</u> if it is valid to reject the data from all the modal branches but one, why should not the irrelevant data from within a given modal branch be rejected except the most relevant datum? This logic, suggests that maximum resolution for a given model parameter is achieved by retaining only one datum, specifically that for which

$$\left(\frac{\partial c(T_{i,n})}{\partial P_j}\right)^{-1} \sigma(T_{i,n}) \qquad (5)$$

is a minimum. In Table 7 to 9 is listed the resolution to be expected under criterion (5) for each of the modal data set branches taken separately, together with the period, in parentheses, at which the extremum occurs. Thus, for example, if <u>only</u> the datum of fundamental mode Rayleigh wave phase velocity at <u>a</u> period of 40 sec were measured and used in the inversion, the thickness of the lid could be better determined than with any other datum or combination of data, under this assumption of the values of $\sigma$.

A casual assessment of the above discussion would lead the reader to the tantalizing, but erroneous conclusion that the determination of optimum values for, let us say, N mantle parameters (and their uncertainties) is best accomplished by taking only N measurements at certain selected mode-frequency pairs and ignoring all other measurements. The fallacy in this statement arises because the curves of partial derivatives of phase velocity are not delta functions. Another way of stating this is that the conclusions represented by criterion (5) are valid if all other parameters are fixed, i.e. known, during the exploration for the optimum value of the jth parameter. But since none of the N parameters is known, it must be taken into account that the partial derivative curves often have large wings remote from the peaks. In this case, the variation of values of phase velocities with respect to one parameter in the

Table 9. Density Resolution (g/cm$^3$) (criterion (5)) (Knopoff and Panza, 1977)

| EARTH \ MODE | 0 | 1 | 2 | 3 | 4 | 5 |
|---|---|---|---|---|---|---|
| LID | 0.26(225) | 0.88(50) | 0.48(40) | 0.41(50) | 0.52(40) | 0.99(40) |
| CHANNEL | 0.26(100) | 0.25(150) | 0.47(35) | 0.40(45) | 0.51(25) | 0.59(30) |
| SUBCHANNEL | 0.41(225) | 0.31(100) | 0.36(80) | 0.42(50) | 0.58(25) | 0.60(25) |
| "SPINEL"LAYER | 0.26(250) | 0.28(150) | 0.28(70) | 0.49(60) | 0.58(35) | 0.86(25) |

neighborhood of, let us say a peak in the partial derivative curve, influences the value of the variation in phase velocity in the neighborhood of a peak of a partial derivative with respect to another parameter. The worst possible case of two coupled parameters occurs when they both have a peak at the same period-mode number; a case in point arises in the effort to determine both the velocity and thickness of the lid; the partial derivative curves for the two parameters both have their peaks at 40 sec for the fundamental mode. Thus these two parameters are not independent under criterion (5) and the resolution of each is certainly poorer than the values quoted in Tables 7 and 8, although they are probably better than the values indicated in Tables 4 and 5. The full problem requires the determination of the period-mode pairs for which the quantities $\delta P_j$ are minima subject to the condition:

$$\sum_j \frac{\partial c(T_{i,n})}{\partial P_j} \delta P_j = \sigma(T_{i,n}) \qquad (6)$$

This is a linear programming problem of some complexity. It can be shown that often less than N solutions to the problem exist in the case of strongly correlated model parameters.

It is possible to conclude that, in an absolute sense, densities cannot be resolved to better than $\pm$ 0.25 or $\pm$ 0.3 gm/cm$^3$ with the use of higher mode data up to the 5th order, even with the assumption that densities are uncorrelated model parameters. However, to determine the properties of the lid, fundamental mode data probably suffice to give resolution to no better than 15 km in thickness and 0.08 km/sec in velocity; the actual case is worse than this since the variables are correlated. In the case of the lid, addition of higher mode data probably does not represent an improvement. These estimates, and in general those of Tables 7-9, probably represent optimal bounds on the resolution; the actual values depend on the correlation of the model parameters used in the inversion.

### 3.3 The case of apparent resistivity curves

Resistivity measurements indicate the true rock resistivity only when they are carried over homogeneous and isotropic areas. In field work this condition is generally not satisfied and the simplest approximation which can be made is to assume the presence of a multilayered medium formed by homogeneous and isotropic layers. In field work an apparent resistivity can therefore be measured which depends upon the thickness and resistivity of the individual layers. Thus, formally, the problem of inversion of apparent resistivity curves is analogous to the problem of inversion of dispersion data. To invert these data a set of combinations of the unknown parameters (i.e. a volume in the space of unknown parameters) has to be found for which the discrepancy between observations and

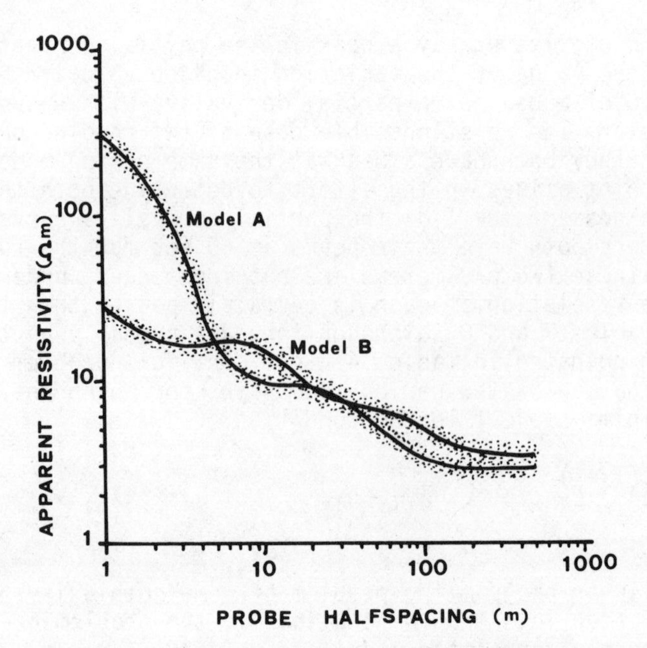


Fig. 3 Apparent resistivity curves computed for the models listed in Table 10, after Calcagnile et al.(1979)

computed properties of the apparent resistivity curves is sufficiently small. An exact inversion of apparent resistivity curves through methods like hedgehog is rather expensive. For this reason here are reported two examples of the application of an approximate inversion procedure which is substantially based on the method proposed by Knpoff and Chang (1977) and Knopoff and Panza (1977), for surface waves, and extended to the case of apparent resistivity measurements by Calcagnile et al.(1979a). The two models reported in Table 10 have been considered. For these models the apparent resistivity curves $\varrho_a$ have been computed using standard techniques. The results of computations are reported in Fig.3, where the shaded area corresponds to 0.05 $\varrho_a(\Omega.m)$. The two curves have been digitized with a sampling interval $\triangle x$ = ln 10/10. In such a way 27 values of apparent resistivity have been obtained for each curve. Assuming as resolution criteria the quantity:

$$\frac{1}{N}\left[\sum_{1=j}^{N}\left(\frac{\partial\varrho_{aj}}{\partial P_i}\right)^2 (\sigma_j)^{-2}\right]^{-\frac{1}{2}} \qquad N=27 \qquad (4')$$

where $P_i$ is the i-th parameter, $\sigma_j = k\ \varrho_{aj}$, k is a percentage error assumed constant, Table 11 and 12 can be constructed.

Table 10. Starting models used in the computation (Calcagnile et al., 1979a)

| Model A | | | Model B | | |
|---|---|---|---|---|---|
| Layer N. | h (m) | $\varrho$ ($\Omega$.m) | Layer N. | h (m) | $\varrho$ ($\Omega$.m) |
| 1 | 0.50 | 39. | 1 | 1.00 | 390. |
| 2 | 1.45 | 13. | 2 | 18.0 | 10.5 |
| 3 | 3.95 | 24.9 | 3 | $\infty$ | 3.70 |
| 4 | 9.15 | 4.80 | | | |
| 5 | 9.15 | 17.3 | | | |
| 6 | $\infty$ | 3.45 | | | |

From a comparison of these two tables it is evident that the same number of experimental data allows a more precise determination of the resistivity layering when simpler models are used. This is a rather obvious conclusion but the application of criterion (4') allows to establish quantitatively the level of detail contained in a given data set. For instance it may be deduced that in model A the presence of two low resistivity layers (layers n.2 and n.4) is required but that their thickness cannot be determined better than within a 40% uncertainty. Now, considering the apparent resistivity value for which

$$( \partial \varrho_{aj} / \partial P_i)^{-1} \sigma_j \qquad (5')$$

is minimum, Tables 13 and 14 can be constructed. In the Tables the entries in parenteses indicate the value of the electrode half spacing, s, in meters, for which $( \partial \varrho_{aj} / \partial P_i)^{-1}$ reaches the minimum value. The results of Table 13 and 14 indicate the more optimistic resolution which can be obtained with the available data set. In fact the conclusions which can be deduced under criterion (5') are valid only when one parameter is inverted keeping all the others fixed. In other words if all parameters are known except one then criterion (5') is valid. When more parameters are

Table 11. Layers thickness resolution (m)(criterion (4') ) (Calcagnile et al.,1979a)

| Layer N. | Model A | Model B |
|---|---|---|
| 1 | 0.10 | 0.05 |
| 2 | 0.60 | 2.7 |
| 3 | 0.75 | ---- |
| 4 | 4.0 | ---- |
| 5 | 2.8 | ---- |

Table 12. Layers resistivity resolution ( $\Omega$.m)(criterion (4') ) (Calcagnile et al.,1979a)

| Layer N. | Model A | Model B |
|---|---|---|
| 1 | 9.5 | 41.0 |
| 2 | 2.1 | 1.0 |
| 3 | 3.6 | 0.2 |
| 4 | 1.0 | ---- |
| 5 | 4.5 | ---- |
| 6 | 0.3 | ---- |

Table 13. Layers thickness resolution (m)(criterion (5') ) (Calcagnile et al.,1979a)

| Layer N. | Model A | Model B |
|---|---|---|
| 1 | 0.05(1.6) | 0.02(5.0) |
| 2 | 0.35(5.0) | 1.40(63.1) |
| 3 | 0.37(20.0) | ---------- |
| 4 | 1.9 (39.8) | ---------- |
| 5 | 1.5 (100) | ---------- |

variable at the same time, criterion (5') becomes more and more inadequate as the coupling between different parameters increases. A typical example of severe coupling is given by layer n.4 of model A, for which the best resolution of thickness and resistivity corresponds to a half spacing s=39.8 m. Thus these two parameters are not independent under criterion (5') and their resolution is certaily worse than the values reported in Tables 13 and 14 but probably better than the values reported in Table 11 and 12. An experienced operator is usually capable of estimating some of the resistivities characterizing the investigated area. In these cases it is possible to fix some resistivity values, thus removing the thickness-resi-

Table 14. Layers resistivity resolution ( $\Omega$.m)(criterion (5') ) (Calcagnile et al., 1979a)

| Layer N. | Model A | Model B |
|---|---|---|
| 1 | 2.7 (1.3) | 19(1.3) |
| 2 | 0.9 (3.2) | 0.50(15.9) |
| 3 | 1.7 (15.9) | 0.13(501) |
| 4 | 0.40(39.8) | ---------- |
| 5 | 0.21(79.4) | ---------- |
| 6 | 0.16(501) | ---------- |

stivity coupling and then the use of criterion (5') turns out to be appropriate. On the other side criterion (4') can be used, for practical purposes, as an upper limit for the estimate of the level of uncertainty at which a model can be resolved.

## 4. INVERSION OF CRUSTAL PARAMETERS

It has been shown that surface waves in the period range 20-250s can be used to resolve the average crustal thickness to within 3-5 km once the distribution of elastic properties is assumed to be known. In this section an analysis is presented of the detail which can be inferred about crustal structures using phase and group velocities of Rayleigh waves in the period range 12-80 sec. To simulate the inversion phase and group velocities (see Fig.4) for the structure given in Table 15 have been computed at the periods 12(2) 24(4) 32,40,50,80 sec. The values in parentheses represent the period interval between adjacent "digitized" values. The structure given in Table 15 may be considered representative of the Apennine region (Calcagnile and Panza,1979). The inversion experiments have been carried out using the hedgehog procedure, considering variable not only the layer thicknesses and S-wave velocities but also density and P-wave velocity, on account of recent results which indicate that group velocities are also rather sensitive to the last two parameters, mainly in the layers near the Earth surface (Cloeting et al.,1979,1980).

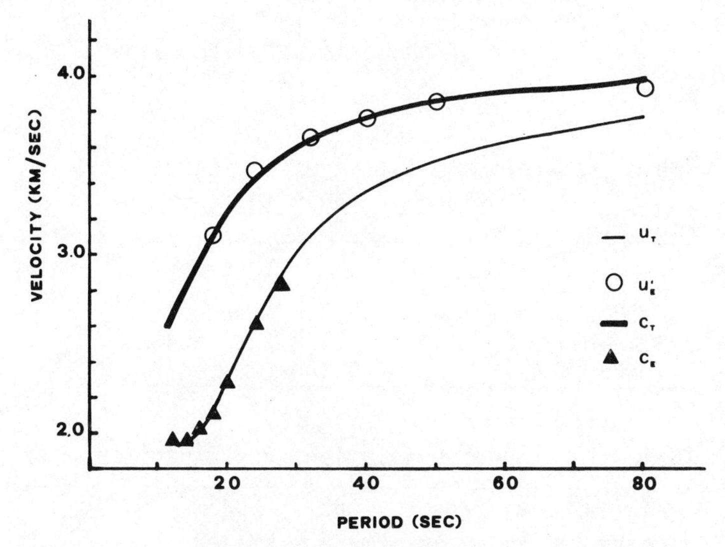


Fig.4 Experimental phase (open circles) and group (full triangles) velocities determined for the Apennines,(Calcagnile and Panza, 1979; Calcagnile et al.,1980). The solid lines give the dispersion relations corresponding to the structure given in Table 15.

Table 15. Average structure for the Apennines used in sections 4 and 5

| Depth (km) | Thickness (km) | $\beta$ (km/sec) | $\alpha$ (km/sec) | $V_P/V_S$ | $\varrho$ (g/cm$^3$) | Layer number |
|---|---|---|---|---|---|---|
| | 1 | 2.00 | 3.8 | 1.90 | 2.2 | 1 |
| 1 | 2 | 2.15 | 3.9 | 1.81 | 2.2 | 2 |
| 3 | 2 | 2.30 | 4.0 | 1.74 | 2.3 | 3 |
| 5 | 9 | 2.60 | 4.5 | 1.73 | 2.6 | 4 |
| 14 | 9 | 3.50 | 6.1 | 1.74 | 2.7 | 5 |
| 23 | 13 | 4.00 | 7.0 | 1.75 | 2.9 | 6 |
| 36 | - - - - | - - - - | - - - - | - - - - | Moho | |
| | 45 | 4.40 | 8.0 | 1.82 | 3.4 | 7 |
| 81 | 290 | 4.50 | 8.2 | 1.82 | 3.6 | 8 |
| 371 | ∞ | 4.90 | 9.0 | 1.84 | 3.7 | 9 |

4.1 Resolution of S-wave velocity distribution.

Experiment N.3. Six crustal parameters are variable, namely the thickness and S-wave velocity of layers n.4,5 and 6 (see Table 15). In fact the number of parameters which vary independently is 5 since it has been chosen to define the thickness of layers n.4 and n.5 through a single parameter. The parameter ranges and steps are given in Table 16. A model is acceptable if it produces dispersion data (phase and group velocity) for which the rms difference, $\sigma$ , and the single point difference, $\varepsilon$ , with respect to the dispersion data of the starting model is less than 0.03 km/sec.

The results of inversion, given in Table 17, clearly indicate the high resolving power of the data set with respect to all parameters

Table 16. Parametrization used with the structure of Table 15

| Parameter Range | Step | |
|---|---|---|
| $3 \leq P1 \leq 15$ | 2 | (km) |
| $5 \leq P2 \leq 21$ | 2 | (km) |
| $2.1 \leq P3 \leq 3.2$ | 0.1 | (km/sec) |
| $3.3 \leq P4 \leq 3.9$ | 0.1 | (km/sec) |
| $3.1 \leq P5 \leq 4.4$ | 0.1 | (km/sec) |

Table 17. Hedgehog solutions for Experiment n.3

| Solution number | P1 km | P2 km | P3 km/sec | P4 km/sec | P5 km/sec |
|---|---|---|---|---|---|
| 1 | 9 | 13 | 2.6 | 3.5 | 4.0 |
| 2 | 9 | 11 | 2.6 | 3.5 | 3.9 |
| 3 | 9 | 11 | 2.6 | 3.6 | 3.9 |

Experiment N.4. Let us consider the same case as before but with $\varepsilon_C = \sigma_C = 0.03$ km/sec and $\varepsilon_U = \sigma_U = 0.06$ km/sec. This experiment is intended to simulate a more realistic case since, for various reasons, two station phase velocities are routinely determined with higher accuracy than one station group velocities. The results of this experiment are contained in Fig.5 and correspond to the value of the S-wave velocity in the lid ($\beta_{LID}$) equal to 4.4 km/sec.

A comparison of these results with the ones of Experiment N.3 indicates clearly that, at the same level of error, group velocities are more resolving with respect to crustal parameters than phase velocities, in agreement with the results obtained in the previous section for the upper mantle. Neverthless it must be mentioned that the thicknesses of layers n.4 and n.5 (see Table 15) and the S-wave velocity in layer n.4 are resolved with the same accuracy as in Experiment N.3. Another relevant consideration to be made is the possibility of reducing the span of possible values for $\beta_5$ and $\beta_6$, if some constraints are placed on the $V_P/V_S$ ratio.

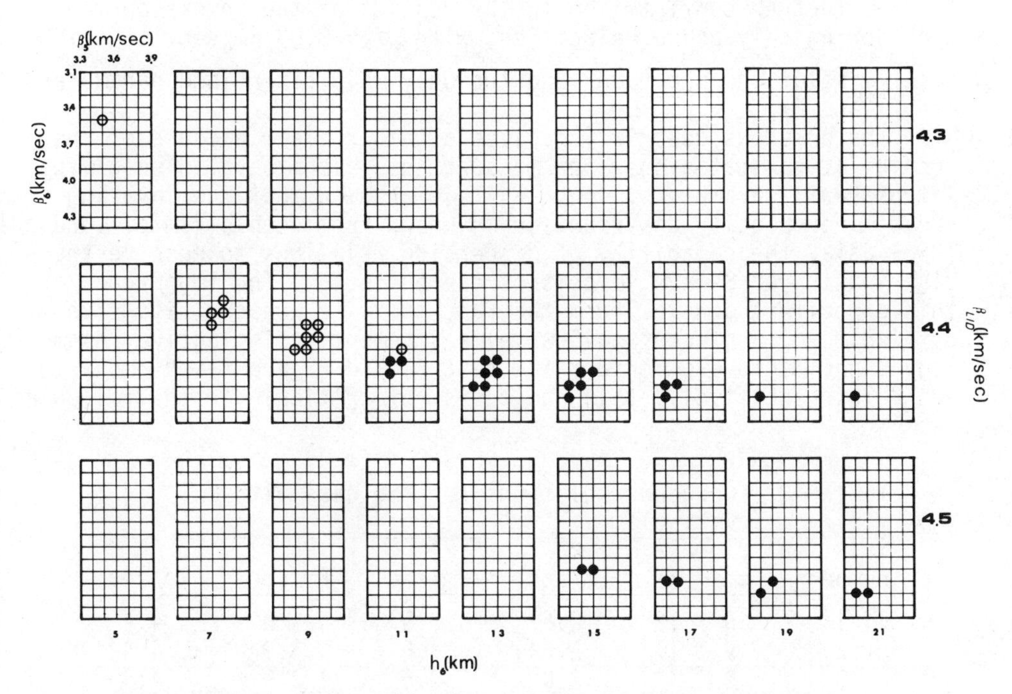


Fig.5 Tic-tack-toe plot for Experiment N.4. Open circles correspond to solutions not satisfying the condition $V_P/V_S=\sqrt{3} \pm 0.05\sqrt{3}$ .

If all solutions for which the $V_P/V_S$ ratio differs from the standard $\sqrt{3}$ value by more than 5% are rejected, the only possible solutions are represented by full dots in Fig.5. In this case the S-wave velocity distribution in layers n.5 and n.6 is rather well resolved since $3.9 \leq \beta_6 \leq 4.2$ km/sec and $3.4 \leq \beta_5 \leq 3.6$ km/sec, while the thickness of layer n.6 ($h_6$) is still affected by an uncertainty of about 10 km. In Fig.5 are reported also solutions for $\beta_{LID}$ different from 4.4 km/sec, which indicate that, with the available data set, it is possible to resolve the S-wave velocity in the lid within about 0.2 km/sec. The span of $\beta_{LID}$ values which has been tested goes from 4.2 km/sec to 4.6 km/sec.

It is interesting to note that a decrease of $\beta_{LID}$ implies a thinning of the crust and, vice-versa, a fast lid ( $\beta_{LID}$=4.5 km/sec) implies a larger crustal thickness. The 39 solutions reported in Fig. 5 are obtained for a $\sigma_c$ = 0.03 km/sec and $\sigma_U$ = 0.06 km/sec level of error. For a level of error $\sigma_c = \sigma_U$=0.03 km/sec only the 3 solutions, listed in Table 18, are obtained. This again indicates that, at the same level of error, group velocities have a resolving power greater than that of phase velocities. This fact finds a further confirmation in the results of the inversion made considering only phase velocities, with $\sigma_c$= 0.03 km/sec. In this case 71 solutions are found and the only well determined parameters are P1 and P3.

The previous description gives a rather clear picture of the possibilities that dispersion measurements alone have in resolving lithospheric parameters. Now, if it is assumed that the layering and P-wave velocity distribution is known from deep seismic soundings (DSS) data, the capability of dispersion relations to resolve the distribution of S-wave velocities can be tested. The importance of this experiment derives from the fact that, at present, it is very difficult to have accurate S-wave measurements using active seismology experiments. To perform the test P1 and P2 are fixed equal to 9 and 13 respectively (this choice is not critical for the following

Table 18. Hedgehog solutions for the case with $\beta_{LID}$ variable

| Solution number | P1 km | P2 km | P3 km/sec | P4 km/sec | P5 km/sec | P6 " km/sec |
|---|---|---|---|---|---|---|
| 1 | 9 | 13 | 2.6 | 3.5 | 4.0 | 4.4 |
| 2 | 9 | 11 | 2.6 | 3.5 | 3.9 | 4.4 |
| 3 | 9 | 11 | 2.6 | 3.6 | 3.9 | 4.4 |

" P6 = $\beta_{LID}$

conclusions and is suggested by the fact that 9 and 13 are the more popular values obtained for the two parameters in the previous experiments), while $\beta_4$, $\beta_5$, $\beta_6$ and $\beta_{LID}$ are assumed variable.

The tests have been made for two cases: a) both phase and group velocities have been inverted at a level of error $\sigma_c = 0.03$ km/sec; $\sigma_U = 0.06$ km/sec; b) only phase velocities have inverted with $\sigma_c = 0.03$ km/sec. The results of the inversions are listed in Table 19. The results contained in the Table clearly indicate that the sub-Moho velocity ($\beta_{LID}$) can be determined with a rather good accuracy using indifferently phase or group velocity data, while crustal parameters are, once more, better resolved by group velocities even if the level of error considered for group velocities is twice that associated to phase velocities. These results indicate the good level of knowledge which may be obtained about the S-wave velocity distribution when combining DSS data and dispersion relations. In such a way, the resolving power of the data set considered with respect to a preassigned set of model parameters has been covered in some detail and the following conclusions can be made.

1) Assuming known the layering of the first 5 km and the velocity below 80 km, and subdividing the remaining earth lithosphere into 4 layers, 3 for the crust and 1 for the lid, in the case $\varepsilon_c = \sigma_c = \varepsilon_U = \sigma_U = 0.03$ km/sec it is possible to resolve: a) crustal thickness within 2 km, that is within about 6% of its average value, b) S-wave velocity in the crustal layers within: less than 0.1 km/sec for layer n.4, about 0.1 km/sec for layer n.5, about 0.1 km/sec for layer n.6, c) S-wave velocity in the lid within: less than 0.1 km/sec. Furthermore, if the layering is known in all crust, then all S-wave velocities are determined within less than 0.1 km/sec.

Table 19. Resolution of S-wave velocity distribution in the lithosphere

| Experiment(a) | | | | Experiment(b) | | | |
|---|---|---|---|---|---|---|---|
| $\beta_4$ | $\beta_5$ | $\beta_6$ | $\beta_{LID}$ | $\beta_4$ | $\beta_5$ | $\beta_6$ | $\beta_{LID}$ |
| km/sec | | | | km/sec | | | |
| 2.6 | 3.5 | 4.0 | 4.4 | 2.6 | 3.5 | 4.0 | 4.4 |
| 2.6 | 3.6 | 4.0 | 4.4 | 2.6 | 3.6 | 4.0 | 4.4 |
| 2.6 | 3.5 | 3.9 | 4.4 | 2.6 | 3.5 | 3.9 | 4.4 |
| 2.6 | 3.5 | 4.1 | 4.4 | 2.6 | 3.5 | 4.1 | 4.4 |
| 2.6 | 3.4 | 4.1 | 4.4 | 2.6 | 3.4 | 4.1 | 4.4 |
| 2.6 | 3.6 | 3.9 | 4.4 | 2.6 | 3.6 | 3.9 | 4.4 |
| | | | | 2.5 | 3.7 | 3.9 | 4.4 |
| | | | | 2.7 | 3.3 | 4.1 | 4.4 |
| | | | | 2.7 | 3.3 | 4.2 | 4.4 |

2) With the same assumptions as before, but in the case $\varepsilon_C = \sigma_C = 0.03$ km/sec $\varepsilon_U = \sigma_U = 0.06$ km/sec it is possible to resolve a) crustal thickness within about 14 km, that is within about 40% of its average value. b) S-wave velocity in the crustal layers within: less than 0.1 km/sec for layer n.4, about 0.3 km/sec for layer n.5, about 0.8 km/sec for layer n.6, c) S-wave velocity in the lid within: about 0.2 km/sec.

3) A considerable reduction in the uncertainty can reasonably be achieved by imposing that the $V_P/V_S$ ratio varies within 5% of the standard $\sqrt{3}$ value. In this case it is possible to resolve: a) crustal thickness within about 10 km, that is within about 27% of its average value, b) S-wave velocity in the crustal layers within: less than 0.1 km/sec for layer n.4, about 0.2 km/sec for layer n.5, about 0.3 km/sec for layer n.6, c) S-wave velocity in the lid within: about 0.1 km/sec.

If the crustal layering is assumed to be known, for a total thickness of 36 km, then the resolution with respect to S-wave velocity in the different layers is less than 0.1 km/sec for layer n.4, about 0.2 km/sec for layer n.5, about 0.2 km/sec for layer n.6 less than 0.1 km/sec for the lid.

From these results we can see how powerful is the study of group and phase velocities in resolving lithospheric structures once combined with crustal data obtained from DSS. Experiments made to determine the resolving power of phase velocities versus group velocities clearly indicate that, if the errors allowed are the same, group velocities are more resolving than phase velocities, at least for crustal parameters.The parametrization used can be criticized since the variation of the thickness of layers n.4 and n.5 is made through the use of a single parameter. To avoid this limitation the following experiment was made.

4) The thickness of layer n.4 was fixed at 9 km, that is the first 14 km of crust have been assumed known. The results of this experiment, made using only group velocities with $\varepsilon_U = \sigma_U = 0.03$ km/sec, can be summarized as follows: a) crustal thickness can be resolved within 10 km, that is within about 27% of its average value, b) S-wave velocity in the crustal layers can be resolved within: less than 0.1 km/sec for layer n.4, about 0.3 km/sec for layer n.5, about 0.7 km/sec for layer n.6. No test was made on lid velocity. The uncertainty in the S-wave velocity of layer n.6 is rather high and can be reduced only by considering the $V_P/V_S$ ratio or assuming known the crustal layering.

5) Constraining the $V_P/V_S$ to vary within 1.65-1.82 the resolution can be summarized as follows: a) crustal thickness within about 8 km, that is about 22% of its average value, b) S-wave velocity in the crustal layers within: less than 0.1 km/sec for layer n.4, about 0.2 km/sec for layer n.5, about 0.3 km/sec for layer n.6.

6) Imposing a total crustal thickness of 36 km the resolution of S-wave velocity in the crustal layers is within: less than 0.1

km/sec for layer n.4, about 0.2 km/sec for layer n.5, about 0.4 km/sec for layer n.6.

As a concluding remark it may be mentioned that the CPU time necessary to perform experiment n.4, with $\beta_{LID}$ = 4.4 km/sec, was slightly less than 14 minutes, using an IBM 370/58. Porter et al. (1979) have estimated relative computer speeds for surface-wave dispersion computation. From their results it follws that, for instance, the same computation could be carried on a CDC 6600 only within about 3 minutes. These values clearly indicate that inversion computations can be made in a relatively fast way also using a non linear approach. Furthermore the results of the tests described in this section can be used to optimize the efficiency of the computations since they can be used as a powerful guide in the choice of the a priori parametrization, which can be realistically resolved with the available data.

## 4.2 Resolution of P-wave velocity distribution

Use is made of the structural model reported in Table 15, and the variation allowed is for the layering and for the P-wave velocity in layers n.4,5 and 6 ($\alpha_4, \alpha_5, \alpha_6$). The P-wave increments are $\Delta\alpha_4 = \Delta\alpha_5 = \Delta\alpha_6$ = 0.2 km/sec, that is twice the ones used to increment $\beta$ in the previous experiments. The results of this test for a level of error $\sigma_C = \sigma_U$ = 0.03 km/sec are reported in Fig.6. The figure clearly illustrates the poor resolving power of the data set with respect to P-wave velocities. The better resolution is limited to the upper crustal layers, but the uncertainty is more than twice than in the case of S-waves. It is obviously meaningless to perform experiments with a higher level of error.

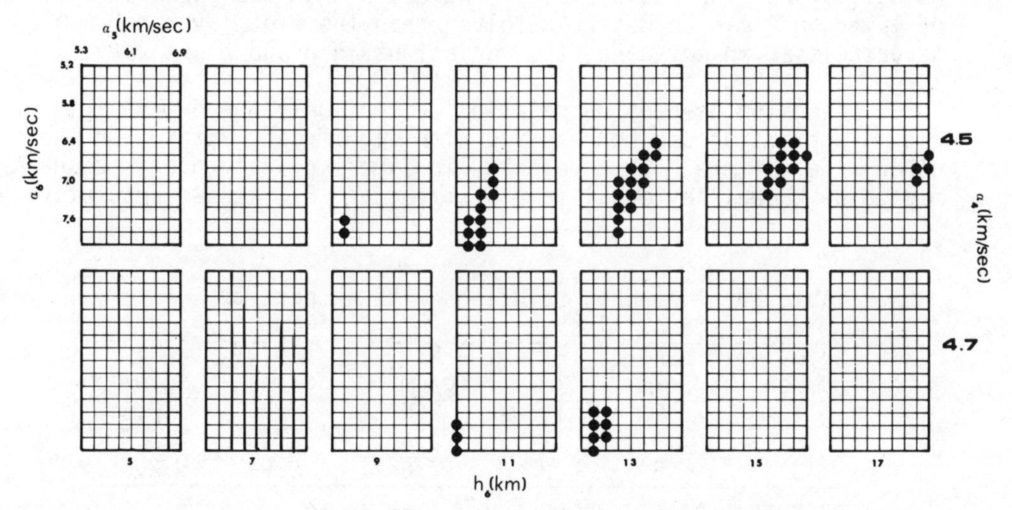


Fig.6 Tic-tack-toe plot for the inversion of P-wave velocity data.

### 4.3 Resolution of density distribution

Usually, in the inversion of surface waves, little attention is paid to P-wave velocity and density distribution. It has been shown that P-waves have little influence on dispersion properties. The situation is rather different when the variations of density are considered. Using the same inversion scheme reported in Table 15 but now allowing variations in the density of layers n.4,5 and 6 ( $\varrho_4$, $\varrho_5$, $\varrho_6$, ), with increments $\Delta\varrho_4 = \Delta\varrho_5 = \Delta\varrho_6 = 0.2$ g/cm$^3$, at a level of error $\sigma_c = \sigma_U =$ 0.03 km/sec the three solutions listed in Table 20 are found. As can be see from the table the largest resolving power is limited to the upper part of the crust, as was already suggested by Cloeting et al.(1979), on the basis of a linear inversion approach. It is worth mentioning that solution n.3 in Table 20 gives for $\varrho_6$ a value of 2.5 g/cm$^3$ which is rather difficult to explain by physical arguments. If this kind of argument is accepted, then the resolution of densities is of the order of 0.2 g/cm$^3$ over the whole crust. An inversion experiment made considering only phase velocities showed that also for densities, at the same level of error, group velocities are much more resolving than phase velocities. Actually phase velocities turn out to be useless for density distribution determination.

### 4.4 Simultaneous inversion of $\varrho$, $\alpha$ and $\beta$ in two crustal layers

Starting from the structure given in Table 15, the thicknesses of all layers have been kept fixes to their starting values and variations have been allowed in $\varrho$, $\alpha$ and $\beta$ in layer n.4 and n.5, in the ranges given in Table 21. The results of inversion are given in table 22 for a level of error $\sigma_c = \sigma_U = 0.03$ km/sec. An analysis of the table clearly indicates that the parameters of layers n.4 can be better resolved than those of layer n.5, nevertheless, in any case, the variations in $\varrho$ and $\alpha$ are very large.

Thus using dispersion relations alone, only the S-wave velocity distribution can be resolved with significant detail. On the other side, if the P-wave velocity in layers n.4 and n.5 is assumed to be known, say equal to 4.5 km/sec and 6.1 km/sec respectively,

Table 20. Hedgehog solutions with respect to densities

| Solution number | P1 km | P2 km | P3 g/cm$^3$ | P4 g/cm$^3$ | P5 g/cm$^3$ |
|---|---|---|---|---|---|
| 1 | 9 | 13 | 2.6 | 2.7 | 2.9 |
| 2 | 9 | 11 | 2.6 | 2.9 | 2.7 |
| 3 | 9 | 9 | 2.6 | 2.9 | 2.5 |

P3 = $\varrho_4$; P4 = $\varrho_5$; P5 = $\varrho_6$ (see text)

Table 21. Parameters used in hedgehog inversion of density P-and S-wave velocity in two crustal layers

| | | |
|---|---|---|
| $2.0 \leq \varrho_4 \leq 2.9$ | $\Delta \varrho_4 = 0.3$ | (g/cm3) |
| $2.1 \leq \varrho_5 \leq 3.0$ | $\Delta \varrho_5 = 0.3$ | (g/cm$^3$) |
| $3.9 \leq \alpha_4 \leq 5.1$ | $\Delta \alpha_4 = 0.3$ | (km/sec) |
| $5.5 \leq \alpha_5 \leq 6.9$ | $\Delta \alpha_5 = 0.3$ | (km/sec) |
| $2.1 \leq \beta_4 \leq 3.2$ | $\Delta \beta_4 = 0.1$ | (km/sec) |
| $3.3 \leq \beta_5 \leq 3.9$ | $\Delta \beta_5 = 0.1$ | (km/sec) |

Table 22. Experiment with $\varrho, \alpha, \beta$ , variable in two layers of given thickness

| $\varrho_4$ (g/cm$^3$) | $\varrho_5$ (g/cm$^3$) | $\alpha_4$ (km/sec) | $\alpha_5$ (km/sec) | $\beta_4$ (km/sec) | $\beta_5$ (km/sec) |
|---|---|---|---|---|---|
| 2.6 | 2.7 | 4.5 | 6.1 | 2.6 | 3.5 |
| 2.6 | 3.0 | 4.5 | 6.1 | 2.6 | 3.4 |
| 2.3 | 2.7 | 4.5 | 5.8 | 2.7 | 3.4 |
| 2.6 | 3.0 | 4.5 | 6.4 | 2.6 | 3.4 |
| 2.6 | 2.7 | 4.8 | 5.8 | 2.6 | 3.4 |
| 2.3 | 2.4 | 4.5 | 5.8 | 2.7 | 3.6 |
| 2.3 | 3.0 | 4.5 | 6.1 | 2.7 | 3.3 |
| 2.6 | 3.0 | 4.8 | 5.8 | 2.6 | 3.3 |
| 2.0 | 2.7 | 4.5 | 5.5 | 2.8 | 3.4 |
| 2.6 | 2.7 | 4.8 | 5.5 | 2.6 | 3.4 |
| 2.0 | 2.4 | 4.5 | 5.5 | 2.8 | 3.5 |
| 2.3 | 3.0 | 4.2 | 6.7 | 2.7 | 3.5 |
| 2.6 | 2.4 | 4.8 | 5.5 | 2.6 | 3.6 |
| 2.6 | 2.4 | 4.5 | 6.1 | 2.6 | 3.7 |
| 2.0 | 2.1 | 4.5 | 5.5 | 2.8 | 3.7 |
| 2.0 | 3.0 | 4.2 | 6.4 | 2.8 | 3.5 |
| 2.3 | 2.1 | 4.5 | 5.8 | 2.7 | 3.8 |
| 2.3 | 2.4 | 4.2 | 6.4 | 2.7 | 3.8 |
| 2.6 | 2.1 | 4.8 | 5.5 | 2.6 | 3.8 |
| 2.6 | 2.1 | 4.5 | 6.1 | 2.6 | 3.9 |
| 2.6 | 2.4 | 4.2 | 6.7 | 2.6 | 3.8 |
| 2.6 | 2.4 | 4.2 | 6.7 | 2.6 | 3.9 |
| 2.6 | 2.7 | 4.2 | 6.7 | 2.6 | 3.7 |

| Allowed initial variations | Variations after inversion |
|---|---|
| $2.0 \leq \varrho_4 \leq 2.9$ | $2.0 \leq \varrho_4 \leq 2.6$ (g/mc$^3$) |
| $2.1 \leq \varrho_5 \leq 3.0$ | $2.1 \leq \varrho_5 \leq 3.0$ (g/cm$^3$) |
| $3.9 \leq \alpha_4 \leq 5.1$ | $4.2 \leq \alpha_4 \leq 4.8$ (km/sec) |
| $5.5 \leq \alpha_5 \leq 6.9$ | $5.5 \leq \alpha_5 \leq 6.7$ (km/sec) |
| $2.1 \leq \beta_4 \leq 3.2$ | $2.6 \leq \beta_4 \leq 2.8$ (km/sec) |
| $3.3 \leq \beta_5 \leq 3.9$ | $3.3 \leq \beta_5 \leq 3.9$ (km/sec) |

than the number of possible solutions reduces to five. Nevertheless the uncertainty in $\varrho_5$ and $\beta_5$ remains rather large. It is interesting to note that low density values correspond to high S-wave velocities and vice-versa. This compensating effect clearly indicates that the two parameters are highly correlated.

The main conclusion of these experiments is that group velocities in the period range 12-80 sec are useful in resolving mainly the upper part of the crust. Nevertheless the combined use of DSS and gravity data is strongly recommended to obtain a good resolution of all crustal parameters. Another remark which has to be made is that the previous conclusions are based on a level of error of 0.03 km/sec, which is a rather optimistic estimate when using analog records, which are manually digitized. This makes it extremely important to consider the possibility of using magnetic tape records which can be digitized automatically. The financial effort which has to be made seems to be worthwhile since, up to now, valid alternative techniques to determine the distribution of S-wave are not available. Another remark in favour of the intensification of dispersion measurements through digital recording system is that this analysis can be performed at low cost (the method does not require field work), either as part of a reconnaissance study or to complement the evidence gathered with other geophysical methods. In general use is made of earthquakes (no need of explosions and related problems) and existing stations with long period seismographs can be easily updated with magnetic recording. To obtain wave propagation paths in areas of special interest, not covered by instruments, one may install temporary stations which do not require special care.

### 4.5 The case of inconsistent data

The inversion of real data sometimes may give rise to the erroneous conclusion that the data have indeed a larger resolving power that predicted by the previous considerations. In general this is due to the fact that the data inverted are inconsistent. To show in detail this fact the following experiments were performed. Considering the higher resolving power of group velocities, these values have been contaminated by a systematic error which was of variable size in each experiment. The inversion scheme was the same as in experiment n.4 of section 4. In the first test we have lowered the group velocity by an amount corresponding to an increase in travel time $\Delta t_U = 0.5$ sec, for a station-epicenter distance of 500 km. With respect to the solutions obtained with consistent data (see Fig.4 with $\beta_{LID} = 4.4$ km/sec) the solutions 9 km; 15 km; 2.6 km/sec; 3.6 km/sec; 4.0 km/sec; and 9 km; 9 km; 2.6 km/sec; 3.7 km/sec; 3.6 km/sec are lost while an extra solution is found, namely 9 km; 11 km; 2.6 km/sec; 3.5 km/sec; 3.8 km/sec. Thus the general picture does not change,but it is already possible to see an

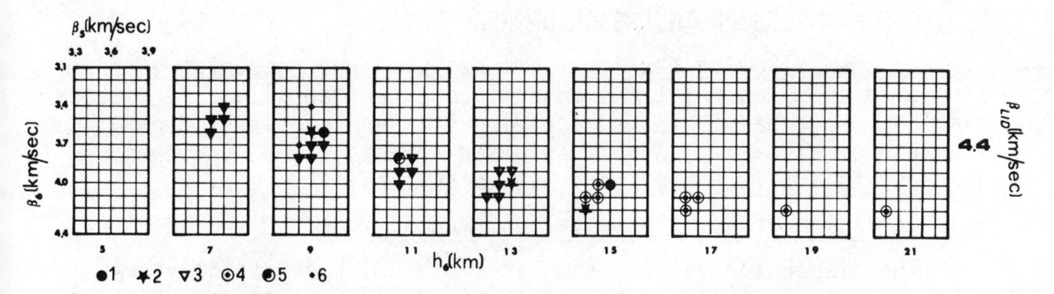

Fig.7 Tic-tack-toe plot for the case of inconsistent data:
1: solutions satisfing only the set of consistent data (see Fig.5 for $\beta_{LID}$ = 4.4 km/sec);
2: solutions common to Experiment N.4 and to the set of inconsistent data with $\Delta t_U$ = 0.5 sec;
3: solutions common to experiment N.4 and to the two sets of inconsistent data corresponding to $\Delta t_U$ = 0.5 sec and $\Delta t_U$ = 1.0 sec;
4: solutions common to experiment N.4 and to the three sets of inconsistent data corresponding to $\Delta t_U$ = 0.5 sec; $\Delta t_U$ = 1.0 sec and $\Delta t_U$ = 2.0 sec;
5: new solutions satisfing only the two sets of consistent data with $\Delta t_U$ = 0.5 sec and $\Delta t_U$ = 1.0 sec;
6: new solutions satisfing only the set of inconsistent data with $\Delta t_U$ = 1.0 sec.

apparent increase in the resolving power. The second test has been performed doubling the increase in travel time ($\Delta t_U$=1.0 sec). In this case a total of five old solutions are lost but at the same time three new solutions are found. The third test was carried out with $\Delta t_U$ = 2.0 sec. In this case only 8 old solutions survive and no new solutions are found. Thus a definitely higher resolving power is obtained but it is indeed only apparent. An inspection of Fig.7 indicates that in this case the S-wave distribution can be apparently resolved within about 0.2 km/sec for the whole crust. The final test, made with $\Delta t_U$ = 4.0 sec, gives no solutions. A synthesis of all these experiments is given in Fig.7. Thus, in general, if the result of the inversion procedure of dispersion data alone gives a very high resolving power with respect to the results described in the previous sections, some care has to be taken before accepting the solutions and a detailed check of the data has to be made.

## 5. SOME EXAMPLES OF INVERSION OF REAL DATA

In the following sections some examples are presented of inversion of dispersion relations determined experimentally. The dispersion data have been obtained applying standard single-station (Levshin et al.,1972 ) and two-station ( Panza, 1976 ) techniques in order to obtain phase and group velocities.

### 5.1 Inversion of Rayleigh-wave fundamental mode

The dispersion values shown in Table 23, representative of North-Central Italy, have been determined and inverted by Calcagnile et al.(1979b). In the model of the cross-section it has been assumed that the region has a crust,an upper mantle, as well as a deeper structure consisting of two jump discontinuities in physical properties at 451 and 671 km depth (Table 24). Below the latter depth, the region is taken to be a half-space. The results described in section 3 indicate the possibility of resolving the structural parameters, mainly S-wave velocity distribution with depth, above the half-space. This is not true in this case since the errors on phase velocities, $\varepsilon_c$, are larger than the ones considered in the theoretical case, mainly at the extremities of the curve (see Table 23), and $\sigma_c$ is equal to 0.06 km/sec and not to 0.03 km/sec as in section 3. Thus for depths larger than 300 km, any reasonable structure is an adequate platform for the determination of the structure of the upper parts of the upper mantle. The interval between the two deeper jump discontinuities a 451 and 671 km is broken into two layers which is intended to model a small gradient in physical properties. The parametrization is fixed for all regions with the exception of upper mantle layers. The numerical values of the fixed parameters are given in Table 24. Since the shortest period available is 25 sec, small to moderate fluctuations in crustal properties do not have a significant influence on the results of inversion for upper manthe properties. Indeed, as has been shown in section 4, also short period surface waves are unsensitive to crustal thickness variations of the order of 5 km.

Table 23. Phase velocities(Calcagnile et al.,1979b)

| T (sec) | c (km/sec) | ε (km/sec) | T (sec) | c (km/sec) | ε (km/sec) |
|---|---|---|---|---|---|
| 250.0 | 4.97 | 0.11 | 69.4 | 3.89 | 0.05 |
| 208.3 | 4.64 | 0.08 | 54.3 | 3.83 | 0.05 |
| 178.6 | 4.39 | 0.08 | 44.6 | 3.81 | 0.05 |
| 156.3 | 4.26 | 0.06 | 35.7 | 3.76 | 0.05 |
| 125.0 | 4.08 | 0.05 | 31.3 | 3.70 | 0.05 |
| 96.2 | 4.01 | 0.05 | 27.8 | 3.62 | 0.07 |
| 83.3 | 3.96 | 0.05 | 25.0 | 3.47 | 0.15 |

Table 24. Fixed parameters in the inversion (Calcagnile et al.,1979b)

| Depth (km) | Layer thickness (km) | $\beta$ (km/sec) | $\alpha$ (km/sec) | $\varrho$ (g/cm$^3$) |
|---|---|---|---|---|
| 0 - 0.2 | 0.2 | 0 | 1.52 | 1.03 |
| 0.2- 4.0 | 3.8 | 1.62 | 2.81 | 2.00 |
| 4.0- 7.0 | 3.0 | 2.00 | 4.00 | 2.20 |
| 7.0- 20.0 | 13 | 3.52 | 6.10 | 2.77 |
| 20.0- 30.0 | 10 | 3.76 | 6.68 | 2.96 |
| 30.0 | | | | |
| | upper mantle | | | |
| 451.0 | | | | |
| 451.0- 571.0 | 120 | 5.30 | 9.76 | 3.80 |
| 451.0- 671.0 | 100 | 5.4 | 9.97 | 4.10 |
| $\geq$ 671.0 | $\infty$ | 6.2 | 11.15 | 4.40 |

Hedgehog is an increasingly expensive program as the number of model parameters increases. The cost of testing the surroundings of each net point increases approximately threefold for each additional model parameter. To reduce costs, as a first step, it is possible to explore the ability of the data to resolve the deeper parts of the upper mantle by assuming that the upper mantle is composed of only two layers as shown in Table 25. The exploration in this case takes place over a three-dimensional parameter space.The three parameters are the thickness of the layers and the S-wave velocities of both. The P-wave velocities in the two layers are coupled to the S-wave velocities for each test model through the formula:

$$\alpha = 7.8 + \frac{4}{3}(\beta - 4.3) \text{ km/sec} \tag{7}$$

Table 25. Hedgehog model 1 for upper mantle(Calcagnile et al.,1979b)

| Depth | | Layer thickness | $\beta$ (km/sec) | $\alpha$ (km/sec) | $\varrho$ ( g/cm$^3$) |
|---|---|---|---|---|---|
| 30 | ----------- | | | | |
| | upper upper mantle | P1(km) | P2 | (P2) | 3.40 |
| 30+P1 | ----------- | | | | |
| | lower upper mantle | 421-P1 | P3 | (P3) | 3.60 |
| 451 | ----------- | | | | |

confidence limits $\sigma = 0.06$ km/sec
single pt.rejection if $|\Delta c| > \varepsilon$ (Table 23).

| Parameter | Range | Starting value |
|---|---|---|
| P1 (km) | 55(30)415 | 265 |
| P2 (km/sec) | 4.2(0.1)4.8 | 4.3 ( $\alpha$ = 7.8) |
| P3 (km/sec) | 4.35(0.1)5.15 | 4.85( $\alpha$ = 8.53) |

Table 26. Acceptable models from Hedgehog inversion of model 1 (Calcagnile et al.,1979b)

| Model number | P1 km | P2 km/sec | P3 km/sec |
|---|---|---|---|
| 1 | 265 | 4.3(7.8) | 4.85(8.53) |
| 2 | 235 | 4.3(7.8) | 4.85(8.53) |
| 3 | 265 | 4.3(7.8) | 4.95(8.67) |
| 4 | 265 | 4.3(7.8) | 5.05(8.80) |
| 5 | 265 | 4.3(7.8) | 5.15(8.93) |

A model is considered acceptable if it passes two criteria. First, the rms difference between the phase velocities computed for the model and the observed values is less than 0.06 km/sec. Second, the absolute difference between the phase velocity predicted at a specific period and the observed values is less than $\varepsilon(T)$, a quantity listed in Table 23. Within the net spacing, the three parameter hedgehog search gave only the five acceptable models listed in Table 26. The entries in parentheses are the values of P-wave velocities determined from equation 7. Thus with very little computer time it is possible to know that the available data are perfectly in agreement with any S-wave velocity in the lower layer ranging at least from 4.85 to 5.15 km/sec. In fact values larger than 5.15 km/sec are not excluded by the inversion results but have been not explored, as a consequence of the a priori choice of the parametrization. This indicates that the data set is not capable of resolving the structure at depths larger than 250 km. As predicted by the analysis of the resolving power made in section 3, it will be shown that the deeper part of the structure can be resolved when dispersion data of higher modes are available. Obviously a better resolution could be reached if more accurate long period data for the fundamental mode are available, but this task is extremely difficult to carry out using the standard long period instruments and when phase differences are measured between relatively near stations. As a consequence of the observation regarding lack of resolution in the lower upper mantle a second inversion can be made fixing the velocity of the lower layer. This reduces the multiplicity of models and the computing expenses. For each model in the new parametrization that is found acceptable, it is reasonable to expect that other models with different velocities in the lower layer will also be found to be acceptable, near the bounds 4.85 to 5.15 km/sec. This is not precisely true due to variations from model to model, but it is a satisfactory approximation for practical purposes. In view of these remarks, the velocity in the lower part of the upper mantle is fixed at 4.85(8.53) km/sec, the upper layer used in the first inversion is broken up into two layers, thus the region is parametrized by four parameters (Table 27), namely the thicknesses and velocities of each of the lid and the channel layers. The P- and S-velocities are again

Table 27. Hedgehog model 2 for upper mantle(Calcagnile et al,1979b)

| Depth | | Layer thickness | $\beta$ (km/sec) | $\alpha$ (km/sec) | $\varrho$ (g/cm$^3$) |
|---|---|---|---|---|---|
| 30 | ---------- | | | | |
| | lid | P1(km) | P3 | (P3) | 3.40 |
| 30+P1 | ---------- | | | | |
| | channel | P2(km) | P4 | (P4) | 3.40 |
| 30+P1+P2 | ---------- | | | | |
| | subchannel | 421-P1-P2 | 4.85 | 8.53 | 3.60 |
| 451 | ---------- | | | | |

confidence limits $\sigma$ = 0.06 km/sec
single pt.rejection if $|\Delta c| > \varepsilon$ (Table 23)

| Parameter | Range | Starting value |
|---|---|---|
| P1 (km) | 15(30)135 | 45 |
| P2 (km) | 80(40)280 | 280 |
| P3 (km/sec) | 4.2(0.1)4.8 | 4.3 |
| P4 (km/sec) | 4.2(0.1)4.8 | 4.3 |

coupled through equation 7. The grid of parametric exploration, and the acceptance criteria are specified at the bottom of Table 27. The hedgehog search delineated nineteen acceptable models,

Table 28. Acceptable models for Hedgehog inversion of parameters model 2 (Calcagnile et al.,1979b)

| Model number | P1 (km) | P2 (km) | P3 (km/sec) | P4 (km/sec) |
|---|---|---|---|---|
| 1 | 225 | | | |
| 2 | 235 | | 4.3(7.8) | |
| 3 | 245 | | | |
| 4 | 255 | | | |
| 5 | 45 | 200 | 4.3(7.8) | 4.4(7.93) |
| 6 | 45 | 240 | | |
| 7 | 105 | 160 | | |
| 8 | 135 | 120 | | |
| 9 | 15 | 240 | 4.4(7.93) | 4.3(7.8) |
| 10 | 45 | 200 | | |
| 11 | 75 | 160 | | |
| 12 | 135 | 80 | 4.3(7.8) | 4.2(7.67) |
| 13 | | | 4.5(8.07) | |
| 14 | 15 | 240 | 4.6(8.2) | 4.3(7.8) |
| 15 | | | 4.7(8.33) | |
| 16 | 15 | 280 | 4.2(7.67) | 4.4(7.93) |
| 17 | 45 | 160 | 4.5(8.07) | 4.2(7.67) |
| 18 | 105 | 120 | 4.4(7.93) | 4.2(7.67) |
| 19 | 135 | 160 | 4.3(7.8) | 4.5(8.07) |

which are listed in Table 28. The solutions can be divided into three groups. In the first twelve cases listed in Table 28, the velocity differences between lid and channel are very small, indicating that the data are incapable of resolving the two regions; these models may essentially be taken to indicate the presence of a more-or-lessuniform layer with S-wave velocity of 4.3 km/sec with small variations around this velocity. The thickness of this layer ranges from 215 to 285 km. The second category of solutions (13-16) are those for which the upper layer is only 15 km thick. In these cases, the lid is only a thin veneer over the thicker channel. Accordingly with the results of the previous section, 15 km is the more optimistic uncertainty affecting lid thickness, thus the main implication of the first two classes of solution is the detection of a more-or-less homogeneous thick zone, starting almost immediately below the Moho, with velocities between 4.3 and 4.4 km/sec. The third group of solutions (17-18) is characterized by a clear lid-to-channel contrast, even if the lid velocities are lower than those measured for young stable continents. The final subcategory of hedgehog solutions is the single case of model 19 which has a positive gradient of velocity with S-wave velocities of 4.3 km/sec in the mantle, below the crust, extending to considerable depth. Below this there is a relatively thick layer of material with velocity 4.5 km/sec.

### 5.2 Simultaneous inversion of fundamental and higher Rayleigh-wave modes

Panza and Scalera (1978) have determined for the area consired in section 5.1 the dispersion relations for the first two higher modes of Rayleigh waves (Table 29). Scalera et al.(1980) have carried out a simultaneous inversion of the fundamental and higher modes, in order to resolve the two main questions left open by the inversion of the fundamental mode, that is the velocity in the lower upper mantle and the presence, or not, of a clear lid-to-channel velocity contrast. The inversion was made using seven variable parameters, namely lid, channel and subchannel thickness and lid, channel, subchannel and "spinel" layer S-wave velocity (see Table 30). The results of inversion are given in Table 31.

Table 29. Phase velocities of Rayleigh higher modes (Panza and Scalera,1978)

| I Higher mode | | | II Higher mode | | |
|---|---|---|---|---|---|
| T sec | c km/sec | ε km/sec | T sec | c km/sec | ε km/sec |
| 30 | 4.45 | 0.11 | 30 | 4.90 | 0.09 |
| 35 | 4.55 | 0.11 | -- | ---- | ---- |
| | | | 40 | 5.24 | 0.09 |
| | | | 50 | 5.43 | 0.09 |
| | | | 60 | 5.77 | 0.09 |
| | | | 70 | 6.16 | 0.11 |

From this Table it can be seen that the lid must have a rather low velocity. In fact only a very thin (lees than 10 km) high velocity veener can not be excluded by the data and this is in agreement with the previous consideration on the resolving power. Furthermore the data allow a rather good definition of the thickness and velocity in the lower upper mantle, which were unresolved using only fundamental mode dispersion data. The more interesting results is the definition of the depth of the so called "olivine-spinel" transition which turns to be around 300 km instead of the usual 400-450 km. This results is in perfect agreement with previous findings in adjacent areas (Mayer-Rosa and Mueller, 2973; Nolet, 1977; Mseddi, 1976).

As a concluding remark it can be observed that the use of the criteria given in section 3.2 has allowed the use of a 7 parameter

Table 30. Structure used in the simultaneous inversion of fundamental and first two higher Rayleigh modes(Scalera et al.1980)

| Depth (km) | | Layer thickness | $\beta$ (km/sec) | $\alpha$ (km/sec) | $\varrho$ (g/cm$^3$) |
|---|---|---|---|---|---|
| | | 0.2 | 0. | 1.52 | 1.03 |
| | | 1.8 | 1.60 | 2.80 | 2.00 |
| | CRUST | 3.0 | 2.00 | 4.00 | 2.20 |
| | | 15.0 | 3.50 | 6.10 | 2.75 |
| | | 10.0 | 3.75 | 6.70 | 2.95 |
| 30 | | | | | |
| | LID | P1(km) | P3 | 7.95 | 3.35 |
| 30+P1 | | | | | |
| | CHANNEL | P2(km) | P4 | 8.05 | 3.40 |
| 30+P1+P2 | | | | | |
| | SUBCHANNEL | P7(km) | P5 | 8.95 | 3.70 |
| 30+P1+P2+P7 | | | | | |
| | "SPINEL"LAYER | 540-P1-P2-P7 | P6 | 9.60 | 3.90 |
| 570 | | | | | |
| | | 100. | 5.40 | 9.95 | 4.10 |
| | | 400. | 6.20 | 11.15 | 4.40 |
| | | 220. | 6.45 | 11.70 | 4.60 |
| | | $\infty$ | 6.65 | 12.00 | 4.70 |

confidence limits $\sigma$ = 0.065 km/sec
single point rejection if $|\Delta c| > \varepsilon$(Table 23 and Table 29)

| Parameter | Range | Starting value |
|---|---|---|
| P1(km) | 15(30)105 | 45 |
| P2(km) | 80(40)320 | 280 |
| P3(km/sec) | 4.00(0.1)4.80 | 4.10 |
| P4(km/sec) | 4.00(0.1)4.80 | 4.10 |
| P5(km/sec) | 4.35(0.1)5.35 | 5.25 |
| P6(km/sec) | 4.90(0.1)5.60 | 5.10 |
| P7(km) | 10(40)90 | 10 |

Table 31. Acceptable models for hedgehog inversion of Fundamental and higher Rayleigh modes(Scalera et al.,1980)

| Solution number | P1 km | P2 km | P3 km/sec | P4 km/sec | P5 km/sec | P6 km/sec | P7 km |
|---|---|---|---|---|---|---|---|
| 1 | 15 | 280 | 4.20 | 4.20 | 5.25 | 5.10 | 10 |
| 2 | 15 | 280 | 4.20 | 4.20 | 5.35 | 5.10 | 10 |
| 3 | 15 | 280 | 4.30 | 4.20 | 5.25 | 5.10 | 10 |
| 4 | 15 | 280 | 4.30 | 4.20 | 5.15 | 5.10 | 10 |
| 5 | 15 | 280 | 4.30 | 4.20 | 5.35 | 5.10 | 10 |
| 6 | 15 | 280 | 4.40 | 4.20 | 5.25 | 5.10 | 10 |
| 7 | 15 | 280 | 4.40 | 4.20 | 5.15 | 5.10 | 10 |
| 8 | 15 | 280 | 4.40 | 4.20 | 5.35 | 5.10 | 10 |
| 9 | 15 | 280 | 4.40 | 4.20 | 5.05 | 5.10 | 10 |

inversion with the use of a very limited amount of CPU time, which for a CDC 6600 computer is of about 5 minutes.

### 5.3 Simultaneous inversion of phase and group velocities of Rayleigh-wave fundamental mode

This concluding chapter on the description of the inversion of real data is the natural application of the results obtained in section 4. The set of data reported in Fig.4 was considered. The phase and group velocities were determined by Calcagnile and Panza (1979) and Calcagnile et al.(1980) for the Apennines area and are reported in the Figure as open circles and full triangles respectively. The solid lines represent the curves which have been digitized to perform the experiment described in section 4 and correspond to the starting structure reported in Table 15. The parametrization used is the same one shown in Table 16. The set of solution is shown in Table 32. The acceptance criteria have been $\varepsilon_c = 0.06$ km/sec $\sigma_c = 0.03$ km/sec $\varepsilon_U = 0.12$ km/sec $\sigma_U = 0.08$ km/sec in accordance with the previous observations on the relative accuracy of phase and group velocity measurements. The results listed in Table 32 indicate rather clearly the impossibility of resolving, in a satisfactor way, the properties of the lower crustal layer. This is not surprising and actually confirms experimentally the theoretical results obtained in section 4. A better resolution is obtained for the other two layers, again in agreement with the theoretical results.

Cassinis et al.(1979) have compiled a summary of the available DSS data on the Italian area from which it is evident that the region of Central Apennines is almost completely unexplored, however north and south of this area the total crustal thickness is on the average between 30 and 35 km. If this fact is kept in mind solutions n. 1,2,3,4,5,7 can be easily disregarded since they give a crust of 38 km or more; the same applies to solutions 27,29,30, 31,32,33,34,36,37 since they correspond to a crust of only 30 km. In such a way the uncertainty in the S-wave velocity of the lower crustal layer is considerably reduced, even if the presence or not

Table 32. Acceptable models from Hedgehog inversion of real data (Calcagnile et al.,1980)

| Solution number | km | km | km/sec | km/sec | km/sec |
|---|---|---|---|---|---|
| 1 | 9 | 15 | 2.6 | 3.6 | 4.1 |
| 2 | 9 | 17 | 2.6 | 3.6 | 4.1 |
| 3 | 9 | 15 | 2.6 | 3.5 | 4.1 |
| 4 | 9 | 15 | 2.6 | 3.6 | 4.0 |
| 5 | 9 | 17 | 2.6 | 3.5 | 4.1 |
| 6 | 9 | 13 | 2.6 | 3.6 | 4.0 |
| 7 | 9 | 15 | 2.7 | 3.5 | 4.1 |
| 8 | 9 | 13 | 2.7 | 3.5 | 4.1 |
| 9 | 9 | 13 | 2.6 | 3.5 | 4.0 |
| 10 | 9 | 13 | 2.6 | 3.7 | 4.0 |
| 11 | 9 | 13 | 2.7 | 3.5 | 4.0 |
| 12 | 9 | 13 | 2.6 | 3.4 | 4.1 |
| 13 | 9 | 13 | 2.7 | 3.4 | 4.1 |
| 14 | 9 | 13 | 2.6 | 3.7 | 3.9 |
| 15 | 9 | 11 | 2.6 | 3.6 | 3.9 |
| 16 | 9 | 11 | 2.7 | 3.5 | 4.0 |
| 17 | 9 | 11 | 2.7 | 3.6 | 3.9 |
| 18 | 9 | 11 | 2.6 | 3.7 | 3.9 |
| 19 | 9 | 11 | 2.7 | 3.4 | 4.0 |
| 20 | 9 | 11 | 2.6 | 3.7 | 3.8 |
| 21 | 9 | 9 | 2.6 | 3.6 | 3.8 |
| 22 | 9 | 9 | 2.7 | 3.5 | 3.9 |
| 23 | 9 | 9 | 2.7 | 3.6 | 3.8 |
| 24 | 9 | 9 | 2.7 | 3.5 | 3.8 |
| 25 | 9 | 9 | 2.6 | 3.7 | 3.7 |
| 26 | 9 | 9 | 2.6 | 3.8 | 3.7 |
| 27 | 9 | 7 | 2.7 | 3.7 | 3.7 |
| 28 | 9 | 9 | 2.6 | 3.8 | 3.6 |
| 29 | 9 | 7 | 2.7 | 3.7 | 3.6 |
| 30 | 9 | 7 | 2.6 | 3.6 | 3.6 |
| 31 | 9 | 7 | 2.7 | 3.6 | 3.6 |
| 32 | 9 | 7 | 2.6 | 3.8 | 3.5 |
| 33 | 9 | 7 | 2.6 | 3.7 | 3.5 |
| 34 | 9 | 7 | 2.7 | 3.7 | 3.5 |
| 35 | 9 | 7 | 2.6 | 3.8 | 3.4 |
| 36 | 9 | 7 | 2.6 | 3.9 | 3.4 |
| 37 | 9 | 7 | 2.6 | 3.9 | 3.3 |

of a crustal low velocity layer just above the Moho still remains unresolved. It is evident that, when data on the average crustal layering is available it will be possible to determine with less uncertainty the S-wave velocity distribution, and thus to assess the problem of the existence or not of the crustal low velocity layer, with a consequent relevant contribution to the understanding of the seismic regime of the Apennines

REFERENCES

Backus, G.E., and Gilbert, J.F.,1968, The Resolving Power of Gross Earth Data, Geophys.J.Roy.Astron.Soc., 16:169.

Backus, G.E., and Gilbert, J.F.,1970, Uniqueness in the inversion of inaccurate gross Earth data, Phil.Trans.Roy.Soc., A,266: 123.

Bessonova, E.N., Fishman, V.M., Ryaboi, V.Z., and Sitnikova,G.A., 1974, The tau-method for inversion of travel times. I. Deep seismic sounding data, Geophys.J.Roy.Astron.Soc.,36:377.

Bessonova, E.N., Fishman, V.M.,Shnirman, M.G., and Johnson, L.R., 1976, The tau-method for inversion of travel times.II. Earthquake data, Geophys.J.Roy.Astron.Soc.,46:87.

Brodskii M., and Levshin, A, 1979, An Asymptotic approach to the inversion of free oscillation data, Geophys.J.Roy.Astr.Soc., 58:631.

Calcagnile, G., Canziani, R., Monterisi, R., and Panza, G.F.,1979a, Potere risolutivo di sondaggi dipolari(Schlumberger), Geol. Appl. e Idrogeol.,14:181.

Calcagnile, G., D'Ingeo, F., and Panza,G.F.,1980, The lithosphere in south-eastern Europe: Preliminary results, Proc.Symp.8, EGS-Wien, 11-14 Sept.,1979, in press.

Calcagnile, G., and Panza, G.F.,1979, Crustal and upper mantle structure beneath the Apennines region as inferred from the study of Rayleigh waves, J.Geophys.,45:319.

Calcagnile, G., Panza, G.F., and Knopoff, L.,1979b, Upper mantle structure of North-Central Italy from Rayleigh waves phase velocities, Tectonophysics,56:51.

Cassinis, R., Franciosi, R., and Scarascia, S.,1979, The structure of the Earth's crust in Italy. A preliminary typology based on seismic data, Boll.Geof.Teor.Appl.,21:105.

Cloeting, S., Nolet, G., and Wortel, R.,1979, On the use of Rayleigh wave group velocities for the analysis of continental margins, Tectonophysics,59:335.

Cloeting, S., Nolet, G., and Wortel, R.,1980, Standard graphs and tables for the interpretation of Rayleigh wave group velocities in crustal studies, Proc.Roy.Neth.Ac.Sci.,B83(1):101.

Keilis-Borok, V.I., and Yanovskaya,T.B.,1967, Inverse problems of seismology, Geophys.J.Roy.Astron.Soc.,13:223.

Keilis-Borok,V.I.,1971, The inverse problem of seismology, in: "Proc.Int.School of Physics E.Fermi, course 50", J.Coulomb and M.Caputo, ed., Academic Press, New York.

Knopoff, L., 1977, Relative errors in Group velocity measurements. J.Geophys.,43:509.

Knopoff, L.,1972, Observation and Inversion of Surface wave dispersion, Tectonophysics,13:497.

Knopoff, L., and Chang, F-S,1977, The inversion of Surface Wave Dispersion Data with Random Errors, J.Geophys.,43:299

Knopoff, L., and Panza, G.F.,1977, Resolution of Upper Mantle Structure using higher modes of Rayleigh waves,Ann.Geofis.,

30:491.
Kovach, R.L.,1978, Seismic surface waves and Crustal and Upper Mantle Structure, Rev.Geophys.Space Phys.,16:1.
Levshin, A., Pisarenko, V., and Pogrebinsky, G.,1972, On a frequency time analysis of oscillations, Ann.Geophys.,28:211.
Mitchel, R.G., 1977, The Structure of Western North America from Multimode Rayleigh Wave Dispersion, Ph.D.Thesis,University of California, Los Angeles, U.S.A.
Mayer-Rosa, D., and Mueller, St.,(1973, The gross velocity depth distribution of P and S waves in the upper mantle of Europe from Earthquake observations, Z.Geophys.,39:395.
Mseddi, R., 1976, Determination sismique de la structure du Manteau superieur sous la Mediterraneè, Ph.D.Thesis,-Univ.Pierre et Marie Curie, Paris,France.
Nolet, G., 1975, Higher Rayleigh Modes in Western Europe; Geophys. Res.Letters,2:60.
Nolet, G., 1977, The Upper Mantle under Western Europe inferred from the Dispersion of Rayleigh Modes, J.Geophys., 43:265.
Nolet, G., and Panza, G.F., 1976, Array Analysis of Seismic Surface Waves: Limits and Possibilities, Pure and Appl.Geophys., 114:773.
Panza, G.F.,1976, Phase velocity determination of fundamental Love and Rayleigh waves, Pure and Appl.Geophys.,114:776.
Panza, G.F., Mueller St. and Calcagnile, G., 1978, The gross features of the lithosphere-asthenosphere system in the European-Mediterranean area, ESC-EGS Symposium on "Deep seismic sounding and Earthquakes" Strasbourg 29 August - 5 September 1978.
Panza, G.F., and Scalera, G., 1978, Higher Modes Dispersion Measurements, Pure and Appl.Geophys., 116:1274.
Porter, L.D., Schwab, F.A., Nakanishi , K.K., Weeks, I.F., Panza, G.F., Mantovani, E., McMenamin, D., Smythe, W.D., Liao, A.H., Landoni, J., Biswas, N.N., Chang, D.S., Bor, S.S., and Kausel, E.,1979, Relative computer speeds for surface-wave dispersion computation, Bull.Seism.Soc.Am., in press.
Press, C.L., 1968, Earth models obtained by Monte Carlo inversion, J.Geophys.Res., 73:5223.
Scalera, G., Calcagnile, G., and Panza G.F.,1980, On the "400-km discontinuity" in the Mediterranean area, this volume.
Valyus, V.P.,1972, Determining seismic profiles from a set of observations, in: "Computational Seismology", V.I Keilis-Borok, ed., Consult.Bureau,New York.
Valyus, V.P., Keilis-Borok, V.I., and Levshin, A.,1969, Determination of the upper-mantle velocity cross-section for Europe, Proc.Acad.Sci.USSR,185,n.3(in Russian).

# THE INVERSION OF DEEP SEISMIC SOUNDINGS:

# PROBLEMS AND CONSTRAINTS

Salvatore Scarascia

Istituto per la Geofisica della Litosfera- C.N.R.

Via Bassini I5 - Milano

SUMMARY

Some examples of record-sections obtained in the exploration of the Crustal Structure in Italy (Italian Geodynamics Project) are shown, describing the main features of the recorded events.

On the base of both kinematic and dynamic considerations the conclusion is drawn that the recorded events should represent the effect on surface of "diving waves" that are generated if a continuous variation of velocity with depth is verified.

Some approximate formulas for calculating the velocity-depth function are given, as well as the principles and recurrence formulas for calculating the travel-time curves from a crustal model (direct problem) and the crustal model from the obtained travel--time curves (inverse problem) in the case of horizontal layering.

Ambiguities arise in calculating the "low-velocity layers", but these ambiguities do not affect so much the interpretation of layers in the lower crust.

The principles of the "ray tracing method" are explained, which should be applied for a better refining of the crustal model along a profile where lateral inhomogeneities are present. Some examples of application illustrate this concept.

At the end some representative cross-sections are shown and a general description of the crustal structure in Italy is given, pointing out the large variations of typology evidenced in the different regions and the consequent geodynamic implications.

REFERENCES

Giese P., The determination of the velocity-depth distribution for separated travel-time segments, U.M.P. Committee (1970)

Giese P., Stein A., Versuch einer einheitlichen Auswertung tiefenseismischer Messungen aus dem Bereich zwischen der Nordsee und den Alpen, Z. Geophysis. 37 (1971)

Pavlenkova N.L., The interpretation of refracted waves by the reduced travel time curve method, Izv. Earth Physics No. 8 (1973)

Colombi B., Scarascia S., Sulla interpretazione dei profili sismici crostali. Calcolo diretto della funzione velocità-profondità, Rivista Italiana di Geofisica, vol. XXII, N. 3-4 (1973)

Scarascia S., Evoluzione e ruolo della sismica crostale in Italia, Memorie della Soc. Geologica Ital., Suppl. 2, vol. XIII (1974)

COMBINED REFLECTION AND REFRACTION MEASUREMENTS FOR

INVESTIGATING THE GEOTHERMAL ANOMALY OF URACH

R. Meissner

Institut für Geophysik

Universität Kiel

SUMMARY

After field work and first results from the combined seismic experiments in the Urach area have been described at the previous meeting (Meissner, Bartelsen, 1979), many new information velocities and structure in the Urach area are now available. Velocity data consist of (1) horizontal velocities from the first (refracted) arrivals from the reflection units and MARS stations, reaching depths down to 5 Km (2) velocities from the second and third arrivals from the MARS stations in the wide angle range, down to depths of 29 Km, and (3) near vertical stacking and intervals from the MARS stations in the wide angle range, down to depths of 29 Km, and (3) near vertical stacking and interval velocities down to 29 Km. The structural information is mainly based on the near vertical reflection work, but important additional data have been collected from the stationary refraction (=MARS) stations along the two lines.

Parallel to the interpretation of field data theoretical calculations based on experiments with gneiss and granitic samples in a high pressure - high temperature apparatus (Kern, 1978), (Meissner, Fakhimi, 1977), have been performed in order to distinguish the effect of temperature on the velocity from that of anisotropy and inhomogeneity of material.
Experiment with (1) constant pressure p and variable temperature T, (2) constant T and variable p, and (3) variable p and T for certain p-T function similar to that in the Urach area have been selected.

The experiments based on 3 component velocity measurements of four gneiss samples show the large influence of a possibility inhomogeneity which may reach up to 20% of the mean p-velocity of I4 gneiss samples investigated. The effect of anisotropy may reach $v_p$ values I3% different from the mean. Compared to these figures the pure T- effect is about 50 m/sec for a 50°C difference in T at constant p which is about 1% change in $v_p$. It might be enhanced by certain hydrothermal alterations.

The horizontal velocity of the refracted arrivals from the reflection units can be compared with the velocity of the sonic log of URACH III showing the different degree of resolution and indications of the suspected anisotropy. Velocities from the evaluation MARS stations in the wide angle range give velocity depth functions which have to be compared with the interval velocities of the near angle-reflection work. In general velocities are lower than those in the vicinity of the Urach area. Velocities obtained from the slant ray paths of the wide angle survey tend to be larger than the near-vertical velocities, but a final correlation has not get been established. The first structural interpretation from the stacked records of the near vertical reflection survey of the long profile shown two unusual reflectors at the crust-mantle boundary, dipping slightly to the ENE with an indication of a disturbed region below the center of the heat flow anomaly. The crust-mantle boundary at 28-29 Km depth is slightly shallower than in the vicinity of the Urach area.

In general, the whole Urach area is marked by a shallow crust, by lower crustal velocities and by two strong reflectors in the lower crust, where velocities increase with depth and additional near-vertical reflections indicate a lammellar structure. Locally, near Urach, velocities show a minimum in the upper layers and a disturbed zone at greater depth.

REFERENCES

Meissner R., Bartelsen H., Field work and first results of seismic reflection and refraction measurements in the area of the Urach geothermal anomaly; summary and half yearly report, project No G/B 60 (I979)

Kern H., The effect of high temperature and high confining pressure on compressional wave velocities in quartz-bearing and quartz-free igneous and metamorphic rocks, Tectonophysics, 44, I85-203 (I978)

Meissner R., Fakhimi M., Seismic anisotropy as measured under high-pressure, high-temperature conditions, J.R. astro. Soc. 49, I33-I44 (I977)

# THE APPROACH TO THE INVERSION PROCESS IN REFLECTION SEISMICS

A.A. Fitch and K. Helbig

formerly
Seismograph Service Ltd
Holwood, Keston
Kent BR2 6HD
England

Vening Meinesz Laboratory
Rijksuniversiteit Utrecht
P.O. Box 80.021
3508 TA Utrecht
The Netherlands

The seismogram generated by plane waves falling vertically on a stack of plane parallel layers can be expressed in terms of the sequence of the reflection coefficients corresponding to the interfaces between individual layers. The reflection coefficients can be expressed through the wave impedances, which are parameters characterizing the layers and are related to the elastic constants and the density (and thus also to the velocity) of the layers. A first approximation to the synthetic seismogram is the convolution of the input pulse with the sequence of reflection coefficients. In this approximation, only the simple wave path (surface - reflector - surface) is taken into account. Internal multiples ("reverberants") can be taken into account by algorithms such as that described by the "Burg ladder".

An unambiguous inversion of a (noise-free) synthetic seismogram obtained by the processes just described is not possible. Algorithms can be described for the recovery of the sequence of reflection coefficients from such a seismogram when the pulse shape is exactly known. However, these processes are not stable: any inaccuracy in the data - be it noise, limited accuracy of the digital representation of the seismogram, or limited knowledge of the pulse - leads to errors which, in the further process, are amplified exponentially. Practical inversion is possible if the pulse shape and the sequence of reflection coefficients are constrained so that the problem becomes overdetermined. Under such conditions, approximations satisfying some "least error criterion" can be obtained which are the "best approximations" within these constraints.

# 1. SOME FUNDAMENTAL CONCEPTS

## 1.1. The seismic wave impedance

If a plane compressional wave propagates through a homogeneous medium in the z-direction, the balance of forces on a volume $dx.dy.dz$ centered at z can be expressed as

$$(1)\quad \ddot{u}.\rho.dx.dy.dz = dx.dy(\sigma(z)+\frac{\partial\sigma}{\partial z} - (\sigma(z) - \frac{\partial\sigma}{\partial z}\cdot\frac{dz}{2})),$$

$$\ddot{u}.\rho = \frac{\partial\sigma}{\partial z},$$

where $\ddot{u}$ is the acceleration, $\rho$ the density and $\sigma$ the normal stress acting on the surface element $dx.dy$ (fig. 1).

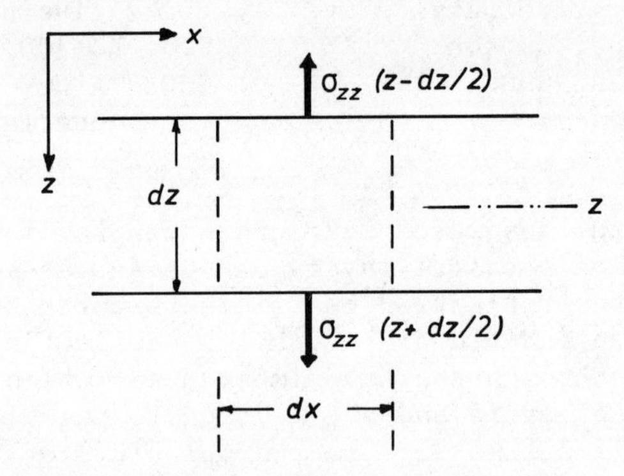


Fig. 1. Stresses acting on the cube $dx.dy.dz$ during the passage of a plane compressional wave in z-direction.

The stress is related to the strain (the space derivative of the displacement $u$) by the one dimensional form of Hooke's law:

$$(2)\quad \sigma = a.\frac{\partial u}{\partial z},$$

where $a$ is the elastic constant corresponding to the strain propagated in plane compressional waves. Since we have assumed homogeneity, $\frac{\partial a}{\partial z} = 0$. Thus (1) and (2) can be combined to produce the onedimensional wave equation.

$$(3)\quad \ddot{u} = \frac{\partial^2 u}{\partial z^2}$$

This wave equation is solved by *any* twice differentiable function of the argument $(z\mp\sqrt{\frac{a}{\rho}}t)\equiv(z\mp vt)$, where $v=\sqrt{\frac{a}{\rho}}$ is the velocity of propagation for plane compressional waves. If we express the displacement as

$$u = u_o.F(z\mp vt) \ ,$$

where $F$ is an arbitrary twice differentiable function of its argument, we obtain for the displacement velocity

$$(4) \quad \dot{u} = \mp u_o.\sqrt{\frac{a}{\rho}} . F'(z-vt)$$

and for the stress

$$(5) \quad \sigma = a.\frac{\partial u}{\partial x} = u_o.a.F'(z-vt) \ .$$

From (4) and (5) follows: $\dot{u}$ and $\sigma$ have the same space-time dependence. In other words: The quotient of stress and particle velocity in a plane progressive wave is a constant quantity characteristic of the material in which the wave is travelling. The quotient of stress and particle velocity in any wave is called the "(wave) impedance" (in electromagnetic waves it is the quotient of the electric component and the magnetic component, dimension [V/m]/[A/m] = [Ω]). For plane propagating compressional waves it is obtained from (4) and (5) as

$$(6) \quad Z = \frac{\sigma}{\dot{u}} = \mp\sqrt{a.\rho} = \mp \frac{a}{v} = \mp\rho.v.$$

The negative sign in (6) corresponds to wave travelling in the positive $x$-direction (frequently one defines the plane wave impedance as the quotient of *pressure* and particle velocity, which makes the positive sign correspond to the propagation in the positive $x$-direction).

The plane wave impedance is closely related to the power density propagating in a plane wave. The total energy of the wave is the sum of kinetic and potential energy:

$$(7) \quad dE_{kin} = \tfrac{1}{2}\rho.dx.dy.dz.\dot{u}^2, \ dE_{pot} = \tfrac{1}{2}dx.dy.dz.\sigma.\frac{\partial u}{\partial x} = \frac{1}{2a} dx.dy.dz \ \sigma^2$$

$$= \tfrac{1}{2}\frac{Z^2}{a} dx.dy.dz.\dot{u}^2 \ .$$

It follows from (6) that $Z^2/a=\rho$, thus the two terms are equal, and the *energy density* is

$$(8) \quad \frac{dE}{dx.dy.dz} = \rho.\dot{u}^2$$

The *power density* is the energy flowing in the time increment $dt$ through the differential cross section $dx.dy$. With $v=dz/dt=\mp\sqrt{\frac{a}{\rho}}$ one has

$$(9) \quad \frac{dP}{dx.dy} = \frac{dE}{dt.dx.dy} = \mp\rho.v.\dot{u}^2 = Z\,\dot{u}^2 \,(= \frac{\sigma^2}{Z})$$

The wave impedance can thus also be defined as the ratio of the power density to the square of the particle velocity, with the convention that a negative power density corresponds to energy lost (moving to +∞) and a positive power density corresponds to energy gained.

In standing waves the wave impedance is purely imaginary (no net energy flow), for curved wave fronts - e.g. close to sources - it is complex (only part of the energy imparted travels in the wave, the rest is "blind" energy remaining in the source area).

The above derivation applies - after a few obvious changes - equally well to shear waves.

## 1.2. Reflection at a single interface at vertical incidence

A plane wave falling on a plane interface with vertical incidence is partially reflected back into the first medium, partially transmitted into the second medium (fig. 2). We define the reflection coefficient $r$ as the ratio of particle displacement velocity in the reflected and incident wave, respectively, and the transmission coefficient $t$ as the corresponding ratio for transmitted and incident waves. Accordingly, $r$ is positive if a downward directed displacement velocity in the incident wave causes a downward directed displacement velocity in the reflected wave. Alternatively, one could define the reflection coefficient as a stress ratio. The derivations would not change, but the sign of the reflection coefficient would be opposite.

At the interface, both displacement velocity and stress must be continuous, thus we have with the definition of $r$ and $t$ and with (6):

$$(10) \quad 1 + r = t \qquad \text{and}$$

$$(11) \quad -Z_1 + Z_1 r = -Z_2 t \qquad \text{, thus}$$

$$(12) \quad r = \frac{Z_1 - Z_2}{Z_1 + Z_2} \qquad \text{and}$$

$$(13) \quad t = \frac{2Z_1}{Z_1 + Z_2}$$

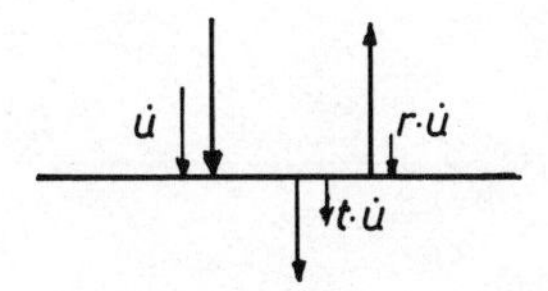


Fig. 2. Reflection and transmission of a plane wave at vertical incidence - continuity of displacement velocity: $(1+r).\dot{u}=t.\dot{u}$

To be observed at the surface, every signal that has been transmitted downward through an interface has to cross this interface a second time in the opposite direction. If we call this upwards transmission coefficient $t'$, we have with $1+r' = t'$ and $r' = -r$ the two-way transmission coefficient.

(14) $t.t' = 1-r^2$.

Equation (12) can be written in a few alternative forms:
With $\bar{Z} = (Z_1+Z_2)/2$ and $\Delta Z = Z_2 - Z_1$ we get

(12a) $r = \Delta Z/(2\bar{Z})$.

If one uses instead $Z_2 = (Z_2/Z_1)^{1/2}(Z_2 Z_1)^{1/2}$ and

$Z_1 = (Z_2/Z_1)^{-1/2}(Z_2 Z_1)^{1/2}$, one obtains with

$q = (Z_2/Z_1)^{1/2}$

$$r = -\frac{q-1/q}{q+1/q} = \frac{\exp(\ln q)-\exp(-\ln q)}{\exp(\ln q)-\exp(-\ln q)}$$

(12b) $r = -\mathrm{tgh}(\frac{1}{2}\ln\frac{Z_2}{Z_1}) \approx -\frac{1}{2}\ln\frac{Z_2}{Z_1}$ (Peterson et al., 1955).

In practical applications, information about velocities is often more easily available than information about densities. Therefore, it is important to resolve the contributions of velocity changes and density changes to the reflection coefficient. With

$$v_1 = \bar{v} - \tfrac{1}{2}\Delta v\ , \quad \rho_1 = \bar{\rho} - \tfrac{1}{2}\Delta\rho$$

$$v_2 = \bar{v} + \tfrac{1}{2}\Delta v\ , \quad \rho_2 = \bar{\rho} + \tfrac{1}{2}\Delta\rho$$

and the abbreviations

$$r_v = -\frac{\Delta v}{2\bar{v}}, \quad r_\rho = -\frac{\Delta \rho}{2\rho}$$

one obtains

$$\text{(12c)} \quad r = (r_v + r_\rho)\left(1 - \frac{r_v r_\rho}{1 + r_v r_\rho}\right)$$

## 2. THE INTERPRETERS REQUIREMENTS

For the complete seismic understanding of a part of the earth the interpreter should have either the acoustic impedance values - or the reflection coefficient - at every point. With this information he has knowledge of structure and of the shapes of sedimentary features; and limited knowledge which bears on lithology, fluid content, load on the rocks, over-pressure or under-pressure and temperature. Geological considerations indicate which of these factors has the greatest effect on change of impedance, and thus the impedance values can be interpreted in terms of this particular factor.

The complete model in the interpreter's mind is thus a three-dimensional array of acoustic impedance values, derived largely from surface seismic observation, but with important, high precision values from well geophysics, and estimated values from lithological logs and surface and subsurface exposures. Associated with this last group of data are, of course, other geological observations that provide the key to interpretation.

The interpreters requirements are met if he has at every place the log of reflection coefficients, since from this he can derive the log of impedances. The log of reflection coefficients can be calculated from velocity and density logs observed in a well (see 3.1), but it cannot be observed directly by surface seismic methods:

Imagine a system in which the input signal to the earth is a unit pulse, and each reflector returns just a unit pulse reflection - an extension of the simple concept of echo sounding at sea. If this were the real earth response, then the surface seismogram would be similar to the log of reflection coefficients the interpreter requires.

However, the earth response is much more complicated, and the recorded seismogram does not resemble very closely the log of reflection coefficients.

It is a principal objective of seismic data processing to produce, as nearly as it can be achieved, a log of reflection coefficients from the recorded seismic data. It is this "undoing" of the additional effects of the propagation through the earth and the effects of data recording which constitutes the "inversion" of the seismic trace.

The subject of inversion will be treated by first describing the "forward" processes, the earth and recording processes. Next will be examined the inverse processes as they would appear in a noise-free environment. Finally, the paper will deal with the inverse processes in the noisy environment which is the practical situation.

## 3. THE FORWARD PROCESS: THE SYNTHETIC SEISMOGRAM

Three assumptions are made in this discussion and persist through much of the analysis of the inversion process:

- i The earth consists of horizontal, isotropic layers.
- ii All observations are made with normally incident plane waves.
- iii These layers are of such a thickness that the two-way travel time in each is one sample time interval.

Assumption (i) is very far from the real earth as the geologist knows it. One must always be aware of this, and be prepared to reexamine inverse processes if the real earth departs far enough from this model to invalidate the assumptions.

Assumption (ii) does not represent usual seismic practice: the generation of plane waves is difficult and costly. However, rays observed close to the source ("zero offset") have been reflected at nearly normal incidence. Nevertheless, the actual wave fronts will always show some curvature. Again one must either correct for this effect, or make sure that the conclusions are not invalidated by wave front curvature or oblique incidence.

Assumption (iii) is a specification of assumption (i) chiefly made in the interest of observational convenience:
Continuous velocity logs taken in boreholes indicate that the velocity changes over distances as small as 30 cm - the current resolution of the velocity log - and might even change over still smaller distances. We therefore have to deal with a great number of layers. Since the seismogram is to be expressed in time, it is convenient to choose as the model a so-called *Goupillaud medium*, in which the travel time through each layer is equal to $\Delta t$. Since this approach leads to a sampled version of a seismogram, it is further convenient to make the *two-way travel time* through each of the layers equal to the sampling time used in actual data acquisition.

### 3.1. The "primaries-only" approximation to the unit pulse response

At each interface, part of the wave is reflected with reflection coefficient $r$, the remainder is transmitted. The actual seismogram is the superposition of arrivals from many different paths most of them reflected several times (figure 3). Since reflection coefficients are typically less than 0.1 (and quite often less than 0.03), it appears to be reasonable to disregard in a first approximation all arrivals that have been reflected more than once. Further, the reflection coefficients quoted above lead according to (14) to two-way transmission factors of 0.99 and 0.999, respectively. Thus, with the same order of approximation the transmission losses can be neglected.

If the wave falling vertically on the stack of layers consists of a "unit pulse" - i.e. if its sampled version is (1,0,0...), the response is - under these simplifying assumptions - simply the sequence of reflection coefficients ($r_1$, $r_2$, $r_3$, .......), as mentioned in section 2.

This "primaries-only unit pulse response" - or more simply the "reflection coefficient log" - can be approximately obtained from a log of wave impedances constructed from a velocity log and a density log re-sampled for double travel time equal to the sample time used in surface recording. If $Z_n$ and $Z_{n+1}$ are the $n$th and $(n+1)$th sample, respectively, and $Z_o$ an arbitrary normalization impedance, (12b) can be written as

$$(15)\quad r_n = -\,\mathrm{tgh}\left(\tfrac{1}{2}\left(\ln\frac{Z_{n+1}}{Z_o} - \ln\frac{Z_n}{Z_o}\right)\right) = -\,\Delta\mathrm{tgh}\left(\tfrac{1}{2}\ln\frac{Z}{Z_o}\right)$$

$$\simeq -\tfrac{1}{2}\,\Delta_n \ln\frac{Z}{Z_o}\,.$$

In other words, the $n$th reflection coefficient is approximately equal to half the negative difference between the logarithm of the $(n+1)$th and the $n$th sample of the impedance.

Generally the density changes much less than the velocity, thus the dimensionless ratio $Z/Z_o$ in (15) can be approximated by $v/v_o$, and the reflection coefficient log can be estimated on the basis of a velocity log alone. However, a further improvement is possible: if the sedimentary sequence in question has reasonably constant lithology, the increase of both density and velocity can be assumed to be due to the same diagenetic processes. Under such circumstances one can expect a relationship of the form

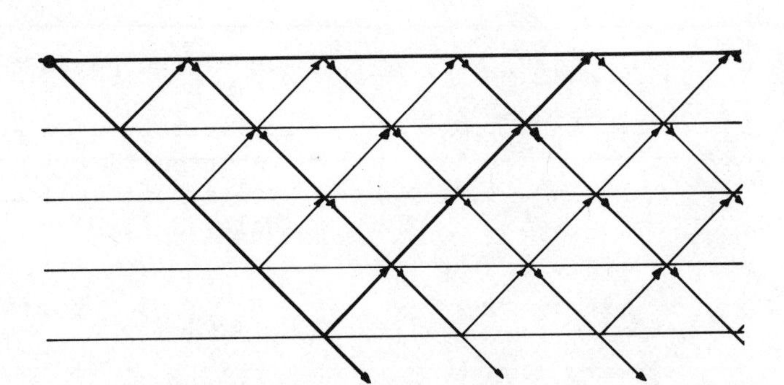


Fig. 3. Generation of complete seismogram (including internal multiples in a "Goupillaud" medium at vertical incidence. Heavy line represents primary reflection from fourth interface.

$(\rho/\rho) = (v/v_o)^{\alpha}$ corresponding to

$(Z/Z_o) = (v/v_o)^{1+\alpha}$ with $\alpha>0$ (and generally $\alpha<\frac{1}{2}$).

With this we get

(15b) $r_n = -\text{tgh}\ (\frac{1+\alpha}{2}\ \ln \frac{v}{v_o}) \approx -\frac{1+\alpha}{2}\ \ln \frac{v}{v_o}$ .

The "primaries only" approximation to the response to an arbitrary pulse given by its sampled values $(p_1, p_2, \ldots\ldots, p_N)$ is the convolution of the pulse with the reflection coefficient log:

(16) $s = p*r$ defined by $s_m = \sum_{n=1}^{N} p_n\ r_{m+1-n}$.

For instance, for the pulse $p_1$, $p_2$, $p_3$ and the reflection coefficient log $r_1$, $r_2$, $r_3$, $r_4$ one has the scheme

| reflection coefficients | | | | | | | | |
|---|---|---|---|---|---|---|---|---|
| | $p_1$ | $p_2$ | $p_3$ | | | | | undelayed pulse |
| $r_1$ | | $r_1p_1$ | $r_1p_2$ | $r_1p_3$ | | | | pulse reflected at first interface delayed $1.\Delta t$ |
| $r_2$ | | | $r_2p_1$ | $r_2p_2$ | $r_2p_3$ | | | pulse reflected at 2nd delayed $2.\Delta t$ |
| $r_3$ | | | | $r_3p_1$ | $r_3p_2$ | $r_3p_3$ | | |
| $r$ | | | | | $r_4p_1$ | $r_4p_2$ | $r_4p_3$ | |
| | | $s_1$ | $s_2$ | $s_3$ | $s_4$ | $s_5$ | $s_6$ | response |

$$\equiv r_1p_1, r_1p_2+r_2p_1, r_1p_3+r_2p_2+r_3p_1, r_2p_3+r_3p_2, r_3p_3+r_4p_2, r_4p_3$$

Convolution of sampled functions becomes a simple multiplication if the sequence of samples is expressed as polynomials in z (i.e. if the sequences are z-transformed): the reflection coefficient log

$$r(z) = \sum_{n=1} r_n z^{n-1}$$

convolved with the pulse

$$p(z) = \sum_1 p_n z^{n-1}$$

gives the response

$$s(z) = \sum_1 s_n z^{n-1}; \text{ thus}$$

(16a) $s(z) = p(z).r(z)$.

If one substitutes these definitions into (16a), carries out the multiplication, and then compares coefficients of equal powers of z one finds that (16a) is indeed equivalent to (16).

## 3.2. The complete response to a unit pulse

The actual response, taking into account the "reverberants" (that is the internal multiples between the layers), is much more complicated than the "primaries-only" response. In fact, neglecting the internal multiples was somewhat rash: it is correct that their amplitudes are very small, but since there is such a large number of layers, there is a large number of internal multiples with the same delay. The sum of these internal multiples can be considerable:

Assume for instance, a sequence of reflection coefficients of equal magnitude $r$ but alternating sign. The direct (undelayed) signal undergoes at each interface a transmission loss of $(1-r^2)$. The total transmission loss for sufficiently large $n$ is

$$(1-r^2)^n \simeq e^{nr^2}, \text{ because } \lim_{q\to\infty} (1+\frac{1}{q})^q = e$$

Each of the first order reverberants experiences the same transmission loss. In addition, amplitudes of first order reverberants are multiplied by $r^2$ due to the two reflections it undergoes at the bottom and the top of a layer (since one is a reflection from above and the other from below and signs of reflection coefficients alternate, this product is always positive). Each of the first order reverberants that had as their deepest point the $n$th interface has thus an amplitude proportional to

$$r^2\ (1-r^2)^n \simeq r^2 e^{nr^2}.$$

There are $2n-1$ such first order reverberants; thus the contribution of the $n$th interface to the complete unit pulse response begins with

$$z^n\ (1-r^2)^n\ (1+(2n-1)r^2 z+\ldots.).$$

For example, it follows that for $r= 0.03$ only 550 layers are necessary to make the first order reverberants as large as the direct wave. Though the example is somewhat artificial, it is by no means completely unrealistic (see O'Doherty and Anstey, 1971). One thus has to take the reverberants into account.

In figure 3 the path of the signal defined as the primary reflection from the fourth layer is indicated. It is obvious that the contribution of this layer consists of this primary reflection and a long tail of multiple reflections.
As an approximation one can regard the *downgoing* wavetrain observed at layer $n$ as a model of the contribution of layer $n$ to the total response.

Figure 4 illustrates the downgoing waves observed at 30 m intervals in a well. It is seen that the waveform is changing from one level to the next because of the loss and the delayed gain described at each interface. These changes are quite small, and the spike series has almost the property of stationarity.

With the assumption of a constant pulse shape the actual seismogram thus can be approximated by the convolution

(17) $s \approx p * r * m$ .

where $s$, $p$, $r$, $m$ are the sampled representations (or the corresponding polynomials in z) of seismogram, pulse, reflection coefficient log, and the train of internal multiples generated by a unit pulse, respectively.

An algorithm to calculate the complete unit pulse response can be based on the analysis of the network of figure 3: one only has to find out all possible paths contributing to the $n$th sample of the seismogram and sum the corresponding amplitudes. Algorithms based on this concept are efficient for the calculation of synthetic seismograms, but not so well suited for the inverse process, i.e.

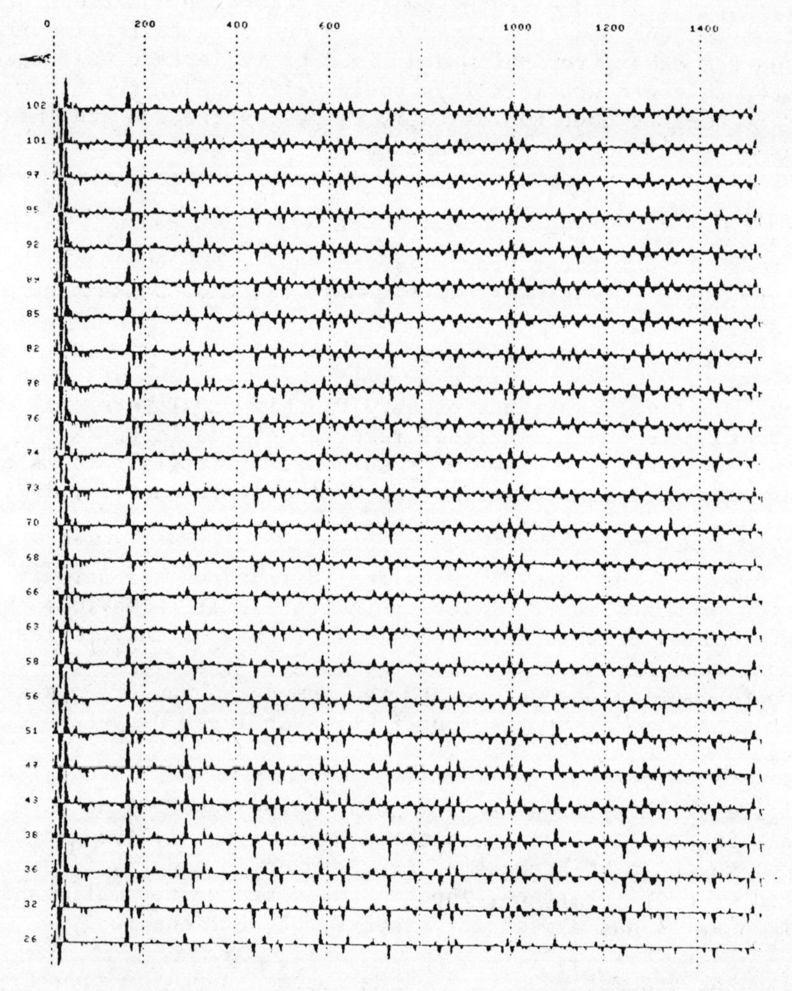


Fig. 4. Downgoing wave trains observed in a borehole at 30 m (100 ft) intervals.

the estimation of the reflection coefficient log from the seismogram. Below we therefore describe a method to calculate the wave field at an interface if the field at an adjacent interface and the corresponding reflection coefficient is known.

Figure 5 describes the relationship between the waves travelling towards and away from an interface. Continuity - as in (10) - leads to

$$(18)\quad u_1 = u_2\ (1-r)+d_1 r = u_2 + r.(d_1 - u_2)$$

$$(19)\quad d_2 = d_1\ (1+r)-u_2 r = d_1 + r.(d_1 - u_2)$$

or, resolved for $d_2$ and $u_2$

$$(18a)\quad d_2 = -\frac{1}{1-r}\,d_1 - \frac{r}{1-r}\,u_1$$

$$(19a)\quad u_2 = \frac{-r}{1-r}\,d_1 + \frac{1}{1-r}\,u_1\ .$$

Equation (18) and (19) are generalizations of (12) and (13): with $u_2 = 0$ one obtains reflection and transmission for a downward incident wave, and with $d_1 = 0$ for an upward incident wave. (18) and (19) must hold for all times (i.e. for all powers of z in the polynomial representation).

We denote the downward and upward wave trains immediately below the top of layer $n$ by

$$\mathcal{D}_n = \Sigma d_{n,k}\, z^{\frac{n}{2}+k-1} \quad \text{and} \quad \mathcal{U}_n = \Sigma u_{n,k}\, z^{\frac{n}{2}+k+1}\ ,$$

respectively. The wave trains immediately above the bottom of layer $n$ and the top of layer $n+1$ are denoted by $\mathcal{D}'_n$ and $\mathcal{U}'_n$,

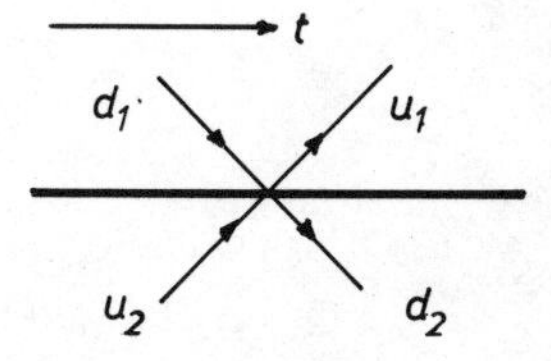


Fig. 5. Continuity of up- and downgoing waves at an interface.

respectively, and it follows that $\mathcal{D}'_n = \mathcal{D}_n . z^{\frac{1}{2}}$ and $\mathcal{U}'_n = \mathcal{U}_n . z^{\frac{1}{2}}$. Substitution into (18a) and (19a) yields

$$(20)\quad \mathcal{D}_{n+1} = \frac{z^{\frac{1}{2}}}{1-r_n}\mathcal{D}_n - \frac{r_n . z^{\frac{1}{2}}}{1-r_n}\mathcal{U}_n \quad \text{and}$$

$$(21)\quad \mathcal{U}_{n+1} = \frac{r_n . z^{\frac{1}{2}}}{1-r_n}\mathcal{D}_n + \frac{z^{-\frac{1}{2}}}{1-r_n}\mathcal{U}_n \quad ,$$

which can be written as

$$(22)\quad z^{\frac{1}{2}}(1-r_n)\mathcal{D}_{n+1} = \mathcal{D}_n . z - r_n \mathcal{U}_n \quad \text{and}$$

$$(23)\quad z^{\frac{1}{2}}(1-r_n)\mathcal{U}_{n+1} = -r_n . \mathcal{D}_n . z + \mathcal{U}_n \quad .$$

The corresponding expressions for upward continuation are

$$(22a)\quad z^{-\frac{1}{2}}(1+r_{n-1}) . \mathcal{D}_{n-1} = (\mathcal{D}_n + r_{n-1}\mathcal{U}_n) z^{-1} \quad \text{and}$$

$$(23a)\quad z^{-\frac{1}{2}}(1+r_{n-1}) . \mathcal{U}_{n-1} = (r_{n-1}\mathcal{D}_n + \mathcal{U}_n) .$$

The essential part of (22) and (23) - i.e. after dropping common terms on the l.h.s. - can be described by the Burg-ladder (figure 6). The polynomials calculated for layer $n$ with the ladder have to be multiplied by

$$z^{-n/2} . \frac{1}{\Pi^n (1-r_i)} .$$

As an example, we calculate the complete unit pulse response for a single layer above a half space. The surface has the reflection coefficient $r<0$, the interface between layer and half space the coefficient $r_1$. The response $R$ can be regarded as twice the up-going wave train $\mathcal{U}_1$: thus we have in the layer

$$\mathcal{D}_1 = 1 + \frac{R}{2} \quad \text{and} \quad \mathcal{U}_1 = \frac{R}{2} ,$$

and in the half space

$$(1-r_1) . z^{\frac{1}{2}} . \mathcal{D}_2 = \mathcal{D}_1 z - r_1 \frac{R}{2} \qquad\qquad (1-r_1) . z^{\frac{1}{2}}\mathcal{U}_2 = -r_1 z \mathcal{D}_1 + \mathcal{U}_1 .$$

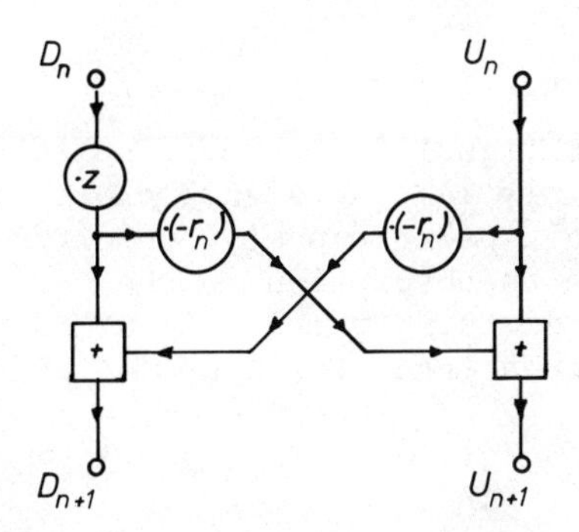


Fig. 6. The "Burg-ladder" symbolizes the algorithm for the downward continuation of wave trains through a Goupillaud medium (except for the common factor $z^{\frac{1}{2}}.(1-r_n)$:
$D_{n+1}$ is the sum of $z.D_n$ and $(-r_n).U_n$, and
$U_{n+1}$ is the sum of $U_n$ and $z.(-r_n).D_n$.

Since $U_2=0$, one has

$$-r_1.z(1+\frac{R}{2}) + \frac{R}{2} = 0 \text{ and}$$

$$r_1 z = \frac{R}{2}(1-r_1 z).$$

Thus $R = \frac{2r_1 z}{1-r_1 z} = 2(r_1 z + r_1^2 z^2 + r_1^3 z^3 + \ldots).$

The signal transmitted into the halfspace is

$$T = D_2 = \frac{1}{1-r_1} \cdot z^{-\frac{1}{2}}((1+\frac{R}{2}) - r_1 \frac{R}{2}) = \frac{1}{1-r_1} {}^{\frac{1}{2}}(1+\frac{r_1 z}{1-r_1} - \frac{r_1^2}{1-r_1 z}$$

$$= (1+r_1).z^{\frac{1}{2}} \frac{1}{1-r_1 z} = (1+r_1).z^{\frac{1}{2}} (1+r_1 z + r_1^2 z^2 + \ldots).$$

## 4. COMPLICATIONS: DIFFERENCES BETWEEN ACTUAL AND SYNTHETIC SEISMOGRAMS

Even if the assumptions made at the beginning of chapter 4 are satisfied in a particular geological situation there would be differences between an actual seismogram and the synthetic seismogram.

4.1. Pulse broadening

The ideal input signal is the unit pulse 1, 0, 0, ... (or, in polynomial form, simply 1). The signal close to an explosion approximates this unit pulse closely. Unfortunately, the loss of high frequencies due to absorption in the earth and the loss of low frequencies in recording results in a broadening of the pulse. This means in particular that *the shape is not constant* throughout the seismogram.

4.2. Other signals and noise

These are best classified as (i) those generated by the seismic source, and (ii) those having other causes.

(i) Most of what is generated by the source - other than the compressional wave which is wanted - can be described as either unwanted transmission paths or unwanted modes of vibration. Unwanted paths are refractions and diffractions. These arrivals are of significance to the interpreter, but they impede the process of inversion. Included with the diffractions are the waves from "scatterers" - such things as large boulders near the surface, or headlands on the coast.

The unwanted modes of vibration include shear waves, and a range of surface waves - Rayleigh waves, Love waves etc. They include also waves at interfaces in the subsurface - pseudo-Rayleigh waves on the sea bottom, seam waves in coal etc. These contributions to the actual seismogram are reproducible: if one repeats the observation, exactly the same kind of disturbances will result.

The source may also produce irregular seismic signals - rocks blown from a shot-hole may land near a geophone, a vibrator dropped to the ground will input a spike, and so on.

(ii) Any natural or artificial sound will be recorded on the seismogram if it is loud enough at the geophone. Wind and storm sounds, trees and plants moving, animals or insects on or close to geophones, traffic on the ground or in the air, and other man-made noises.

The irregular source effects and the noise listed under (ii) are not reproducible: a repetition of the observations results in a different realization of noise. The *relative* contribution of these effects thus can be reduced by summation of repeated observations.

### 4.3. Departures from the simple earth model

Most geological environments depart from the horizontally layered model, sometimes by angles sufficiently large to invalidate an inversion process based on horizontal assumptions. The presence of dip in the rocks results in two important changes.

(a) The partition of energy by reflection and transmission is changed, i.e. the reflection coefficient is changed, if the incident ray is not normal. Moreover, there is then mode conversion at the surface of reflection, and a primary compressional wave gives both compressional and shear components in the reflected and transmitted waves. The relationships are expressed in the Zoeppritz equations.
(b) The time intervals between the multiple elements are changed.

## 5. THE INVERSE PROCESS FOR NOISE-FREE DATA

If a time series $s = s_1 s_2 s_3 \ldots$ is known to be the convolution of a unknown unit pulse response $m = m_1 m_2 \ldots$ with an exactly known pulse $p = p_1 p_2 \ldots p_n$, the response $m$ can be, at least theoretically, recovered.

This recovery can be done stepwise: it follows from direct inspection of fig. 3 that $m_1 = s_1/p_1$. Thus the contribution due to $m_1$ is known and can be subtracted from the series. After this subtraction, the series is

$$s^{(1)} = 0, (s_2 - \frac{s_1}{p_1} \cdot p_2), (s_3 - \frac{s_1}{p_1} \cdot p_3), \ldots$$

$$= 0, \quad s_2^{(1)} \quad , \quad s_3^{(1)} \quad , \ldots$$

The series begins now with the contribution due to $m_2 = s_2^{(1)}/p_1$. The process described is nothing but the *deconvolution* of $s$, with $p$ (or the *convolution* of $s$ with $p^{-1}$, where $p^{-1} * p = 1,0,0 \ldots$):

$$s = m * p \Leftrightarrow s * p^{-1} = m * p * p^{-1} = m.$$

If one uses the z-transform to represent the time series, *convolution* becomes *multiplication of polynomials* in z, and *deconvolution* becomes a *division of polynomials*:

with $s = \sum_1 s_i z^{i-1}$, and $p = \sum_1^n p_j . z^{j-1}$

one has

(24) $m = \sum_1 s_i z^{i-1} / \sum_1^n p_j z^{j-1}$

If the seismogram $s$ is regarded as the convolution of the reflection coefficient log $r = r_1\ r_2\ \ldots$, the polynomial division (24) is already sufficient for the inverse process: we have then $m \equiv r$, and with (15) we obtain

(25) $Z_{n+1} = Z_n . e^{2\mathrm{tgh}^{-1}(-r_n)}$.

With known $z_1$ we could construct the sequence of impedances.

We have seen, however, that the unit pulse response $m$ is *not* the reflection coefficient log, but that the internal multiples (the "reverberants") have to be taken into account. In this case the recovery of $r_n$ from $s$ and $p$ is - formally - still possible: assume that in layer $n$ below the source the downgoing wavetrain

$$D_n = \sum_{k=1} d_{n,k} . z^{\frac{n}{2}+k-1}$$

and the upgoing wavetrain

$$U_n = \sum_{k=1} U_{n,k} . z^{\frac{n}{2}+k+1}$$

are known. The first sample of the upgoing wavetrain is only due to the downgoing wavetrain

$$u_{n,1} = r_n . d_{n,1}, \text{ thus}$$

$$r_n = \frac{u_{n,1}}{d_{n,1}} .$$

Once $r_n$ is known, the wavetrains $D_{n+1}$ and $U_{n+1}$ in layer $(n+1)$ can be calculated using the algorithm symbolized by the Burg ladder (fig. 6). This gives us

$$r_{n+1} = \frac{u_{n+1,1}}{d_{n+1,1}}$$

Iteration of this process leads to the complete log of reflection coefficients below layer $n$, and (25) could, in principle, be applied to recover the impedances. Such a calculation can be initiated in a well, where the up- and downgoing wavetrains have been separated. With some assumptions the calculation can be initiated from surface seismograms.

Unfortunately, algorithms of this type do not work in practice: the pulse $p$ is not exactly known, the seismogram $s$ is not the convolution of unit pulse response $m$ and pulse shape $p$. In reality we have

$$s = p * m + n$$

where $n$ is a stochastic time series representing the noise. An additional difficulty is that the basic assumptions - plane waves incident on plane parallel layers - cannot be realized. And even if all this could be overcome difficulties would remain due to the finite accuracy of the digital representation of the data.

Algorithms such as those described here are unstable: at some point in the deconvolution the subtracted pulse is too large. Thus the next sample in the unit pulse response is negative and too large. This process quickly blows up rendering the resulting series meaningless, even if the arithmetic is correct. The "undoing" of the generation of the reverberants - sometimes called "dynamic predictive deconvolution" - is beset by the same difficulties, only more so, since the process is highly non-linear.

Thus one can summarize the nature of the seismic inversion problem:

(a) The log of reflection coefficients is under-determined from the observations. There are not enough equations to determine both the pulse and the log.

(b) If the pulse is known or estimated, complete inversion, first for the pulse broadening and then for the reverberant system, can be carried out arithmetically.

(c) In the presence of noise - the practical environment - both these calculations go unstable, and the output oscillates with increasing amplitude.

(d) The approach which is used is to place some constraints on the form of the pulse and on the nature of the log of reflection coefficients so that they are now over-determined by the observations; these statistical methods can be applied to obtain the "best" estimates of pulse and log within these constraints, according to some least error criterion.

(e) The further study of inversion thus resolves itself into several different fields of study:
 - i The inversion process
 - ii The seismic and geological evidence for the constraints or the models which are postulated for these calculations
 - iii Comparison of the output from the inversion process with the reflection coefficient log calculated from velocity logs in wells and with impedances calculated on the basis of velocities and densities of core samples from the well.

A constrained inversion method is described in this volume by Lailly.

## 6. REFERENCES

Lailly, P., 1980, The inverse problem in one space dimension reflection seismics (this volume).

O'Doherty, R.F., and Anstey, N.A., 1971, Reflections on amplitudes, Geophysical Prospecting 19: 430-458.

Peterson, R.A., Fillipone, W.R., and Coker, F.B., 1955, The synthesis of seismograms from well log data, Geophysics 20: 516-538.

Robinson, E.A., 1975, Dynamic predictive deconvolution, Geophysical Prospecting 23: 779-797.

Schneider, W.A., Larner, K.L., and Burg, J.P., 1964, A new data processing technique for the elimination of ghost arrivals on reflection seismograms, Geophysics 29: 738-805.

# THE INVERSE PROBLEM IN 1-D REFLECTION SEISMICS

Patrick Lailly

I.N.R.I.A.
BP. 105
78150 Le Chesnay
FRANCE

This lecture explains some results obtained by A. BAMBERGER, G. CHAVENT, P. LAILLY.

This study has been done at INRIA (Institut National de Recherche en Informatique et Automatique), and has been supported by IFP (Institut Français du Pétrole) and SNEA (Société Nationale Elf Aquitaine).

## I - INTRODUCTION

We deal with the inverse problem in 1-D reflection seismics assuming a horizontally stratified medium and a plane (horizontal) wave excitation, the problem is to find some parameters (only function of the depth) of the substratum from surface observations.

The previous assumptions make the forward problem quite simple since the model reduces to a 1-D model.

But the inverse problem appears as a non linear inverse problem (due to multiple reflections) and this is the main difference with some inverse problems studied in reflection seismics.

The inverse problem has been studied first by Kunetz[8] a major difficulty appeared : the result was found to depend in a very unstable way on the seismic data.

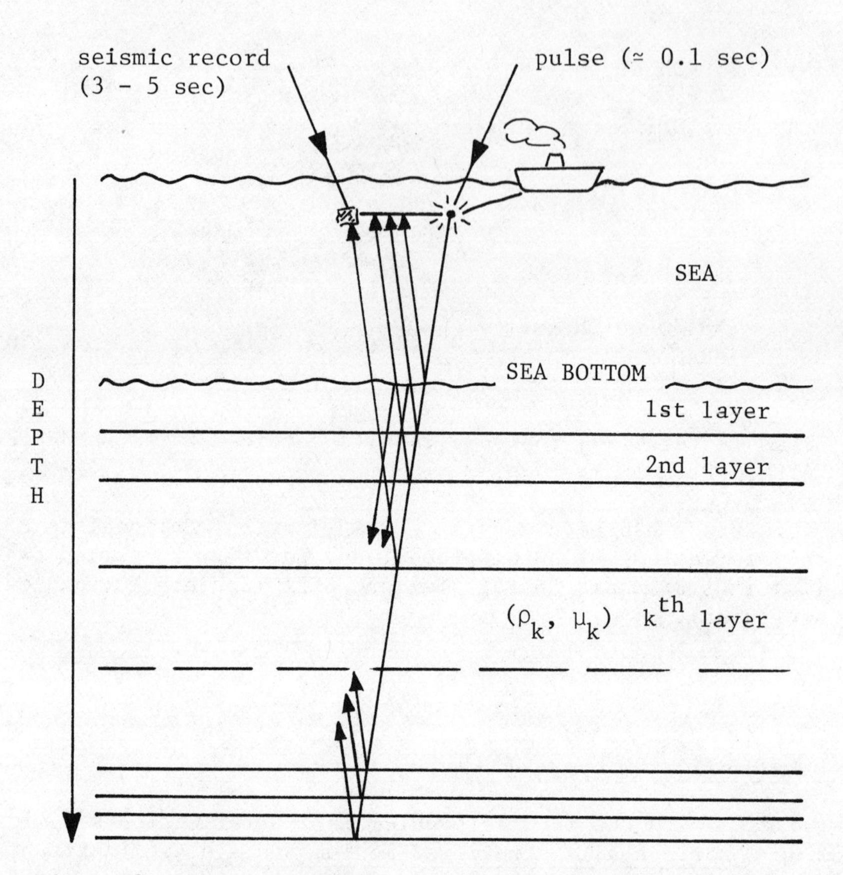


Figure 0. The inverse problem of reflection seismics.

KNOWN: pulse $g(t)$.
UNKNOWN: some parameters of the substratum (functions of depth).
OBSERVATION: seismogram $Y_d(t)$ (measurement of the displacement at the surface of the sea, for instance).

Bamberger[1] used a regularization procedure to (partially) overcome this difficulty : he characterized the undetermination for the solution and gave a stable method leading to a regularized result which is stable when noise corrupts the data.

These two points will be the main results of the first (somewhat mathematical) part of this lecture.

In a second part, we shall describe the numerical methods that we have used for the computations.

At last, we shall illustrate by some numerical results, the interest of the suggested method with emphasis on the stability with respect to the noise and the importance of the multiple reflections.

## II - STATEMENT OF THE PROBLEM

In this section we give the mathematical model that will be used in the following and we recall the main properties of the solutions. Then we define, in a general form, the inverse problem we shall be interested in.

### 2.1 - The 1-D wave equation

We have assumed that the medium is horizontally layered and that it is excited by a plane wave propagating vertically. We assume in addition that:

- the layers are constituted of a homogeneous, isotropic, linearly elastic solid,
- the second Lame's parameters of these solids are zero (*)

Owing to these assumptions, we can use, as a model, the 1-D wave equation :

$$\begin{cases} (1) \quad \rho(z) \dfrac{\partial^2 y}{\partial t^2} - \dfrac{\partial}{\partial z}\left(\lambda(z) \dfrac{\partial y}{\partial z}\right) = 0 & z \in ]0,L[ \quad t \in ]0,T[ \\ (2) \quad -\lambda(0) \dfrac{\partial y}{\partial z}(0,t) = g(t) \quad \text{(B.C)} & t \in ]0,T[ \\ (3) \quad y(z,0) = \dfrac{\partial y}{\partial t}(z,0) = 0 \quad \text{(I.C)} & z \in [0,L] \end{cases}$$

(*) This assumption is not necessary if the excitation is a P wave or a S wave.

where the following notations have been used :

$$
(4)\begin{cases}
z : \text{depth} \quad 0 \le z \le L & L : \text{large enough} \\
t : \text{time} \quad 0 \le t \le T & T : \text{duration of the observation} \\
y(z,t) : \text{displacement at depth } z \text{ and time } t \\
g(t) : \text{(pressure) pulse} \\
\lambda(z) : \text{1st Lamé's parameter of the medium at the depth } z \\
\rho(z) : \text{density of the medium at the depth } z
\end{cases}
$$

This model is the simplest one (1 space dimension, linear equation, no damping term), we recall in the next paragraph the mean properties of the solutions :

- propagation without deformation of the wave in a homogeneous medium,
- reflection and transmission of the incident wave at an interface between 2 layers.

## 2.2 - Properties of the solutions of the model (recall)

We shall use the method of characteristics to obtain the qualitative properties of the solution. The method of characteristics will also be used (cf. § 4.4) to solve numerically the wave equation.

### 2.2.1 - Solutions in an homogeneous medium

Theorem 1 : The solutions of the wave equation

$$
(5) \quad \rho \frac{\partial^2 y}{\partial t^2} - \frac{\partial}{\partial z}\left(\lambda \frac{\partial y}{\partial z}\right) = 0 \qquad z \in \mathbb{R}
$$

when $\rho$ and $\lambda$ are constant are sum of an upgoing and a downgoing wave propagating at the velocity $c = \sqrt{\frac{\lambda}{\rho}}$

Proof :

Let us consider :

$$(6)\quad \begin{cases} u(z,t) = \sqrt{\rho}\,\dfrac{\partial y}{\partial t} + \sqrt{\lambda}\,\dfrac{\partial y}{\partial z} \\ v(z,t) = \sqrt{\rho}\,\dfrac{\partial y}{\partial t} - \sqrt{\lambda}\,\dfrac{\partial y}{\partial z} \end{cases}$$

Since :

$$\begin{cases} \sqrt{\rho}\,\dfrac{\partial u}{\partial t} - \sqrt{\lambda}\,\dfrac{\partial u}{\partial z} = 0 \\ \text{and} \\ \sqrt{\rho}\,\dfrac{\partial v}{\partial t} + \sqrt{\lambda}\,\dfrac{\partial v}{\partial z} = 0 \end{cases}$$

u and v are respectively constant along the (characteristic) lines whose equation is :

$$\text{and} \begin{cases} dz = -\sqrt{\dfrac{\lambda}{\rho}}\; dt \\ dz = \sqrt{\dfrac{\lambda}{\rho}}\; dt \end{cases}$$

So u and v are respectively upgoing and downgoing waves propagating at the velocity $c = \sqrt{\frac{\lambda}{\rho}}$ and the result follows since :

$$(7)\quad \frac{\partial y}{\partial t} = \frac{1}{2\sqrt{\rho}}\,(u + v)$$ ■

REMARK 1

The following inequality :

$$T \leq \int_0^L \frac{dz}{c(z)}$$

is required to have a proper formulation of the problem (1), (2), (3). This means that the waves, emitted at $z = 0$ must arrive at $z = L$ later than T. Otherwise, we should have specified what is the boundary condition at $z = L$. ■

### 2.2.2 - Reflection and transmission (joined half spaces)

Let us consider a medium constituted of 2 layers :

- layer 1 (parameters $\rho_1$, $\lambda_1$)
- layer 2 (parameters $\rho_2$, $\lambda_2$)

Let $y_1$, $u_1$, $v_1$ and $y_2$, $u_2$, $v_2$ denote the restrictions of y, u, v (cf. (5), (6)) respectively to the layers 1 and 2.

We suppose that we know the incident waves $v_1$ and $u_2$ and we want to calculate the waves $u_1$ and $v_2$ coming from the reflection and transmission phenomena of the incident waves (cf. figure 1).

We express the continuity of the (velocity of) displacement and of the pressure at the interface between the two layers :

$$\begin{cases} \dfrac{\partial y_1}{\partial t}(z_0,t) = \dfrac{\partial y_2}{\partial t}(z_0,t) \\ -\lambda_1 \dfrac{\partial y_1}{\partial z}(z_0,t) = -\lambda_2 \dfrac{\partial y_2}{\partial z}(z_0,t) \end{cases} \qquad z_0 = \text{depth of the interface}$$

which can be exprimed in terms of u and v :

$$(8) \quad \begin{cases} \dfrac{1}{\sqrt{\rho_1}} \dfrac{u_1+v_1}{2} = \dfrac{1}{\sqrt{\rho_2}} \dfrac{u_2+v_2}{2} \\ \sqrt{\lambda_1} \dfrac{u_1-v_1}{2} = \sqrt{\lambda_2} \dfrac{u_2-v_2}{2} \end{cases}$$

The solution of the linear system (8) (in $u_1$, $v_2$) is :

$$(9) \quad \begin{cases} u_1 = u_2 + r\,(v_1-u_2) \\ v_2 = v_1 + r\,(v_1-u_2) \end{cases} \quad \text{with } r = \frac{\sqrt{\rho_1\lambda_1} - \sqrt{\rho_2\lambda_2}}{\sqrt{\rho_1\lambda_1} + \sqrt{\rho_2\lambda_2}}$$

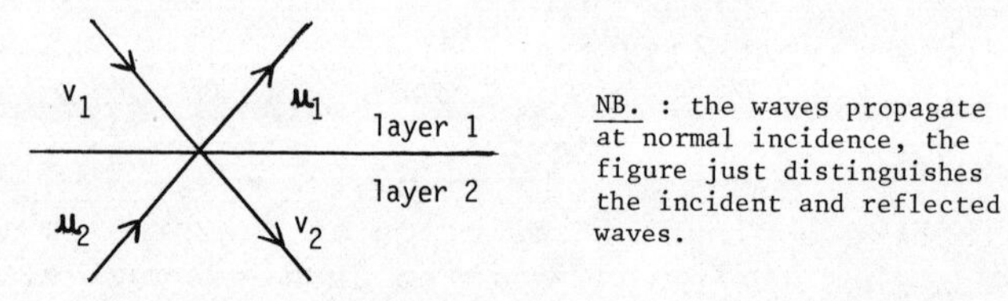


Figure 1. Calculations using formulae (9), of the reflected-transmitted waves $u_1$, $v_2$ in terms of the incident waves $u_2$, $v_1$.

Comments :

- Consider only one incident wave equal to 1, the reflected and transmitted waves will respectively be equal to r and 1 + r : the number r is called the reflection coefficent at the interface.

- The interface gives rise to reflection-transmission phenomena only if the layers are such that the product $\rho(z)\ \lambda(z)$ is discontinuous at the interface. ∎

2.3 - A first formulation of the inverse problem

In the sequel we suppose that we dispose of the following observation :

(10) $Y_d(t) = \frac{\partial y}{\partial t}(0,t) + (b,t)$ ← noise

↑ velocity of the displacement measured at the surface

REMARK 2 :

The noise b(t) can include the difference between our model and the physical phenomenon of wave propagation in the actual medium as well as the noise of measurement. ■

The inverse problem is: assuming that the pulse g(t) is known, to find the parameters $\rho(z)$ and $\lambda(z)$ from the observation $Y_d(t)$.

REMARK 3 :

In offshore seismics the observation is not $\frac{\partial y}{\partial t}(0,t)$ but the pressure at some depth. With some adaptation most of the results and the suggested method can be applied. ■

## III - MATHEMATICAL STUDY OF THE INVERSE PROBLEM

We recall that the pulse g(t) is supposed to be known. Let us now define the operator $H_g$.

(11) $H_g : (\rho, \lambda) \longrightarrow \frac{\partial y}{\partial t}(0,t)$ with y solution of (1), (2), (3)

unknown parameters ⟶ response of the system to the parameters ,

The inverse problem is then to invert the operator $H_g$: so we must ask some questions about this operator :

- is it one to one ?

- which topology on the set of parameters makes the operator $H_g$ continuous ?

### 3.1 - Some properties of the direct operator : unknown parameters $\rightarrow$ response of the system

1rst result : We cannot find both $\rho$ and $\lambda$ ($H_g$ is not one to one.

This result is well known. Let us though recall the proof :

We introduce a new depth variable x which is the travel time :

$$(12)\quad x(z) = \int_0^z \frac{d\xi}{c(\xi)} \qquad \text{with } c(z) = \frac{\lambda(z)}{\rho(z)}$$

and the acoustical impedance $\sigma$ of the medium

$$(13)\quad \sigma(z) = \sqrt{\lambda(z)\rho(z)}$$

If we change the depth variable z into the new depth variable x in equations (1), (2), (3) our model becomes :

$$\begin{cases} (14)\quad \sigma(x)\dfrac{\partial^2 y}{\partial t^2} - \dfrac{\partial}{\partial x}\left(\sigma(x)\dfrac{\partial y}{\partial x}\right) = 0 & x \in ]0,T'[ \quad x \in ]0,T[ \\ (15)\quad -\sigma(0)\dfrac{\partial y}{\partial x}(0,t) = g(t) \quad \text{(B.C.)} & t \in ]0,T[ \\ (16)\quad y(x,0) = \dfrac{\partial y}{\partial t}(x,0) = 0 \quad \text{(I.C.)} & x \in [0,T'] \end{cases}$$

$$(17)\quad \text{with } T' = \int_0^L \frac{d\xi}{c(\xi)}$$

REMARK 4 :

Theorem 1 shows that, in the model using the travel time variable, a wave propagates with a velocity equal to one, which could have been predicted ! ■

The change of variable $z \rightarrow x$ has no effect upon the observation $Y_d(t)(t)$ : the couple of parameters $(\sigma(x), \sigma(x))$ gives the same seismogram as the couple of parameters $(\rho(z), \lambda(z))$. So the operator $H_g$ is not one to one.

We can define, in the set of couples $(\rho,\lambda)$, the equivalence relation $R_T$ :

$$(18)\quad \begin{cases} (\rho_1, \lambda_1) R_T (\rho_2,\lambda_2) \text{ iff these couples define the same impedance} \\ \text{in terms of the travel time over } [0, \frac{T}{2}] \text{ (i.e., give the same} \\ \text{response over } [0,T]). \end{cases}$$

So we shall be interested with the representatives $(\sigma(x), \sigma(x))$ of the equivalence classes and we still denote by $H_g$ the operator

$$(19)\begin{cases} H_g : \{\sigma(x),\ x \in [0, \frac{T}{2}]\} \to \{\frac{\partial y}{\partial t}(0,t),\ t \in [0,T]\} \\ \text{with } y \text{ solution of (14), (15), (16).} \end{cases}$$

Then we have the following theorem (Gopinath and Sondhi[7], Bamberger[1]).

Theorem 2 :

Assuming some regularity properties of the considered functions $\sigma(x)$, the operator $H_g$ is one to one.

($\delta$ = Dirac function)

Comments :

The previous theorem shows that we can expect to find $\sigma$ as any other representative of the equivalence class associated to $\sigma$. We have chosen $\sigma(x)$ as the representative of the equivalence classes since it appears that to find $\sigma(x)$ from $Y_d(t)$ is the natural formulation of the inverse problem as we can see in the following example :

Let us consider the impedance distribution $\sigma$ given on fig. 2 (with a reflection coefficient r at $x = \lambda/2$).

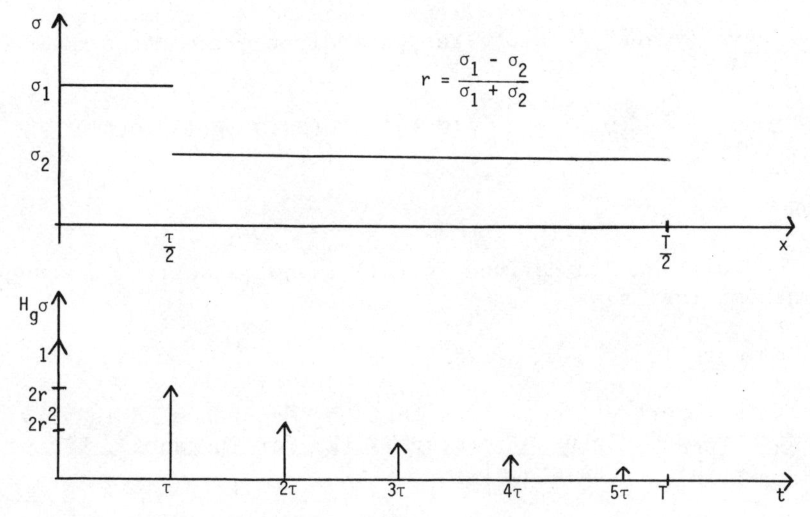


Figure 2. The impulse response of a distribution constituted of two layers.

It can be shown that the impulse response to $\sigma$ is :

$$H_g\sigma = \delta(t) + 2r\delta(t-\tau) + \ldots + 2r^k \ \delta(t-k\tau) + \ldots$$

So we can easily reconstruct the reflection coefficients distribution (and then the impedance distribution) from the impulse response : this is the basis of the Kunetz's method which tries to find, in a progressive way, the reflection coefficients.

2nd result : the operator $H_g$ is continuous for a very weak distance over the set $\Sigma_b$ :

(20) $\Sigma_b = \{\sigma(x),\ x \in [0, \frac{T}{2}]$ s.t. $\quad 0 < \sigma^- \leq \sigma(x) \leq \sigma^+ < \infty$

almost every where in $[0, \frac{T}{2}]\}$

More precisely, we have the following theorem (Gerver[5], Bamberger[1,2]- :

Theorem 3 :

*Assuming a sufficiently regular pulse g(t), the operator $H_g$ is continuous on $\Sigma_b$ into $C^0([0,T])$, for the following "distance" over $\Sigma_b$ :*

$$(21)\begin{cases} d(\sigma,\sigma') = \sup\limits_{x,x'\in[0,\frac{T}{2}]} \left| \int_0^x \sigma(\xi)d\xi - \int_0^{x'} \sigma'(\xi)d\xi \right| \\ \text{with } x \text{ and } x' \text{ linked by the relation } \int_0^x \frac{d\xi}{\sigma(\xi)} = \int_0^{x'} \frac{d\xi}{\sigma'(\xi)} \end{cases}$$

*and we have* (y and y' denote respectively the responses to $\sigma$ and $\sigma'$)

$$(22) \quad \sup_{t\in[0,T]} \left| \frac{\partial y}{\partial t}(0,t) - \frac{\partial y'}{\partial t}(0,t) \right| \leq C(\sigma^-,\sigma^+,g,T)d(\sigma,\sigma')$$

REMARK 5 :

The "distance" d defined in (21) is not exactly a distance in the mathematical sense.

Comments :

It is important to realize that the "distance" d is very weak with respect to usual distances ($L^2$ for instance). Let us illustrate this on an example.

We define (cf. figure 3) the impedance distributions $\sigma$ and $\sigma'$:

$$\sigma = \begin{cases} \sigma_2 & \forall x \quad [2i\Delta x,\ (2i+1)\Delta x[ \\ \sigma_1 & \forall x \quad [(2i+1)\Delta x,\ 2(i+1)\Delta x] \\ 0 & \forall x \ \text{t.q} \quad t_1 = 2N\Delta x \le x \le \frac{T}{2} \end{cases} \qquad i=0,\ldots,N-1$$

(we suppose $\sigma_2 > \sigma_1$).

$$\sigma' = \begin{cases} \sqrt{\sigma_1\sigma_2} & \forall x \in [0,\ \frac{t_1(\sigma_1+\sigma_2)}{2\sqrt{\sigma_1\sigma_2}}[ \\ 0 & \forall x \in [\frac{t_1(\sigma_1+\sigma_2)}{2\sqrt{\sigma_1\sigma_2}},\ \frac{T}{2}[ \end{cases}$$

REMARK 6 :

We can notice that the impedance distributions $\sigma$ and $\sigma'$ give a reflection coefficient equal to one respectively at

$x = t_1$ and $x = \dfrac{t_1(\sigma_1+\sigma_2)}{2\sqrt{\sigma_1\sigma_2}}$. ■

We are going to show that

(23) $d(\sigma,\sigma') \le (\sigma_2-\sigma_1)\Delta x$

so that $\sigma$ and $\sigma'$, very different with respect to usual distances are very close with respect to the "distance" d when $\Delta x$ is small.

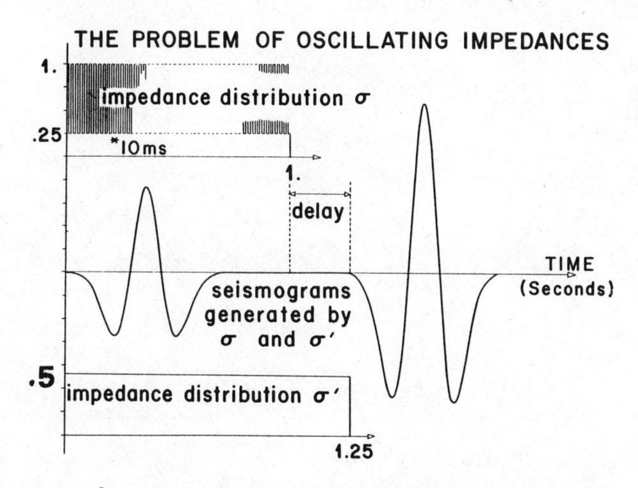


Figure 3. Instability of the inverse problem.

Proof of (23) :

i) let us first look at the case $x \in [0,t_1]$ and

$$x' \in [0,\ t_1 \frac{\sigma_1+\sigma_2}{2\sqrt{\sigma_1\sigma_2}}]$$

We have the following relations (cf. figure 4) :

$$(24)\ \int_0^x \frac{d\xi}{\sigma(\xi)} = (\frac{1}{\sigma_1} + \frac{1}{\sigma_2})\ \frac{x}{2} - \varepsilon(x)$$

(where $\varepsilon(x)$ is a positive continuous function, bounded by $\frac{1}{2}(\frac{1}{\sigma_1} - \frac{1}{\sigma_2})\Delta x$, zero for $x = 2i\Delta x \quad i \in N$)

$$(25)\ \int_0^{x'} \frac{d\xi}{\sigma(\xi)} = \frac{x'}{\sqrt{\sigma_1\sigma_2}}$$

$$(26)\ \int_0^x \sigma(\xi)d\xi = (\frac{\sigma_1+\sigma_2}{2})\ x + \varepsilon_1(x)$$

(where $\varepsilon_1(x)$ is a positive continuous function, bounded by $\frac{\sigma_2-\sigma_1}{2}\Delta x$, zero for $x = 2i\Delta x \quad i \in N$)

$$(27)\ \int_0^{x'} \sigma'(\xi)d\xi = \sqrt{\sigma_1\sigma_2 x'}$$

From (24), (25), we find :

$$(28)\ x' = \frac{\sigma_1+\sigma_2}{2\sqrt{\sigma_1\sigma_2}}\ x - \varepsilon(x)\ \sqrt{\sigma_1\sigma_2}$$

so that :

$$(29)\ \left|\int_0^x \sigma(\xi)d\xi - \int_0^{x'} \sigma'(\xi)d\xi\right| = \varepsilon_1(x) + \varepsilon(x)\sigma_1\sigma_2$$

$$\leq (\sigma_2-\sigma_1)\Delta x$$

ii) For $x = t_1$ the corresponding value of $x'$ is (cf. (28))

$t'_1 = \frac{\sigma_1+\sigma_2}{2\sqrt{\sigma_1\sigma_2}}\ t_1$ so that (cf. (29))

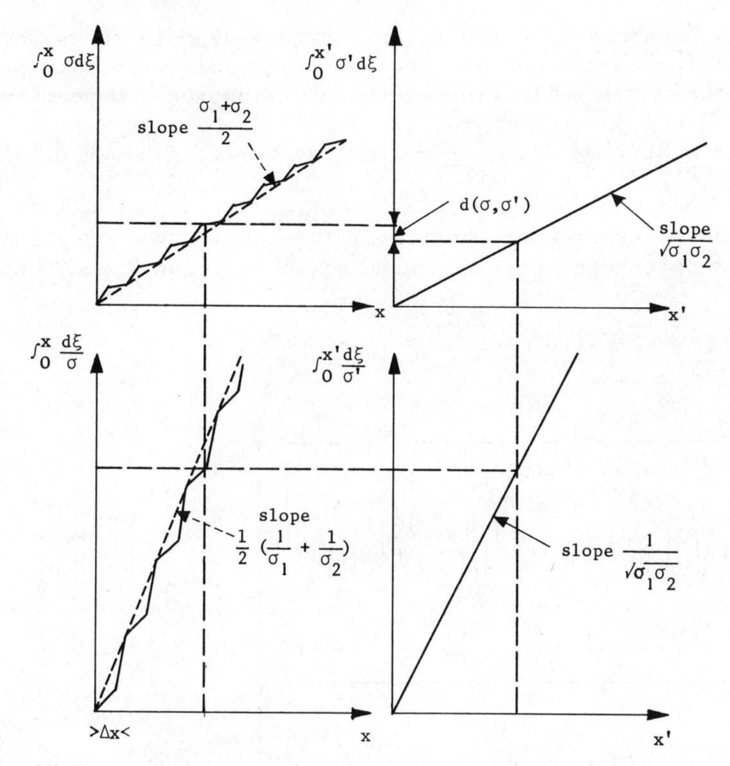


Figure 4. Calculation of $d(\sigma,\sigma')$.

$$\int_0^{t_1} \sigma(\xi)d\xi - \int_0^{t'_1} \sigma'(\xi)d\xi = 0$$

and $\int_0^x \sigma(\xi)d\xi - \int_0^{x'} \sigma'(\xi)d\xi$ remains zero for $x > t_1$ and $x' > t'_1$. ∎

2 Consequences for the inverse problem

1°) We can see (cf. figure 3) that, as theorem 3 had predicted, the impedance distributions $\sigma$ and $\sigma'$ give very close seismograms for a large enough pulse. Of course the impulse seismograms generated by $\sigma$ and $\sigma'$ are very different : the constant C in (22) depends on the pulse g(t) and is as greater as g is narrow. Figure 5 shows how the seismograms generated by a sequence of oscillating impedances go to the seismogram generated by $\sigma'$ as the oscillating step becomes small with respect to the wave length.

So we can see that, for usual pulses, the inverse problem is ill-posed : 2 very different impedance distributions $\sigma$ and $\sigma'$ generate infinitely close seismograms. In this sense we can say that the inverse problem has not a unique solution.

2°) Apparent velocity : The previous example shows that some (oscillating) media can generate apparent delays in wave propagation : the total reflection of the distribution $\sigma$ appears on the seismogram at time $2t_1 \frac{\sigma_1+\sigma_2}{2\sqrt{\sigma_1\sigma_2}}$ instead of $2t_1$. Though $\sigma$ is represented as a function of the travel time, waves do not propagate, for a macroscopic point of view, with a velocity equal to 1 but with an apparent velocity $C_a = \frac{2\sqrt{\sigma_1\sigma_2}}{\sigma_1+\sigma_2} \leq 1$.

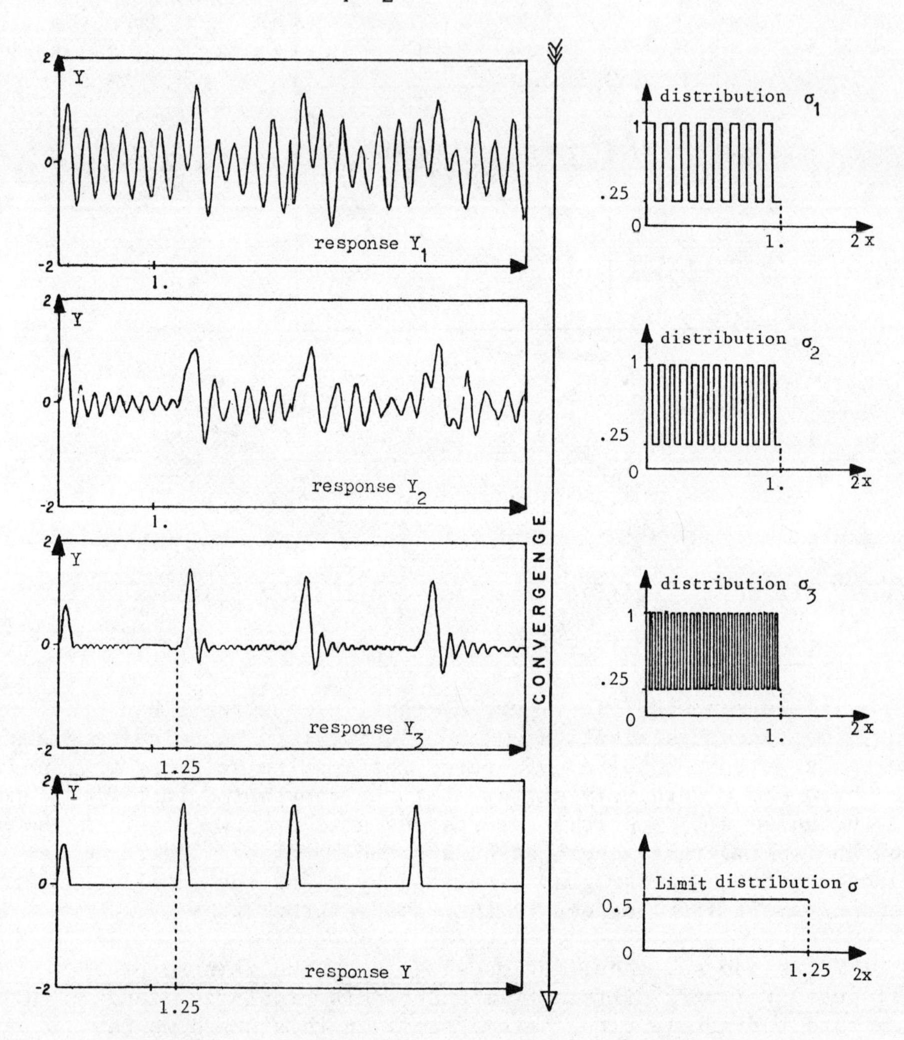


Figure 5. Numerical illustration of the convergence.

The apparent velocity does not depend on the oscillation frequency but only on the oscillation amplitude. We can express the apparent velocity $C_a$ in terms of the ratio $q = \frac{\sigma_1}{\sigma_2}$ or in terms of the reflection coefficient r. Table 1 gives the apparent velocity for different values of q or r. The apparent velocity can lead to problems for adjusting the result of the inverse problem with sonic log.

Table 1 : Apparent velocity in oscillating distributions

| r (reflection coefficient) | 0.82 | 0.67 | 0.54 | 0.43 | 0.33 | 0.25 | 0.18 | 0.11 | 0.05 | 0 |
|---|---|---|---|---|---|---|---|---|---|---|
| $q = \frac{\sigma_1}{\sigma_2}$ | 0.1 | 0.2 | 0.3 | 0.4 | 0.5 | 0.6 | 0.7 | 0.8 | 0.9 | 1.0 |
| $C_a$ | 0.58 | 0.76 | 0.84 | 0.9 | 0.94 | 0.96 | 0.98 | 0.98 | 1.00 | 1.00 |

### 3.2 - A regularized formulation of the inverse problem

In order to eliminate the previous ambiguity in the solution of the inverse problem we shall seek the solution in a set smaller than $\Sigma_b$ but sufficiently large to contain a great number of impedance distributions that can be found in nature.

As the difficulty comes from oscillating impedances it is natural to introduce the set :

$$(30)\quad \Sigma_{var} = \{\sigma \in \Sigma_b \quad \text{s.t.} \quad \text{var}\, \sigma \leq M\}$$

where M is a constant that we choose a priori.

We recall that the variation of a function is defined by

$$(31)\quad \text{var}\, \sigma = \sup_{\substack{\text{subdivisions}(x_i) \\ 0 = x_0 < x_1 < x_i < x_{i+1} = \frac{T}{2}}} \sum_i |\sigma(x_{i+1}) - \sigma(x_i)$$

(it measures the oscillating character of a function).

We can prove the following theorem (Bamberger[1]).

Theorem 4 :

On $\Sigma_{var}$ the "distance" d defines a stronger topology than the one induced by the $L^2$ norm :

(32) $\forall \varepsilon > 0 \ \exists \eta$ s.t. $\forall \sigma, \sigma' \in \Sigma_{var}$

$$d(\sigma, \sigma') < \eta \Rightarrow \left( \int_0^{\frac{T}{2}} (\sigma - \sigma')^2 dx \right)^{\frac{1}{2}} < \varepsilon$$

REMARK 7 :

The set $\Sigma_{var}$ is not the only one to have the property given in theorem 4. We have chosen $\Sigma_{var}$ because we want to take account of impedance distributions with strong discontinuities.

Consequences for the inverse problem :

- the problem "to find $\sigma \in \Sigma_{var}$ from $H_g(\sigma)$" is mathematically a well posed problem,

- unfortunately we cannot identify many physical impedance distributions which are not in $\Sigma_{var}$. We can only find a regular representative of such distributions. Then it is important to keep in mind that, from each regular solution, we can construct a solution as irregular as we want and having a response infinitely near . (Cf. Bamberger[1]).

### 3.3 - Approximation of the inverse problem

In order to solve the inverse problem with a computer an approximation technique is required. We need to replace the function $\sigma(x)$ by a function depending on a finite number of parameters.

Let us define the set $\Sigma_{\Delta x}$ :

(33) $\Sigma_{\Delta x} = \{\sigma^c \in \Sigma_{var}$ which are piecewise constant on equal travel time intervals of length $\Delta x\}$

Then we have :

Theorem 5 : (approximation of $\Sigma_{var}$ by $\Sigma_{\Delta x}$) :

$\forall \varepsilon \in 0\ \exists\ \Delta x$ such that for all $\sigma \in \Sigma_{var}$ there exists $\sigma^c \in \Sigma_{\Delta x}$ such that $d(\sigma, \sigma^c) < \varepsilon$.

Consequence :

If we choose $\Delta x$ sufficiently small we can approximate, as precisely as we wish, the impedance distributions in $\Sigma_{var}$ (for the $L^2$ norm) and their responses.

REMARK 8 :

We have chosen $\Sigma_{\Delta x}$ in order to generate impedances with strong discontinuities. An other reason of this choice (for a practical point of view) will be given in § 4.4. ■

3.4 - Definitive formulation of the inverse problem, stability with respect to the noise

Because of the noise, it may happen that our observation $Y_d(t)$ is not the response to any impedance distribution ; so a natural formulation of the inverse problem is the following :

"To find an impedance distribution $\tilde{\sigma} \in \Sigma_{\Delta x}$ which generates a synthetic seismogram fitting the observed seismogram at best".

or more precisely :

$$(34)\quad \begin{cases} \text{Given } g(t),\ T,\ M,\ \Delta x, \text{ to find } \tilde{\sigma} \in \Sigma_{\Delta x} \text{ such that} \\ J(\tilde{\sigma}) = \min J(\sigma) \text{ where} \\ J(\sigma) = \int_0^T \left(\frac{\partial y}{\partial t}(0,t) - Y_d(t)\right)^2 dt \\ \text{with } y \text{ solution of (14), (15), (16) with the parameter } \sigma \end{cases}$$

Then the inverse problem appears as an optimal control problem in a partial differential equation as studied by Lions[11] and applied to parameter estimation by Chavent[3].

Some mathematical results :

We want to define a projection on the set $H_g(\Sigma_{\Delta x})$ which is closed in $L^2$ but not convex ! So there is some difficulty to give sense to problem (34) and to have a continuity property of the projection operator when it exists. We refer to Bamberger[2] to have a precise statement of the following result

There exists a neighbourhood of the set $H_g(\Sigma_{\Delta x})$ on which we can define a projection operator which is lipschitz continuous.

This shows that the method is stable with respect to perturbations on the observation when they are not too strong.

Advantages of this formulation :

We can expect 2 advantages of this formulation :

- the determination of the solution $\tilde{\sigma}$ is global as opposed to the Kunetz's method which is a progressive method and so accumulates errors as the computation (depth) advances
- it has been designed for noise corrupted observation (by the projection and regularization procedures) this is not the case of the method suggested by Gjevick, Nilsen, Hoyen[6].

Per contra, the method leads to somewhat costly computations.

3.5 - Some remarks about the comparison of the computed solution with sonic logs

1°) The case of oscillating impedances

As indicated in § 3.2, we cannot identify impedance distributions whose variation is greater than the (a priori) chosen bound M. More precisely we can always invert the seismogram generated by these distributions but some cautions are required when we compare the physical impedance distribution $\sigma$ and the computed impedance distribution $\tilde{\sigma}$.

An approach could be to compare the computed distribution to a regular distribution $\sigma^r$ generating a seismogram near the one generated by $\sigma$. This distribution $\sigma^r$ can be computed by the following algorithm :

To find the growing sequence of points whose abscissa is $x'_k$ and the sequence of numbers $\sigma^r_k$ such that :

$$(35)\quad \begin{cases} \int_{x'_k}^{x'_{k+1}} \sigma(x)\,dx = \sigma^r_k\,\Delta x \\ \int_{x'_k}^{x'_{k+1}} \frac{dx}{\sigma(x)} = \frac{\Delta x}{\sigma^r_k} \qquad k = 0,1,\ldots \\ x'_0 = 0 \end{cases}$$

The distribution $\sigma^r$ is defined by :

$$\sigma^r(x) = \sigma^r_k \quad \text{for } k\Delta x \le x \le (k+1)\Delta x$$

REMARK 9 :

The impedance distribution $\sigma^r$ is piecewise constant on intervales of length $\Delta x$ (in travel time) but its variation can be greater than M. ∎

REMARK 10 :

When $\mathrm{var}\sigma \le M$ the algorithm (35) gives the construction of the distribution $\sigma^c$ mentioned in theorem 5. ∎

The usual procedure to regularize an impedance distribution (with a view to compute synthetic seismograms) is to apply a band pass filter to the sonic log. This procedure is satisfactory when we do not take account of the multiple reflections : we just have suppressed high frequencies in the impulse response that do not appear in the seismogram. But this procedure may be no more valid when we want to take account of all the multiple reflections : the high frequencies in the impedance distribution may have an effect on the low frequencies of the impulse response (an oscillating distribution leads to a delay).

2°) Effect of the mean value of the impedance on the seismogram

It can be shown (and this is well known from geophysicists) that, when the pulse $g(t)$ is zero mean, the seismogram scarcely depends on the mean value of the impedance. So, in this case, we cannot expect to find the mean value of the impedance.

## IV - COMPUTATION OF THE SOLUTION

### 4.1 - Solution of optimization problems by gradient methods

The definitive formulation of the inverse problem appears as an optimization problem. We can only solve these problems in finite dimension vectorial spaces : this has lead us to approximate the (infinite dimension) set $\Sigma_{var}$ by the (finite dimension) set $\Sigma_{\Delta x} \subset R^N$ (N = number of layers). We shall see (cf. § V) that $\Delta x$ must be chosen very small ($\simeq$ 4ms) so that $N = \frac{T}{2\Delta x}$ is large (at least 300-400).

So we cannot explore the whole set $\Sigma_{\Delta x}$ in order to see which is the function $\tilde{\sigma}$ which minimizes the cost function $J(\sigma)$. Then we must use an (iterative) descent method :

Starting from a point $\sigma^0 \in \sigma_{\Delta x}$ we construct a sequence $\sigma^n$ of functions in $\Sigma_{\Delta x}$ such that $J(\sigma^{n+1}) < J(\sigma^n)$ just looking whether such a $\sigma^{n+1}$ exists in neighboorhood of $\sigma^n$. We expect that the sequence $\sigma^n$ converges to $\tilde{\sigma}$. (This may not happen if the function $J(\sigma)$ is not convex : $\sigma^n$ may converge to a local minimum for some $\sigma^0$).

An important class of descent methods are gradient methods : we can find a point $\sigma^{n+1}$ such that $J(\sigma^{n+1}) < J(\sigma^n)$ in the direction opposite to the gradient direction. Usually we choose $\sigma^{n+1}$ which minimizes $J(\sigma)$ in this direction : this is an optimization problem in R which is much simpler than the original problem in $\mathbb{R}^N$.

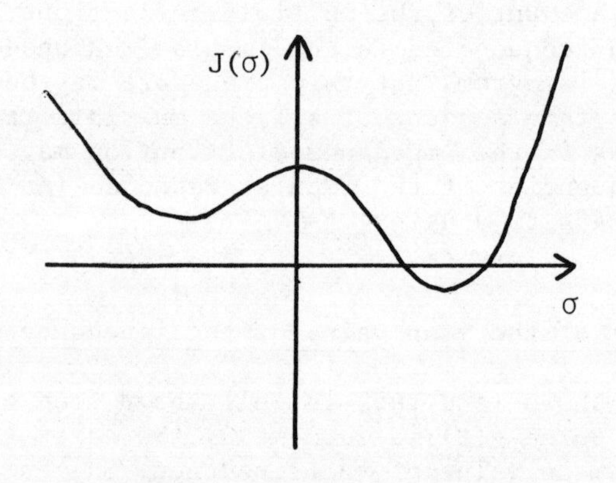


A nonconvex function

4.2 - Calculation of the gradient

There is a difficulty to calculate the gradient : the function $\sigma \to J(\sigma)$ cannot be explicited. Variation calculus can overcome this difficulty. We introduce the solution $p(x,t)$ of the adjoint equations

$$(36)\begin{cases} \sigma^n \dfrac{\partial^2 p}{\partial t^2} - \dfrac{\partial}{\partial x}\left(\sigma^n \dfrac{\partial p}{\partial x}\right) = 0 \\ -\sigma^n(0) \dfrac{\partial p}{\partial x}(0,t) = 2\left(\dfrac{\partial y^n}{\partial t}(0,t) - Y_d(t)\right) \\ p(x,T) = \dfrac{\partial p}{\partial t}(x,T) = 0 \end{cases}$$

($y^n$ = solution (14), (15), (16) with $\sigma = \sigma^n$)

then, in the continuous case ($\sigma^n \in \Sigma_{var}$), the component, associated to one point x, of the gradient at $\sigma^n$ is given by :

$$(37)\quad \int_0^T \frac{\partial y^n}{\partial t}(x,t)\,\frac{\partial p}{\partial t}(x,t) - \frac{\partial y^n}{\partial x}(x,t)\,\frac{\partial p}{\partial x}(x,t)\,dt$$

When $\sigma^n \in \Sigma_{\Delta x}$ the gradient can be obtained by an adequate discretization of (36), (37).

4.3 - Optimization with constraints

The set $\Sigma_{\Delta x}$ is just a part of the vectorial space $\mathbb{R}^N$. The bound constraint :

$$\sigma^- \leq \sigma(x) < \sigma^+ \qquad \forall x \in [0, \tfrac{T}{2}]$$

gives no difficulty : we can just project the descent direction (opposite to the gradient) on $\Sigma_{\Delta x}$ when $\sigma^n$ is on the boundary of $\Sigma_{\Delta x}$.

But the constraint $\text{var}\sigma \leq M$ is very difficult to manage. We have changed the original cost function $J(\sigma)$ (cf. (34)) into (penalization technique)

$$J(\sigma) + \frac{1}{\varepsilon}(\text{var}\sigma - M)^+ \quad \text{where} \quad x^+ = \begin{cases} x \text{ if } x \geq 0 \\ 0 \text{ if } x \leq 0 \end{cases}$$

Then we are lead to an optimization problem without constraints but an other difficulty appears : the cost function is no more differentiable since the mapping : $\sigma \to \text{var}\sigma$ is not differentiable.

This difficulty has been overcome by an adaptation of gradient methods to non differentiable problems (Lemarechal [10]).

REMARK 11 :

The penalization technique is a handy way to introduce constraints in an optimization problem. So we can easily prescribe the computed solution to verify any supplementary information which is available a priori or obtained by other methods. ∎

4.4 - Computational aspects

The following diagram gives the algorithm of the method :

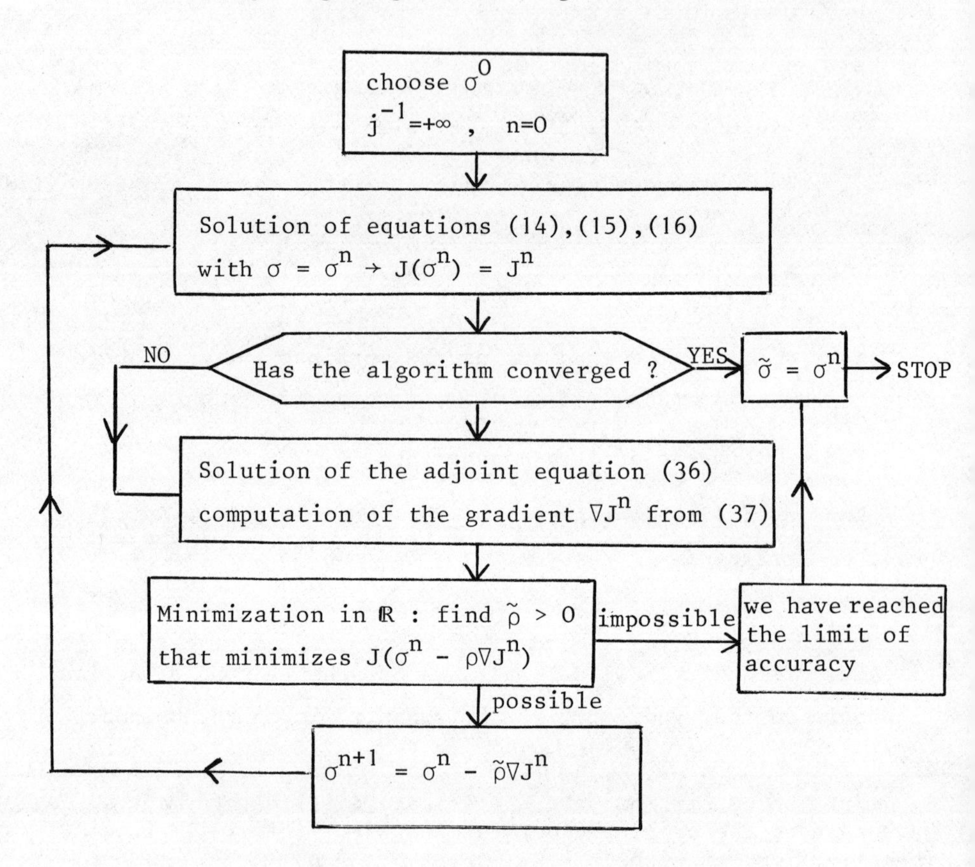


We can see that the size of the problem is large :

- a gradient iteration (minimization in R) requires to solve about 6 times the wave equation (state or adjoint equations)

- the number of iterations will be large (the cost function will appear to be flat and the number of layers N is large).

So a special care has been taken in the choice of the algorithms :

1°) The wave equation is solved by the method of characteristics which gives the exact impulse response in a very quick way when $\sigma \in \Sigma_{\Delta x}$. This impulse is given by :

$$(38) \quad \sum_{k=0}^{k=N-1} \lambda_k \, \delta(t - 2k\Delta x) \qquad (\delta = \text{Dirac function})$$

the numbers $\lambda_k$ being computed applying formulae (9) at the nodes of the net of characteristics (cf. Bamberger[1]).

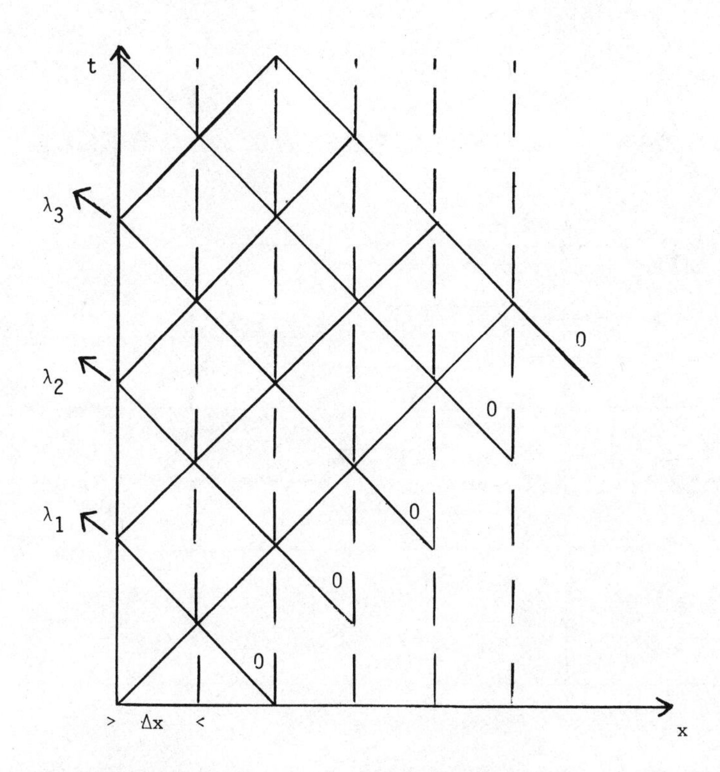


Figure 6. The method of characteristics.

2°) The optimization problem is solved by a fast converging conjugate gradient algorithm (Lemarechal[10]).

These algorithms have considerably reduced the computing time of our first attempts to solve this problem and now, the inversion of one usual seismogram (T = 2.4 sec., N = 300, 300 gradient iterations) requires about 3 minutes for a CDC 7600 computer.

## V - SOME EXAMPLES OF SEISMOGRAM INVERSIONS

In a first serie of runs on simulated data we test the practicability of the method and the stability of the computed result when some noise corrupts the seismogram.

Then we have tried to invert actual recordings. Table 2 gives the characteristics of the runs.

### 5.1 - Runs on simulated data

#### 1°) Principle of the runs

Figure 7 gives the principle of the runs.

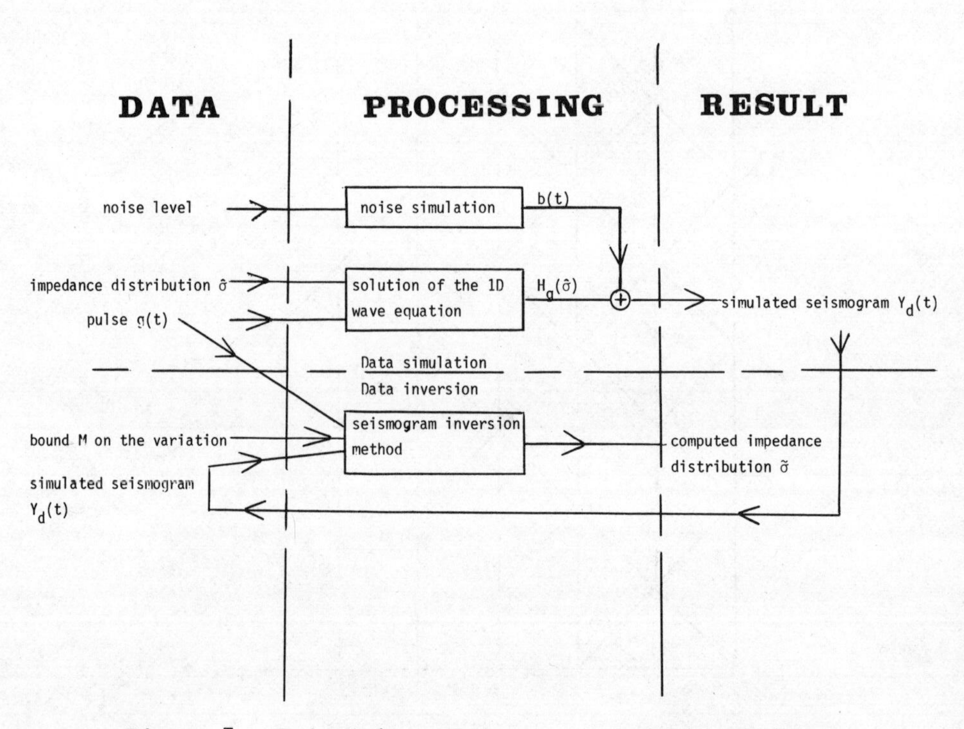


Figure 7. Principle of the runs on simulated data.

| Table 2 | Runs I | Runs II | Runs III |
|---|---|---|---|
| Nature of the data | Simulated seismogram noise levels : 0 % and 50 % | Field data (offshore seismics) | Field data (offshore seismics) |
| Characteristics of the example | Actual impedance distribution | Horizontally stratified substratum deep sea (3 sec.) → plane waves and normal incidence | Horizontally stratified substratum shallow sea (400 ms) → spherical waves and oblique incidence |
| Pulse | Ricker | Flexichoc | Airgun |
| Preliminary data processing | | Seismograms : partial stack (5 traces) without NMO and amplitude correction to compensate the spherical attenuation<br>Pulse : syntesis or estimation | Stack with NMO and Slant stack<br>Estimation of the pulse |
| T (observation duration) | 2.4 sec | 6 sec (but 2.8 sec of water layer) | 5 sec |
| Parameters of the program | $\Delta x$ = 4 ms<br>M = 0 and 28000 | $\Delta x$ = 4 ms<br>M = 15000 (**) | $\Delta x$ = 5 ms (*)<br>M = 15000 (**) |
| Number of layers | 300 | 400 | 500 |
| Number of gradient iterations | 600 | 100 | 30 |
| Computing time for 1 trace | 6 min (CDC 7600) | 80 sec (CDC 7600) | 60 sec (CDC 7600) |

(*) the program has required such a $\Delta x$ so we had to sample again the data.

(**) We have chosen for M the lowest value which allows a fit of the computed seismogram on the recorded seismogram.

Note that the impedance distribution $\hat{\sigma}$ which has been used in the simulation has been obtained from a sonic log recorded in the Sahara. The noise b(t) has been computed by convolution of a white noise with the pulse g(t) so that we are in the worst case where the spectra of the noise and of the signal are the same. The step $\Delta x$ used to compute $\tilde{\sigma}$ is the same as the sampling step of $\hat{\sigma}$. At last the same pulse g(t) has been used of the solution of the inverse problem and the computation of the synthetic seismogram : it is the Ricker pulse which is given on figure 8.

2°) Inversion of synthetic seismograms (no noise)

Figure 9 gives the results obtained after 600 iterations of the gradient algorithm for a non zero mean pulse (top) and for a zero mean pulse (bottom). In this last case, the shape of the impedance distribution is well recovered but not the mean value of the impedance as foretold in § 3.5 2° ; the following iterations do not really improve the result.

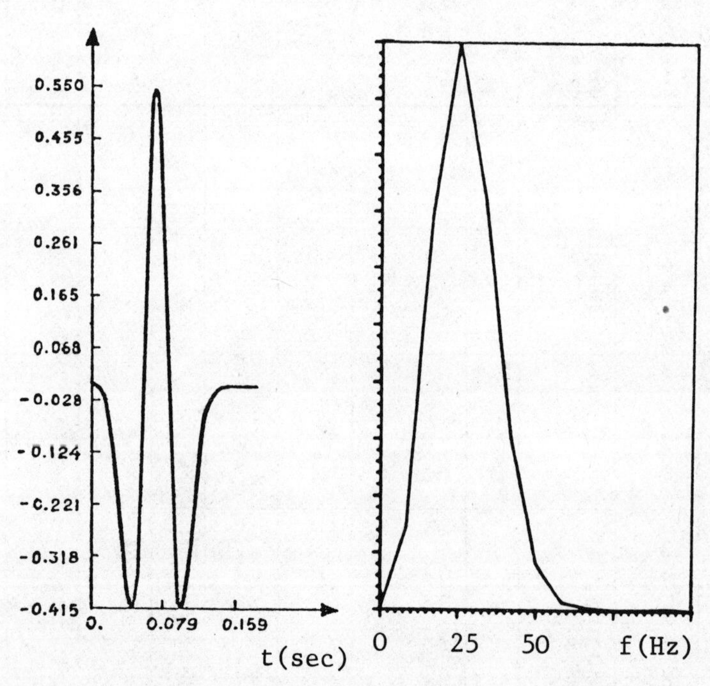


Figure 8. The Ricker pulse (left) and its spectrum (right).

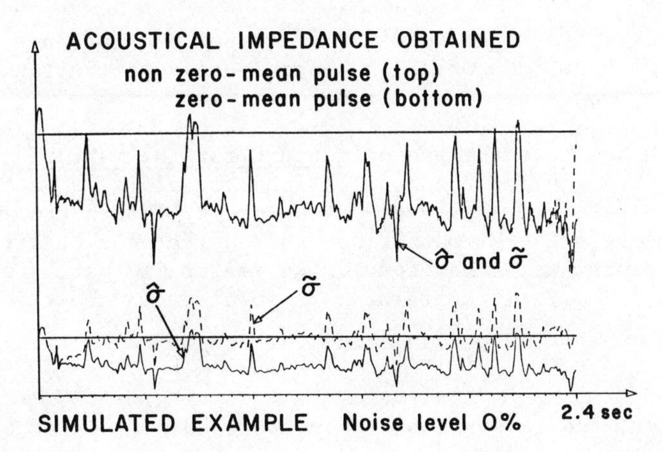


Figure 9. Runs on simulated data (no noise).

3°) Inversion of noise corrupted seismograms, influence of the chosen bound M on the variation

The noise level (*) is 50 % : figure 10 gives the original synthetic seismogram (thin line) and the noise corrupted seismogram (thick line).

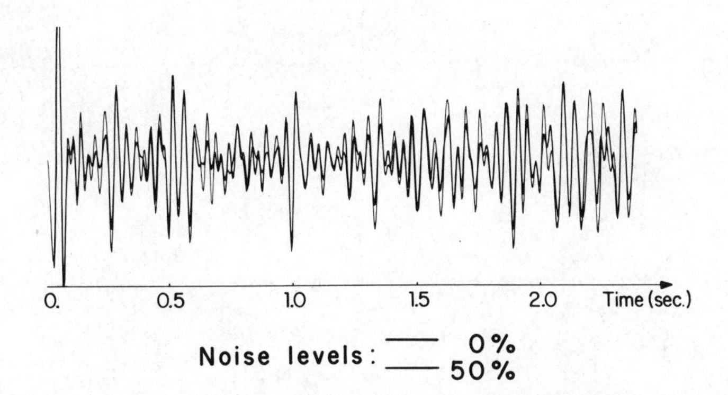


Figure 10. Simulated seismograms.

(*) ratio standard variation of the noise over standard variation of the signal.

Figure 11 gives the impedance distributions obtained with $M=\infty$ (top) and with M=28000 which is approximately the variation of the original impedance distribution $\hat{\sigma}$.

We can see that the use of a bound for the variation improves the result by reducing the oscillations and giving a better recovery of the relative amplitude of the peaks. Though the impedance is computed in a global manner we can see that the quality of the result decreases with the depth ; the reason is that, due to multiple reflections, $Y_d(t)$ contains more information about the shallower reflection coefficients than about the deeper reflection coefficients.

At last, the response of the computed distribution $\tilde{\sigma}$ is superposable to the observation $Y_d(t)$.

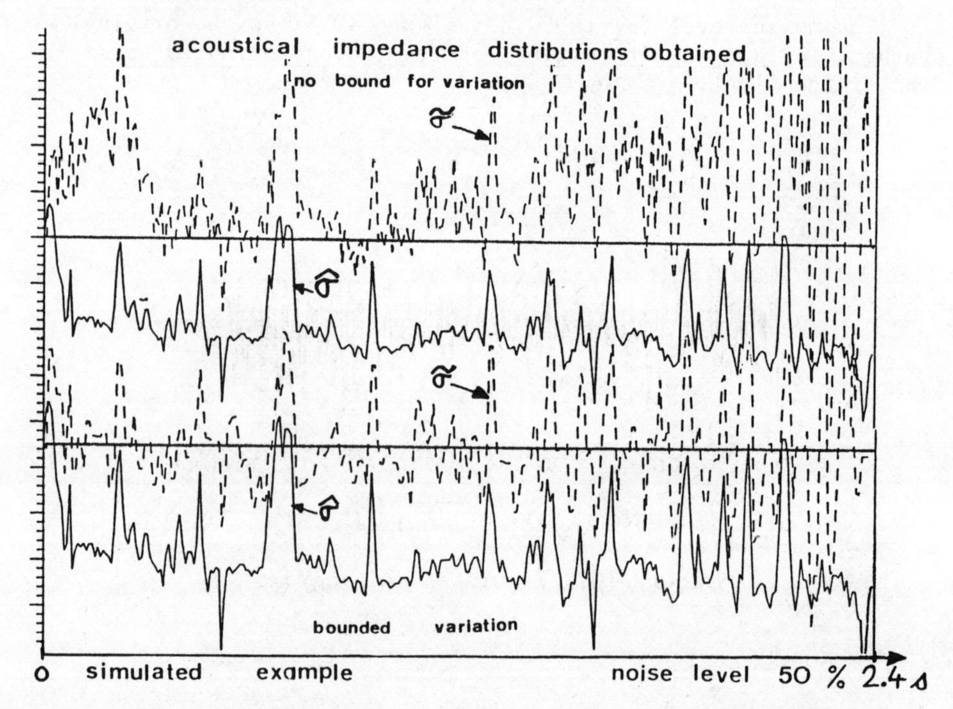


Figure 11. Impedance distributions obtained from a 50% noise corrupted seismogram with (bottom) and without (top) bound on the total variation of the impedance.

4°) Discussion : impedance of the multiple reflections

We have plotted on figure 12, for the (actual) distribution $\hat{\sigma}$ the partial derivatives of the 190th coefficient $\lambda_{190}$ of the impulse response (cf. (38)) with respect to the successive reflections coefficients $r_1,\ldots,r_{300}$.

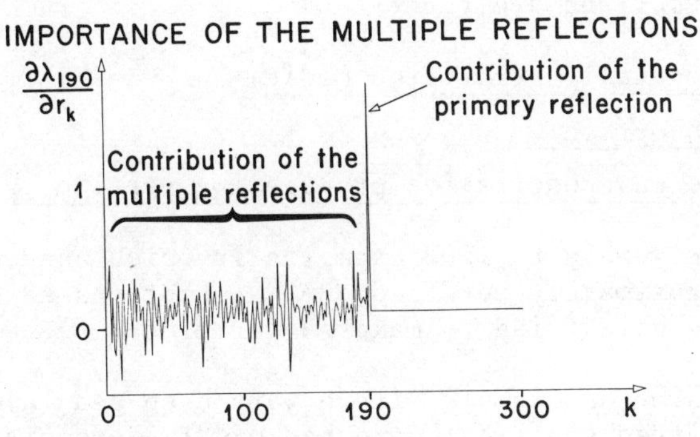


Figure 12. The partial derivatives of $\lambda_{190}$ with respect to $r_1$ up to $r_{300}$.

One checks easily that :

$$(1) \qquad \frac{\partial\lambda_{190}}{\partial r_k} \simeq .2 \quad k < 190 \quad , \quad \frac{\partial\lambda_{190}}{\partial r_k} = 0 \quad k > 190$$

from which we draw two conclusions :

. in the case of the impedance distribution used $\hat{\sigma}$, the multiple reflections cannot be neglected in the determination of the impedance.

. the sequential determination of the $r_k$'s (i.e. first $r_1$, then $r_2$, etc...) from the impulse response vector $\lambda$ -that is the principle of Kunetz's method- is highly unstable : in the case of figure 12, suppose for instance that the reflection coefficients $r_1$ up to $r_{189}$ have been determined within an uncertainty of $\pm$ 0.05; it follows then that $r_{190}$ is determined within an uncertainty of $\pm$ .95, and hence is no more significant !

### 5.2 - Runs on field recordings

We have considered two examples (Runs II and III) of increasing difficulties and interest for exploration.

#### 5.2.1 - Recall of the assumptions required for the validity of the model

These assumptions are :

A1 : horizontally stratified medium

A2 : no absorption

A3 : plane wave excitation propagating vertically

It is fairly easy to find examples in which assumptions A1 and A2 are approximately verified. But assumptions A3 gives some problems. So we have tried to make A3 verified :

- in choosing an example with a very deep sea layer (Runs II) so that the emitted spherical wave has nearly degenerated into a plane wave and the normal incidence information is available since the offset (≃ 300 m) is small with respect to the deepness of the sea. For each shot of the seismic section, a (five traces) stack (without normal moveout corrections) has been done in order to reduce the noise level as well as an amplitude correction to compensate the spherical attenuation.

- by an adequate processing (slant stack) of the data (Runs III).

We refer to table 3 for the description of the examples of the preliminary data processing, of the parameters used in the inversion.

#### 5.2.2 - Runs II (deep sea)

We want to check, on a propicious example, that the method :

- is not too much sensitive about some parameters which are difficult to know and especially about the pulse,

- leads to a result which can be used by geologists (*) (the result must show usual figures such as reflectors, scattering hyperbolas, etc... despite of the noise),

---

(*) the best would be to compare the computed solution with a recorded impedance but no sonic log is available when the sea is so deep.

- makes something more than just a deconvolution (we want to check that the method eliminates multiple reflections).

1°) Influence of the pulse

The "Flexichoc" pulse which has been used is reliable (gives the same excitation from one shot to an other) but it is not recorded. So we must find what is the pulse g(t) which appears in the boundary condition (15). Two methods have been used to solve this problem :

- synthesis of g(t) by convolution of a sequence of four delta functions : the pulse itself and its three ghosts) with a seismic band pass filter. This gives pulse 1 (figure 13).

- a method (cf. Stone[15]) which estimates the pulse from the seismic record. This gives pulse 2 (figure 13).

Figure 13 gives the impedances obtained by inversion of one trace using pulse 1 and pulse 2. Though these 2 pulses are somewhat different the computed impedance distributions are well correlated (but the correlation decreases with the depth).

At last, pulse 1 as well as pulse 2 gives a computed seismogram which is superposable with the recorded seismogram.

2°) Inversion of a seismic section

We have inverted the seismic section (constituted of 80 traces) represented on figure 14. We have plotted the section constituted of the 80 computed distributions of reflection coefficient (figure 15).

We can see that the computed section is a good deconvolution of the original section : it gives very sharp details due to high frequencies that have been allowed in the computed distribution. In particular some reflectors are located very accurately.

The computed section shows each reflector with its own strength (our processing has theoretically preserved the amplitudes) : this may be a good help for the geologist to distinguish the geological interfaces.

The strongly noise corrupted traces of the original seismic section (in the middle and in the left of the section) have not spoiled the quality of the computed section.

At last, we can see that the 2D or 3D phenomenons on the original section (scattering hyperbolas, dipped arrivals) are also conspicuous on the computed section.

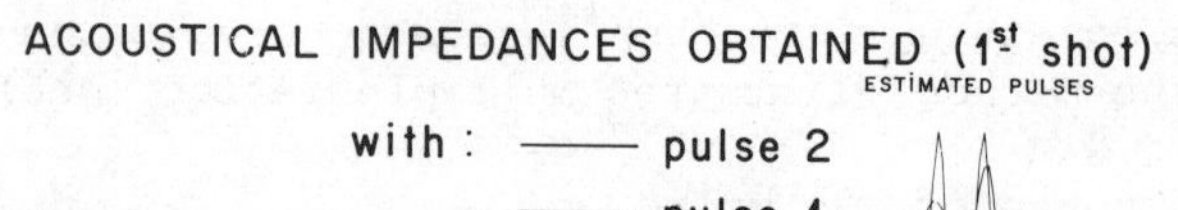


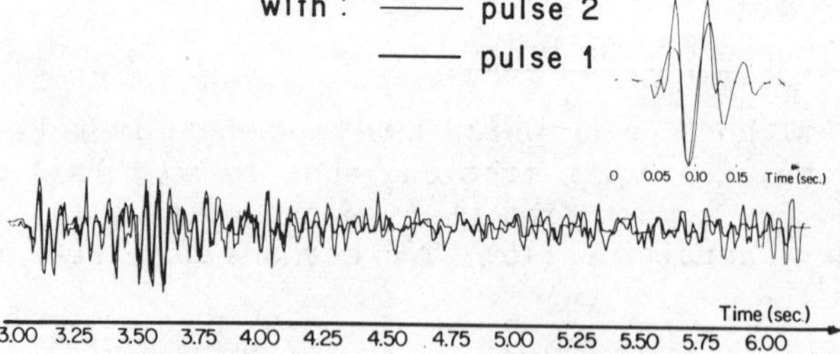


Figure 13. Run 2: Influence of the choice of the pulse on the acoustical impedance obtained.

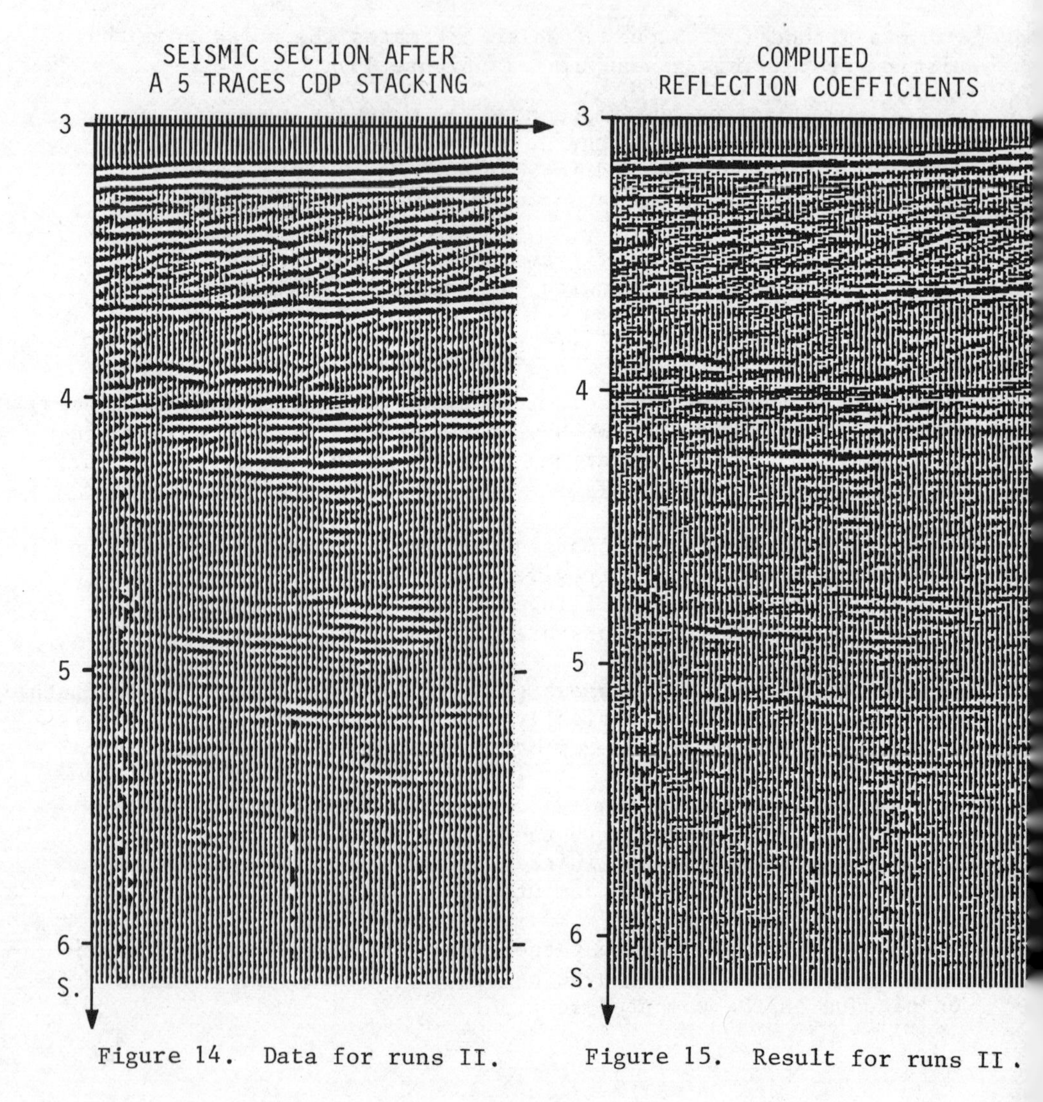


Figure 14. Data for runs II.

Figure 15. Result for runs II.

3°) Importance of the multiple reflections (*)

We want to reply the question : does the method give more than just a deconvolution ?

So we have to compare the computed distribution $\tilde{r}$ of reflection coefficients with the sequence $\tilde{\lambda}$ of coefficients (cf. (38)) of the associated impulse response (this impulse response can be viewed as the deconvolution of the recorded seismogram by the pulse g(t) since the computed seismogram is superposable with the recorded seismogram).

The difference between $\tilde{\lambda}$ and $\tilde{r}$ measures the contribution of all the internal multiple reflections. We have computed the quantity :

$$\frac{\sum_{k=1}^{400} |\tilde{\lambda}_k - \tilde{r}_k|}{\sum_{k=1}^{400} |\tilde{\lambda}_k|}$$

and it appears that this quantity ranges between 20 % and 40 % depending on the trace.

In order to visualize the interpretation error which can be made when multiple reflections are neglected, we have displayed on figure 16 and 17 respectively the reflection coefficients $\tilde{r}$ and the associated impulse responses $\tilde{\lambda}$, with an amplitude equalization function of the depth. We have emphasized on figure 16 some reflectors which are completely or partially hidden on figure 17 by internal multiples. Conversely we have emphasized on figure 17 some reflectors which have been artificially added by the multiples reflections. It can be seen that even in the case of the very flat section of figure 14, the internal multiples may have an influence on the interpretation of the section, especially for the great depths.

---

(*) Notice that the deepness of the water layer is greater than 3 sec (two way travel time) and the seismograms are recorded up to 6 sec so that the only one multiple reflections in the seismic section are internal multiples.

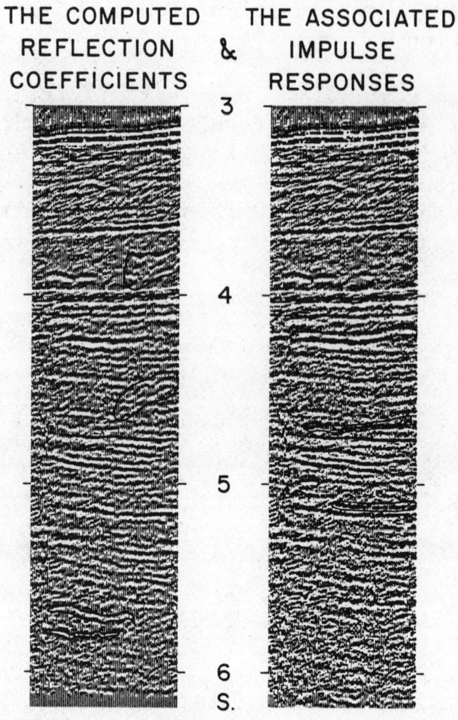


Figure 16.

Figure 17.

5.2.3 - Runs III (shallow sea)

The previous example lacks some comparison of the computed impedance with a sonic log. So we have considered an other example in which the impedance has been measured in a well (figure 18). The seismic section (figure 19) of this example shows that :

- the medium is nearly horizontally layered,
- the water layer is very small (about 0.2 sec. travel time). Hence the recorded seismograms are far from the response to a plane wave excitation.

ACOUSTICAL IMPEDANCE
KFT/SEC*GR/CC

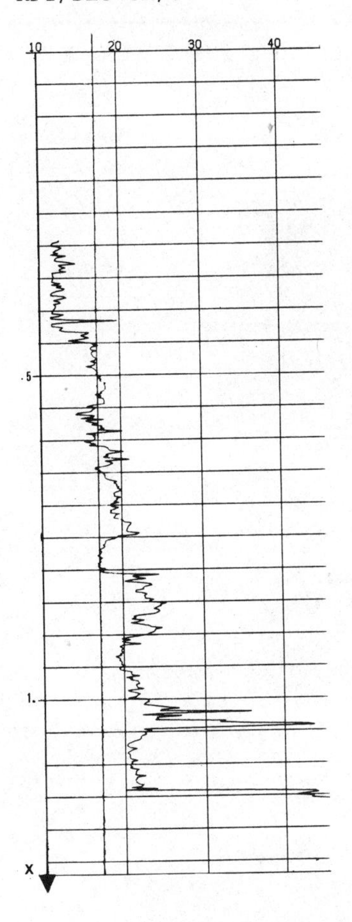


Figure 18. Acoustical impedance distribution recorded in the well.

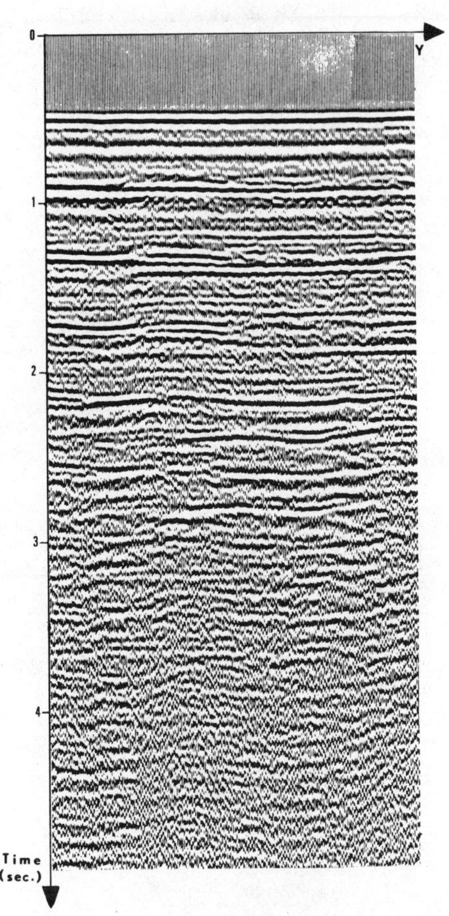


Figure 19. Seismic section for run III (stack with normal move-out correction).

Two ways are often used to reconstruct plane seismograms from the actual recordings :

- the stack with normal moveout corrections which gives, for each value of the lateral coordinate y, an estimation of the response to a plane wave excitation for little depths.
- the stack without normal moveout corrections (slant stack) which gives an estimation of the response to a plane wave excitation for large depths.

Let $Y_{d_1}(t)$ and $Y_{d_2}(t)$ denote respectively the seismograms extracted from the two stacked sections and corresponding to the location of the well. We define a new cost function (instead of the one appearing in (34)) :

$$J(\sigma) = \int_0^T \alpha(t)(Y(\sigma,t) - Y_{d_1}(t))^2 dt + \int_0^T (1 - \alpha(t))(Y(\sigma,t) - Y_{d_2}(t))^2 dt$$

where $\alpha(t)$ is the function :

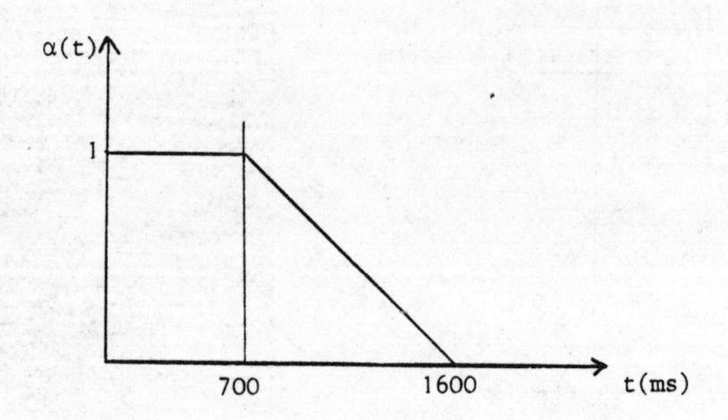


The comparison of the computed and recorded impedance distributions has been made by filtering these two functions by the pulse g(t) since, as we have seen in the previous examples, we can only obtain the components of the acoustical impedance which are in the bandwidth of the pulse g(t).

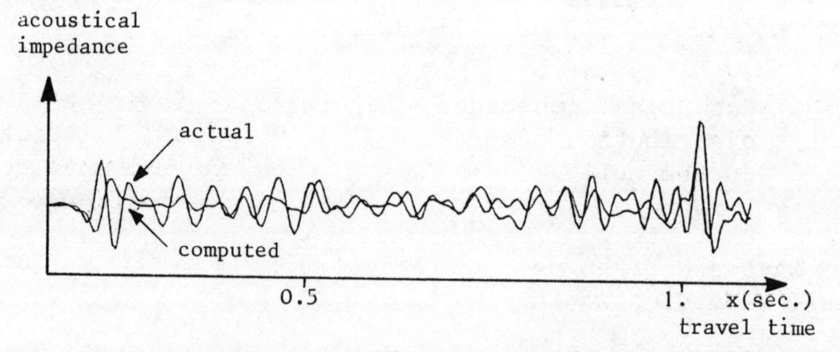


Figure 20. Comparison of the computed impedance distribution to the actual impedance distribution (recorded in the well).

Comparison of the computed impedance distribution to the actual impedance distribution (recorded in the well).

We can see (figure 20) that there is clearly some correlation between the computed and recorded impedance. As a matter of fact a good correlation is very rare for this kind of problem and the difficulties of such a comparison has been examined in a considerable amount of studies (see for instance [9], [13]). Briefly the 3 following problems still remain to solve :

- is the 1D wave equation an accurate enough model to simulate a plane wave propagating vertically in a horizontally layered medium ?
- have we a sufficient knowledge about the pulse g(t) ?
- can we reconstruct the response to a plane wave from the original experiment ?

## VI - CONCLUSION

The optimal control formulation and an adequate choice of the set of admissible parameters has lead to a method which gives a stable result when noise corrupts the data.

This has been checked successfully on simulated data. For inversion of field recordings, the interest of the method relies upon a positive answer to the 3 geophysical and technical problems which have appeared in the shallow sea example.

Though a solution to these problems can be expected, our work at present time is studying 2D models as well for the interest of the direct problem as for a future solution to the 2D inverse problem.

## REFERENCES

1. A. Bamberger, G. Chavent, P. Lailly, "Etude mathématique et numérique d'un problème inverse pour l'équation des ondes à une dimension", Rapport Laboria n° 226, IRIA, BP 105, 78150 Le Chesnay, Rapport Centre de Mathématiques Appliquées n° 14, Ecole Polytechnique, 91128 Palaiseau Cedex, France.

2. A. Bamberger, G. Chavent, P. Lailly, "About the stability of the inverse problem in the 1-D wave equation - Application to the interpretation of seismic profiles", Journal of Applied Mathematics and Optimization, n° 5, p. 1-47, 1979.

3. G. Chavent, "Identification of functional parameters in partial differential equations" in "Identification of parameters distributed systems", Edited by R.E. Goodson and M. Polis A.S.M.E., New York, 1974.

4. D.M. Detchmendy, "Attenuation and inverse problems in wave propagation", Swieeco 1968, Esso production Research Company, Houston, Texas.

5. M.L. Gerver, "The inverse problem for the vibrating string equation", Izv. Acad. Sci. URSS, Physic Solid Earth, 1970-71.

6. B. Gjevick, A. Nilsen, J. Hoyen, "An attempt at the inversion of the reflection data", Geophysical Prospecting, 24, 1976, p. 492-505.

7. B. Gopinath, M.M. Sondhi, "Inversion of the telegraph equation and the synthesis of non uniform lines", Proc. IEEE, 59, 1971, 3.

8. G. Kunetz, "Quelques exemples d'analyse d'enregistrements sismiques", Geophysical Prospecting, II, 1963, 4.

9. M. Lavergne, C. Willm, "Inversion of seismograms and pseudo velocity logs", Geophysical Prospecting, V. 25,1977, p. 231-250.

10. C. Lemarechal, "Non differentiable optimization subgradient and ε-subgradient methods", Lecture Notes in Econ. and Math. Systems, Vol. 117, "Optimization and Operational Research", Springer, Berlin, 1976.

11. J.L. Lions, "Contrôle optimal de systèmes gouvernés par des équations aux dérivées partielles", Dunod, 1968.

12. F. Murat, "Contre exemples pour divers problèmes où le contrôle intervient dans les coefficients", Annali. Mat. ed applicata 4, Vol. 112, p. 49-68, 1977.

13. M. Schoenberger, P.K. Levin, "The effect of subsurface sampling on 1-D synthetic seismograms", Paper presented at the 48th Annual International SEG Meeting, November 1, 1978, San Francisco, U.S.A..

14. S. Spagnolo, "Convergence in energy for elliptic operators", in Numerical solution of partial differential equations, III, Synspade 1975, B. Hubbard ed., Academic Press, New York, 1976.

15. P.G. Stone, "Robust wavelet estimation by structural deconvolution", Paper presented at the 46th Annual meeting of the SEG in Houston, Texas, October 1976.

16. P.C. Wuenschel, "Seismogram synthesis including multiples and transmission coefficients", Geophysics, 25, 1960, 1, 106-129.

## RAY GEOMETRIC MIGRATION IN SEISMIC PROSPECTING

K. Helbig

Vening Meinesz Laboratory, Rijksuniversiteit Utrecht
P.O. Box 80.021, 3508 TA Utrecht, The Netherlands

### INTRODUCTION: DEFINITION OF TERMS

With the term *raw data set* we describe the data obtained in a seismic survey. It is implicitly assumed that they have been pre-processed to minimize the effects of near surface layers, variations in source strength, and variations in source-to-ground and receiver-to-ground coupling. In short, we assume the data to be essentially homogeneous if the deeper subsurface is homogeneous so that all observable imhomogeneities are seismically significant.

The raw data set is a one-valued function depending on two to seven independent coordinates: the spatial coordinates of the source (-group), the spatial coordinates of the receiver (-group), and time (in this context the finite extension of the groups as well as their directivity is neglected). The function value depending on these independent coordinates is particle velocity (or sound pressure in marine observations) at the center of the receiver group. If the data are acquired with coincident source and receiver displaced along a straight line, there is only one spatial coordinate. On the other hand, for measurements in mines with source and receivers in different galleries - or on a strongly irregular surface - one needs all six spatial coordinates. A correction for these elevation differences is possible only if the general direction of the rays is known - a condition that sometimes, for instance when working in mines, is not satisfied.

It is often convenient to regard the independent variable not as a single valued function, but as the real part of a complex (two-valued) function, the analytical signal (Taner et al., 1979). One can then speak of time dependent amplitude and phase of the signal instead of time-dependent value of particle velocity. In constructing the analytic signal one obtains the imaginary part as the Hilbert transform of the real part.

The spatial coordinates assume by necessity discrete values. The time coordinate can be either continuous or discrete, depending on the acquisition technology, but since the continuous data are necessarily band limited they correspond to discrete data with discretization interval determined by the high-frequency limit of the spectral band. Thus the distinction between data that are continuous in time and those that are defined only for discrete values is not significant.

A *seismic event* can be described as a (limited) smooth hypersurface of equal phase connected with an amplitude above that of the immediate temporal vicinity, a "phase line-up cum amplitude increase". If there are only two spatial coordinates, the event-hypersurface is an ordinary surface, for only one space coordinate it is a line, for instance the familiar mark on time sections or individual seismograms. The identification of seismic events in the raw data sets or in any of its sub-sets is called *picking*. Picking requires a minimum density of points not only in the temporal coordinate but also in the source-receiver sub-space. Ideally this density should be commensurate with the spatial band width required. Picking by hand is less prone to errors caused by undersampling than most other processes, thus too low a point density results in a loss of resolution, not necessarily in alias effects. The minimum density is defined by the requirement that the smallest part of an event-hypersurface be occupied by enough points so that its geometrical parameters can be determined with sufficient accuracy.

The causes of seismic events are discontinuities (or smooth changes) in impedance (or velocity) by which the waves coming from the source are bent, reflected, refracted or diffracted into the direction of the receivers (this definition excludes the direct wave). The aim of the seismic data processing steps we discuss here is the conversion of the raw data set into a description of the spatial distribution of the impedance (or velocity). In the light of the foregoing statements it appears that seismic events contain information on the discontinuities of the impedance (or velocity). In view of the band limited nature of the data a smooth change is equivalent to a discontinuity if the gradient is steep enough. In the following the discussion is restricted to acquisition technologies suitable for reflection seismics (i.e. moderate source-receiver separation). Under these conditions most of the events are due to impedance changes and arrive at the receivers by virtue of reflection or diffraction. With the above restrictions the process of conversion of the raw data set (or a processed subset, e.g. a "time section") is called *migration*, though there is no question that the term has a few misleading associations.

The conversion of seismic events to the surfaces of discontinuity can take place either by working on the complete set (or a suitable subset that is quasi-continuous in time) by computer algorithms, or by working on the set of seismic events. The difference between the two approaches is primarily one of the bulk of data: even the most rudimentary quasi-continuous data set is represented by about $10^4$ individual numbers per km, while the corresponding set of seismic events contains not more than 800 numbers per km, which can be further reduced - in view of the inherent smoothness of events - to 100 to 200 per km. One consequence of this drastic reduction is that the *event migration* can be carried out by hand calculation, by mechanical analog computer (i.e., a specialized slide rule), or by programmable pocket calculator. Another consequence is the greater flexibility: since event migration is typically carried out either in the interactive mode or by programmable calculators (with interruptions of the calculation process to plot intermediate results), modifications of program or parameters are easily possible. A typical application where event migration is - in some geometrical aspects - superior to *data set migration* is *ray tracing* (discussed further below). Since event migration is possible only after the events have been identified, it is also called *post-picking migration*.

Data set migration is discussed in another contribution to this volume (Hosken 1980). The processes applied there are inherently simple, the practical difficulties lie primarily in the necessity to access up to $10^6$ data during a typical application. A further difference between the event migration methods discussed here and the data set migration methods is the use of *ray geometry* instead of concepts based on the wave equation. One of the early event migration methods is a notable exception to the use of ray geometry (Hagedoorn, 1954). It is rightly regarded as the forerunner of the Kirchhoff migration method. Since it is conceived as a graphical method to which no simple arithmetic implementation has been published, it is not discussed here.

Migration of seismic events can be regarded as the transformation of the hypersurfaces in the *image space* (the space of the raw data set) into interfaces (i.e. impedance discontinuities) in the *object space* (ordinary 2- or 3-space). This transformation - or mapping - requires knowledge of the distribution of velocity in space since the *event times* (e.g. reflection times) have to be converted to distances. One is faced here with a dilemma recurring frequently in the conversion of seismic data to structural information: the information required for the conversion process may itself only be available as a result of the conversion process. Thus event migration methods can be classified according to the way the velocity distribution is determined before the migration.

Event migration with known constant velocity poses no serious difficulties. Several migration methods are therefore based on the constant velocity model. To apply such methods, the actual (known or unknown) velocity distribution has to be replaced by a suitable *replacement velocity*. This can be determined from known or surmised velocity distributions or from information contained in the raw data set. The replacement velocity depends not only on the actual velocity distribution, but also on the purpose for which it is intended: the replacement velocity for the calculation of the horizontal coordinates of the interface in the object space is different from that for the determination of the vertical coordinate.

Event migration in closed form is also possible if the velocity is an algebraic function of the spatial coordinates. Of practical importance are velocity functions that depend on one coordinate only (generally z, but inclined axis of dependence is also possible). The parameters of a *replacement velocity function* again have to be determined from known or surmised velocity distributions, or from information contained in the raw data set. If a replacement velocity function closely models an actual velocity distribution, it can be regarded as an "all-purpose function". If the approximation is not close, the parameter values chosen again depend on the purpose. Generally the horizontal coordinates are more important than the vertical ones: errors in depth mean that the drill reaches the target later (or earlier), but errors in horizontal position might mean that the target is missed entirely, and that costly deviation operations or even re-drilling become necessary. Therefore parameters for replacement velocity functions (as well as the replacement velocity discussed above) should be adjusted to optimize lateral migration.

The methods mentioned so far are *one-step methods* in the sense that one can determine position and spatial attitude of a late event (a deep interface) without regard to interfaces in the overburden. All effects of the overburden are supposed to be embodied in the replacement velocity (function). This is possible only if no (steeply) dipping interfaces with significant velocity contrast occur in the overburden. A necessary condition for the permissibility of a replacement velocity is that a half-ray (source - reflector or reflector - receiver) that is vertical anywhere is essentially vertical throughout (for the replacement velocity function one has to use "parallel to the axis" instead of "vertical" if inclined axes are to be included). The expression "essentially" is used in the sense "the errors resulting from the approximation are proportional to the deviation from this condition".

When one-step migration methods do not guarantee the required accuracy, one has to use *ray tracing methods*. These are iterative in the sense that before one can migrate a late event, those earlier events that correspond to significant velocity contrasts have to be converted to interfaces in the object space so that the ray path corresponding to the late event can be traced with the help of Snell's law. It is not sufficient to determine the position of only small pieces of the hypersurfaces of "earlier events" that have in the image space the same spatial coordinates as the event to be migrated: ray tracing has to be carried out in the object space, and lateral spatial coordinates of object and image can differ significantly.

## THE CONSTANT-VELOCITY MODEL

The simplification afforded by constancy of velocity allows one to describe a migration algorithm for the most general disposition of sources in space and rather general shape of discontinuities: each event consists of a finite set of septuplets, viz. the six spatial coordinates of source and receiver (by virtue of the principle of reciprocity, source and receiver can be interchanged without affecting the data set) and the event time. Together with the velocity each septuplet defines an ellipsoid of revolution with source and receiver as focal points and the product of event time and velocity as major axis. The physical meaning of this ellipsoid is evident: any reflecting element tangent to the ellipsoid would have resulted in the observation represented by the septuplet (fig. 1).

The envelope to all ellipsoids corresponding to a seismic event contains the impedance discontinuity that has caused this event. According to the topological relationship between ellipsoids and envelope we have to distinguish several cases:

1) The envelope is a two-dimensional manifold (flat or curved surface). This can occur in two different ways: a) the envelope lies outside the ellipsoids (fig. 2). This is the standard case, an "ordinary" reflection. b) The envelope lies inside the ellipsoids

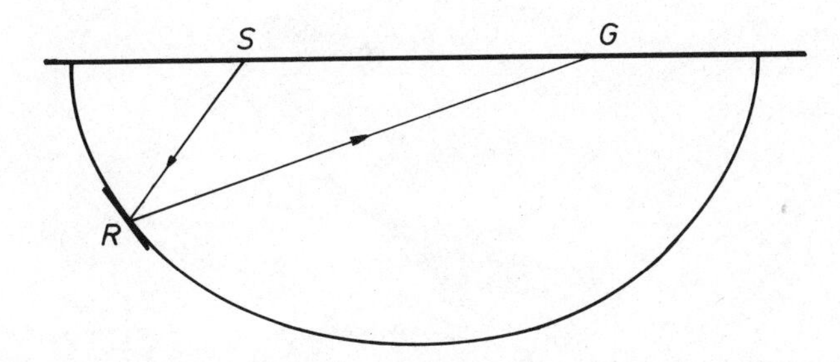


Fig. 1. Axial section of the ellipsoid defined by sources, receiver and length of travel path. The reflecting element must be tangent to the ellipsoid.

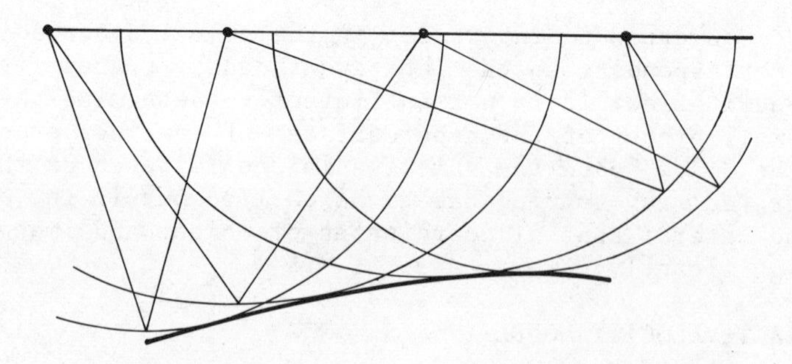

Fig. 2. Discontinuity is external envelope to ellipsoids if seismic event is standard reflection.

(fig. 3). This occurs if the seismic event is the receding branch of a so-called "bow-tie reflection" (fig. 4). The signal on the receding branch is the Hilbert transform of the signal in the two advancing branches (i.e. it is the imaginary part of the analytical signal). Bow-tie reflections occur over synclinal structures, if the center of curvature lies below the plane of observation (fig. 5).

2) The envelope degenerates to a point or line (fig. 6). This corresponds to diffraction by a point shaped scatterer (e.g., a shipwreck on the seafloor) or a linear scatterer (pipeline, trace, of fault plane).

3) The ellipsoids belonging to a seismic event might not intersect at all (fig. 7). This does not represent a physically possible situation but is an indication for an incorrectly

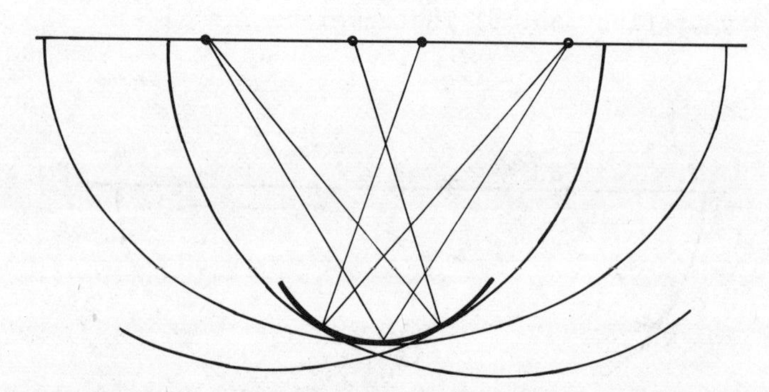

Fig. 3. Discontinuity is internal envelope to ellipsoids if seismic event is receding branch of "bow-tie" (see fig. 4).

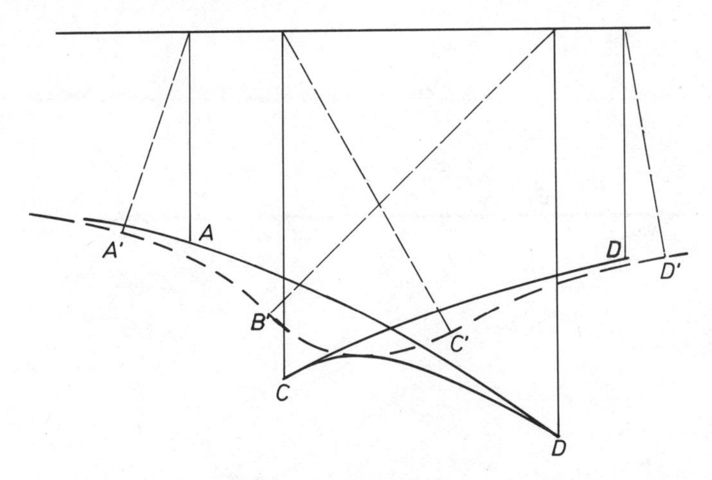


Fig. 4. Generation of "bow-tie" reflection *ABCD* over a shallow syncline *A'B'C'D'* by transmitter-receiver-coincident acquisition technology. *BC* is the receding branch. Solid lines belong to object space, broken lines to image space.

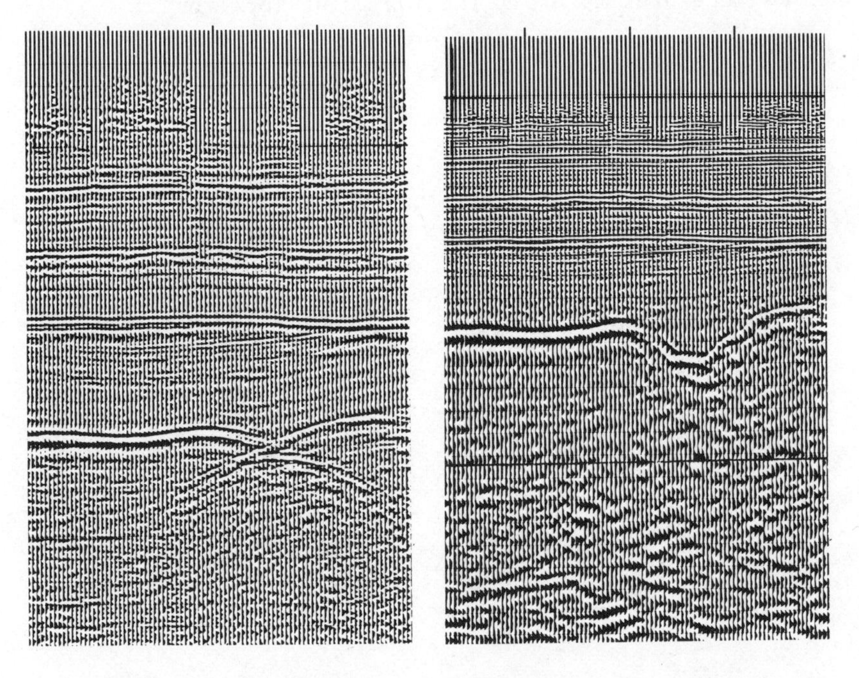

Fig. 5. Example of a "bow-tie" reflection (left). Object space model (right). (courtesy of Prakla-Seismos GmbH)

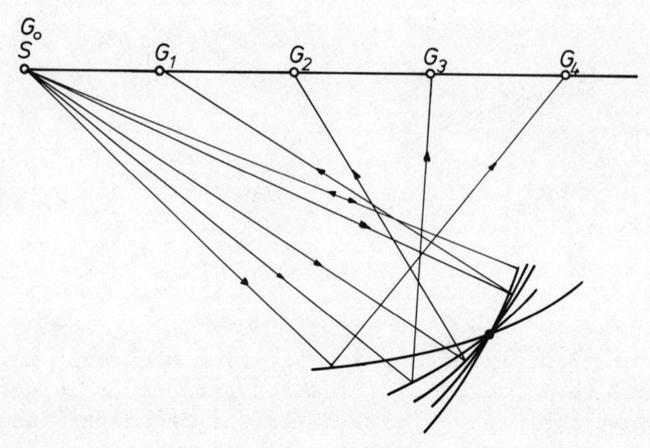


Fig. 6. Five ellipsoids, corresponding to observations with common shotpoint, intersect in a point: discontinuity interface degenerates to a diffracting point (or line).

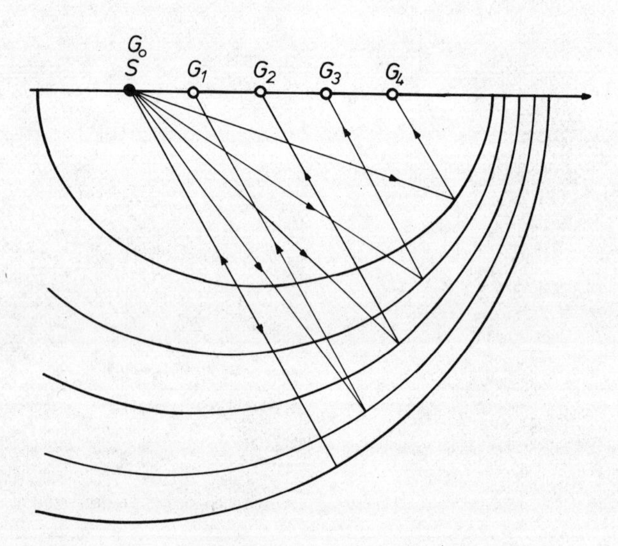


Fig. 7. Five ellipses without a common point. Either the corresponding points in the image space do not belong to one seismic event, or the velocity used for the construction is too high.

identified event: either the points making up the event belong in reality to different events, or, more likely, the signal has travelled with a drastically different velocity. This frequently occurs with shear wave events which are not recognized as such. However, there are other violations of the assumptions that lead to this situation, for instance steeply dipping events in areas where the velocity increases strongly with depth. We assume for the rest of this discussion that events have been correctly identified and that the velocities used are adequate.

## The construction plane

The ellipsoids in the previous illustrations have been drawn as ellipses. The reason for this is not only convenience: data acquisition methods where sources and receivers lie on a straight line do not give any information on cross-dip. The *construction plane* is the plane through source $S$, receiver $G$, and reflection point $R$. The rotational symmetries of all ellipsoids around the line $SG$ forces us to leave the determination of the spatial attitude of this plane until later. One has to keep in mind that these planes are individually determined for each triplet $S,G,R$, that they do not have to be vertical, nor do the planes have to be identical - even for reflecting points belonging to the same discontinuity (and to the same event). The correct spatial attitude can be determined whereever two lines intersect, for all other points one has to interpolate. If sources and receivers are placed in a two-dimensional pattern, determination of spatial attitude is possible nearly everywhere.

## Source-receiver-coincident observations

Source-receiver-coincident observations form a particular simple subset of the total data set. They might be results of single channel observations, they can be the subset of the central traces in split-spread continuous profiling, or they can be simulated from data obtained with a range of separations between source and receiver by the application of the "dynamic correction". Since, in view of the principle of reciprocity, source and receiver are interchangable, these dynamically corrected data are conveniently referred to a *data point* halfway between source and receiver. CDP (Common Data Point) stacked data are the most important type of simulated source-receiver-coincident data. The above process is particularly simple for this kind of data, since the ellipsoids degenerate to circles, so that the akward construction of ellipses is replaced by the much simpler construction of circles.

Actual or simulated source-receiver-coincident observation have the additional advantage that the data set depends on only half as many spatial coordinates as general raw data sets. If sources and receivers are further constrained to the surface, the number of coordinates of the image space is the same as that of the object space. The set of events in the image space quite often has a strong similarity with the discontinuities in the object space. Then one can see many details of the structure without migration. It is this similarity that has caused the term "migration" to be adopted: the image space was regarded as an "uncorrected" version of the object space, and migration was regarded as the correction necessary to move ("migrate") the discontinuities to the correct positions. Care is needed, since the similarities mentioned are quite often deceptive, even misleading.

In data sets corresponding to source-receiver-coincident observations, a point in the object space corresponds to a hyperboloid in the image space (visible in many time sections as diffraction hyperbolae), a point in the image space corresponds to a sphere (the degenerated ellipsoid) in object space. This is a strong reminder that the two spaces are related to each other, but have different internal structure.

Some light is shed on this relationship by the *exploding reflector model* (Loewenthal et al., 1976): a source receiver-coincident time section (zero-offset section) can be thought to be produced by generating at all refractors and diffractors at time zero waves by small explosions. Except for a scale factor the observations at the surface thus produced are indeed a first order approximation of the zero offset section. The object space has the coordinates $x$ and $z$, the image space the coordinates $x$ and $t$. Both can be represented in an $x$-$t$-$z$ system (fig. 8). The space-time-dependent displacement field that is generated in the quarter space $z>0$, $t>0$ by the exploding reflector model is called the *wave field*, image space and object space are boundary surfaces of this wave field (these definitions are easily extended to more general object- and image spaces, but because of the greater number of dimensions these can no longer be directly visualized). A point disturbance at $t=0$ (i.e. in the object space) generates a circular wave with time dependent radius, thus the wave field is a cone with its axis parallel to the time axis and its apex in the object space at the point diffractor. Intersections of this cone with planes that are parallel to the image "space" - in particular with the plane $z=0$, the image space itself - are hyperbolae, intersections with planes parallel to the object "space" are circles. To the familiar superposition of sperical wave fronts from different diffractors (Huygens' principle) in planes parallel to the object space corresponds the superposition of hyperbolae in planes parallel to the image space in *forward modeling* (see the contribution of J.W. Hosken in this volume).

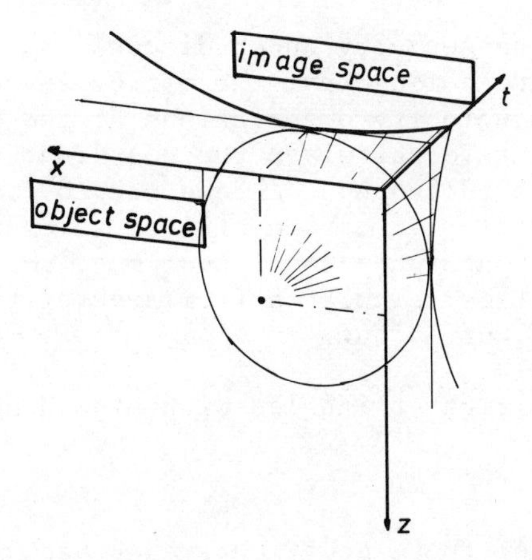


Fig. 8. Wave field generated by a point diffractor in the "exploding reflector model". This model is a first order approximation to a zero offset section (source-receiver-coincident section). The wave field is a cone, intersections of this cone with planes parallel to the object space are circles, with planes parallel to the image space are hyperbolae.

## Plane discontinuities

The methods described so far are geometrical ones - one has to construct ellipsoids and find the envelope. This problem can of course be solved analytically, but unless one restricts the interfaces to simple shapes that can be easily represented in coordinate or parameter form, no closed solution can be given. A restriction of entire discontinuities to simple shapes is unrealistic, but if the reflection points on the discontinuity are close enough to each other, the discontinuity can be locally replaced by its tangent plane. The error introduced by neglecting the local curvature of discontinuities has been discussed by Krey (1965).

*Linear source - receiver array.* We choose a subset of the raw data set where the signal from a source has been observed on several geophones along a straight line through the source (obviously, observations at one geophone position of signals from sources along a line through the geophone are equivalent). This subset corresponds to classical data acquisition methods - from the dip shooting of the early thirties and the continuous profiling of the forties and fifties to the shotpoint- or geophone-gathers of modern CDP surveys. Only recently two-dimensional data acquisition methods have deviated from this basic acquisition geometry.

Fig. 9 shows the geometry under discussion. The rays arrive at $G_i$ as if they were coming from the mirror image $S'$ of the shotpoint $S$. With known velocity observations at two geophones are, in principle, sufficient to calculate the coordinates of $S'$. The arrival time at a third geophone allows one to calculate the velocity. Unfortunately, errors in arrival times influence the velocity obtained in a highly non-linear way, so that simple averaging of velocities calculated from several triplets of observations is not sufficient.

For any of the right triangles with hypothenuse $S'G_i$ we have

$$(x_i-2\xi)^2 + 4\xi^2 - v^2t_i^2 = 0. \tag{1}$$

If we subtract two of these equations we obtain

$$x_i^2 - x_j^2 - 4\xi(x_i - x_j) - v^2(t_i^2 - t_j^2) = 0, \text{ or}$$

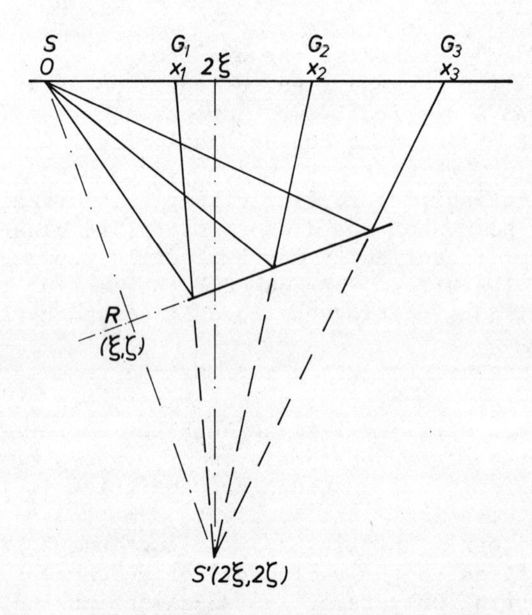


Fig. 9. Geometry for flat reflector: rays arrive at geophones as if they were coming from virtual source $S'$, the mirror image of $S$.

$$4\xi + v^2\frac{t_i^2 - t_j^2}{x_i - x_j} - (x_i + x_j) = 0. \tag{2}$$

If one writes (2) for the combinations $i,j$ and $j,k$ and subtracts, one obtains

$$v^2\left(\frac{t_i^2-t_j^2}{x_i-x_j} - \frac{t_j^2-t_k^2}{x_j-x_k}\right) - (x_i-x_k) = 0,$$

or in symmetrical form

$$\frac{1}{v^2} = \frac{t_i^2}{(x_i-x_j)(x_i-x_k)} + \frac{t_j^2}{(x_j-x_k)(x_j-x_i)} + \frac{t_k^2}{(x_k-x_i)(x_k-x_j)} . \tag{3}$$

From (2) we have

$$\xi = -\frac{t_i-t_j}{x_i-x_j}\, v^2\, \frac{t_i+t_j}{4} + \frac{x_i+x_j}{4} \tag{4}$$

and from (1)

$$\zeta = \tfrac{1}{2}\sqrt{v^2t^2 - (x_i-2\xi)^2} . \tag{5}$$

The coordinates of the actual reflection points $R_i$ are determined as intersection of $S'G_i$ and the reflecting plane:

$$\xi_i = -\frac{\xi^2(2\xi-x_i) + \zeta^2(2\xi+x_i)}{\xi .\ x_i-2\ (\xi^2+\zeta^2)} \tag{6}$$

$$\zeta_i = \frac{2\zeta\ (\xi . x_i-\xi^2-\zeta^2)}{\xi .\ x_i-2\ (\xi^2+\zeta^2)} \tag{7}$$

*Linear source-receiver-coincident data.* The rays corresponding to transmitter-receiver coincident data are reflected at vertical incidence, therefore the coordinates of the reflecting point can be obtained easily (fig. 10):

$$\frac{\xi_1-x_1}{v\ t_1/2} = \frac{\xi_2-x_2}{vt_2/2} = -\frac{v.\Delta t/2}{\Delta x} ,$$

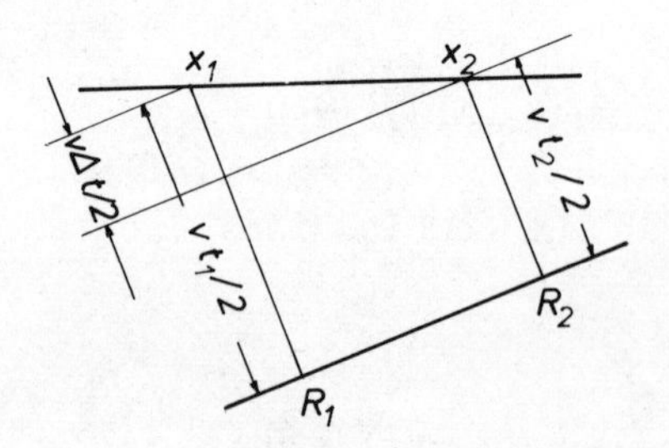


Fig. 10. Construction of reflection points $R_1$ and $R_2$ from zero-offset data over a plane discontinuity.

thus

$$\xi_i - x_i = -v^2 \cdot \frac{t_i}{2} \cdot \frac{1}{2}\frac{\Delta t}{\Delta x} \tag{8}$$

$$\zeta_i = \frac{1}{2}\sqrt{v_2 t_i^2 - 4\,(\xi_i - x_i)^2} \tag{9}$$

Determination of the spatial attitude of the construction plane

It was pointed out above that the construction is not carried out in a vertical plane but in the "construction plane" through source, receiver, and reflection point. The true spatial attitude of this construction plane can be determined if the same reflection point has been observed with two different azimuths of the seismic array or line. Fig. 11 gives an example of two intersecting lines (identified as line I and line II). If the reflection point is the same - i.e. if the two events have been identified correctly as being caused by the same discontinuity - the length $d$ must be the same in both lines, since it is the spatial separation of the common surface point and the common reflection point. If one varies the inclination of a construction plane (i.e. varies the cross dip) the projection of the reflection point on the surface traces out a short line perpendicular to the seismic line. The intersection of the two short lines obtained in this way is the projection of the true reflection point on the surface. Together with the already known spatial separation one can construct in a virtual construction plane true depth and dip of the element of the discontinuity, since this virtual plane is normal to the strike and therefore is vertical.

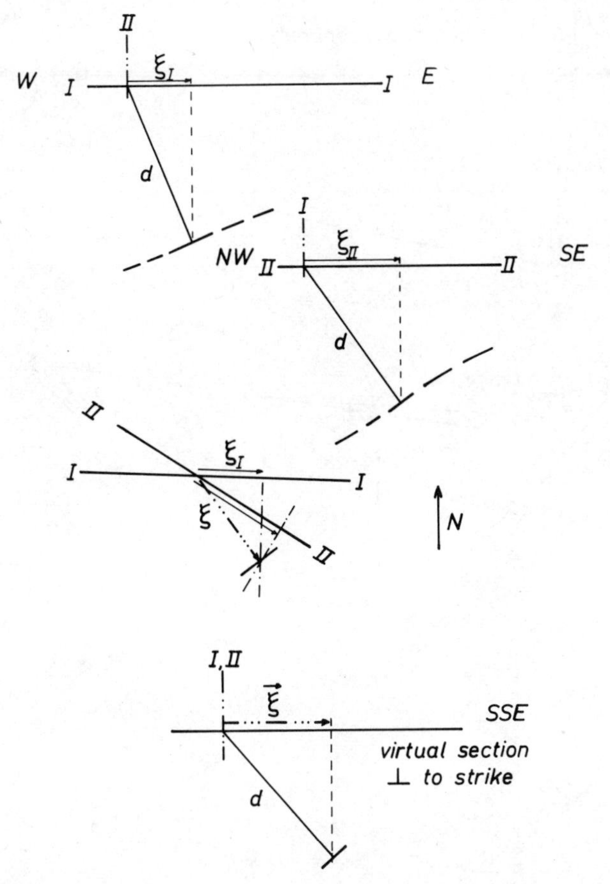


Fig. 11. Construction of the true spatial position and attitude of the reflecting element at the intersection of two lines.

## Compilation of depth contour maps from data in inclined construction planes

The fact that most construction planes are not vertical causes some difficulties in the compilation of depth contour maps: it is easy to determine in a "migrated depth section" those points at the surface for which the length $d$ (fig. 12) has "round" values corresponding to the contour values. However, if the construction plane is not vertical, these are not depth values, nor does the point at the discontinuity to which they refer lie below the observation line at the surface.

The projection of the sequence of reflection points is called the *trace line*. It is generally not known except at intersections

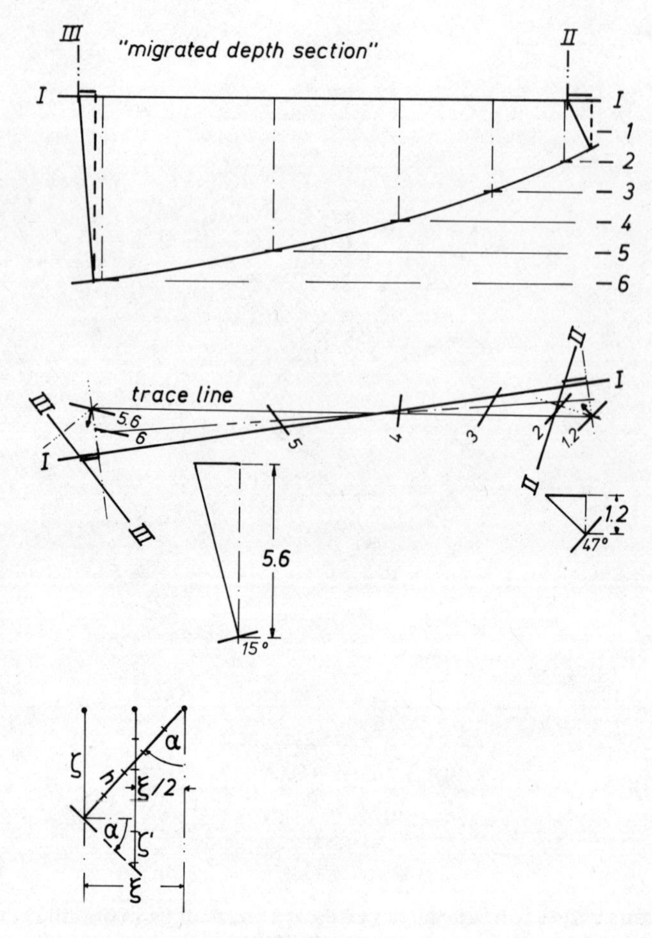


Fig. 12. To compile depth contour maps "round" values $h$ are posted halfway between the surface line and the projection of the trace line (see the cross section at lower left).

of lines. If intersections are close enough, the trace line can be interpolated with reasonable accuracy. One can then read the estimated separation between trace line and surface from the map and calculate the true depths of the trace-line points. These depths have of course no longer "round" depth values and thus do not lie on contours. This difficulty is easily overcome for small cross dips, if one plots the "round" depth values midway between surface line and the projection of the trace line. If the discontinuity is flat over the separation of the trace line and this "half-trace line", one has (see cross section in the lower left of fig. 12)

$$\xi = h \sin\alpha \qquad \zeta = h \cos\alpha$$

$$\zeta' = \zeta + \xi/2.\mathrm{tg}\,\alpha = h\left(\cos\alpha + \tfrac{1}{2}\frac{\sin^2\alpha}{\cos\alpha}\right) = h\,\frac{1-\frac{1}{2}\sin^2\alpha}{\cos\alpha}$$

$$= h\,\frac{1-\frac{1}{2}\left(\frac{x}{h}\right)^2}{\sqrt{1-\left(\frac{x}{h}\right)^2}} \approx h$$

REPLACEMENT VELOCITIES

The constant velocity model discussed in the previous chapter can be extended to more realistic situations if the actual velocity distribution is replaced by a suitable average velocity called the *replacement velocity*. If the velocity is a function of spatial coordinates, the average also depends on position. Aside from the small reduction in complexity, the problem remains to determine this average of the velocities from the data. The problem can be solved approximately if the velocity depends on depth only.

Relations between $v(z)$ and the replacement velocity $v_r(\zeta)$

If the velocity is known as an (arbitrary) function of $z$, the coordinates of the reflection point can be determined. The construction is shown below for source-receiver coincident data. The situation is illustrated in fig. 13. If $x$ and $z$ are the coordinates of a point on the ray and $i$ the deviation from the vertical, we have

$$dz = ds \cos i \text{ and } dx = ds \sin i$$

The time increment corresponding to the line element $ds$ along the ray is $d\tau = ds/v$. We also define the time element corresponding to a vertical segment of the ray as $d\tau' = dz/v$, so that we can convert $v(\tau)$ to $v(\tau')$. Snell's law can be written in the form

$$\frac{\sin i}{v} = p = -\tfrac{1}{2}\, dt/dx_{z=0} \tag{10}$$

The differential quotient in this expression has to be taken in the image space along the seismic event. With (10) we get

$$\xi = p\int_S^R v(z)ds = p\int_S^R v^2(\tau')d\tau = p\int_0^\zeta \frac{v(z)}{\cos i}\,dz = p\int_0^{t'/2}\frac{v^2(\tau')}{\cos i}\,d\tau' \tag{11}$$

In (11) $t'$ is defined by

$$\int_0^{t'/2} v(\tau')\,d\tau' = \zeta\,, \tag{12}$$

i.e. $t'$ is the reflection time to a *horizontal* reflector at the depth $\zeta$.

By comparing (11) with (8) we find that the replacement velocity that allows one to calculate the horizontal coordinate of the reflection point with the expressions valid for the constant velocity case is

$$v_r^2 = \frac{\int_0^{\zeta} v(z)\,dz}{t_0/2} = \frac{\int_0^{t'/2} \frac{v^2(\tau')}{\cos i}\,d\tau'}{t_0/2} \tag{13}$$

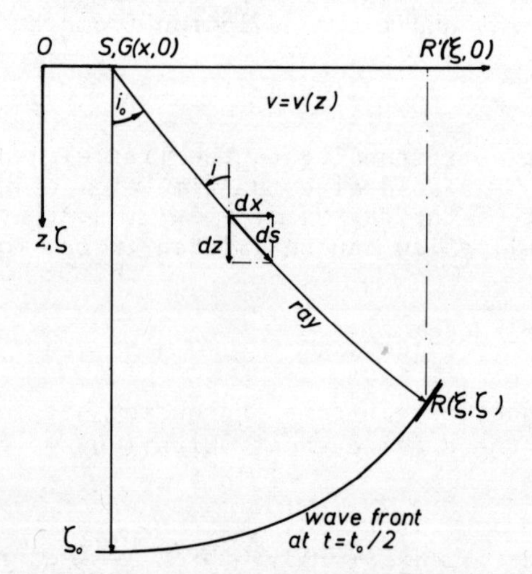


Fig. 13. Ray corresponding to source-receiver-coincident data acquired over halfspace with velocity depending on depth only.

This integral can not be evaluated, since neither $\zeta$ nor $t'$ are known before the construction. For small angle $i$ one can replace cos $i$ by 1 and $t'$ by $t$, the actual event time. With that we obtain an approximation

$$v_r^2 \approx \frac{\int_0^{t/2} v^2(\tau')d\tau'}{\int_0^{t_0/2} d\tau'} = (v_{rms})^2 \text{ , or} \tag{14}$$

$$v_r^2 \approx \frac{\int_0^{\zeta} v(z)dz}{\int_0^{\zeta} \frac{dz}{v(z)}} = (v_{rms})^2 \text{ .} \tag{15}$$

The expression on the r. h. s. of (14) is the square of $v_{rms}$ (*root-mean-square velocity*), the square root of the time average of the square of the velocity. In arriving at (14) we replaced cos $i$ by 1 and $t'$ by $t$. The first approximation means that the correct integrand is larger than the one used, the second approximation means that the correct upper limit is smaller than the one used. The effects of the two substitutions thus cancel to some degree, though to what degree they cancel can only estimated if more is known about $v(z)$.

## Approximation of the rms-velocity by the stacking velocity

If $v(z)$ is known from measurements in boreholes or other investigations, the rms-velocity can be obtained by evaluating the integral in (15). However, such data quite often do not exist. One can in such cases use the *stacking velocity* (if the data have been obtained with CDP or an equivalent technology). The stacking velocity is an experimentally determined quantity: one dynamically corrects data belonging to a common data point (and common azimuth) with a whole range of velocities. The velocity that best removes the curvature of the event is accepted as the stacking velocity $v_s$.

In determining $v_s$ one uses criteria like the minimum fluctuation around a straight line or the maximum sum of amplitudes after dynamic correction. For general investigations these experimental definitions of the stacking velocity are not too well suited (see, however, Al Chalabi, 1973; 1974) since one optimizes the parameters of a hyperbola. A simpler way to investigate the relationship between rms and stacking velocity is via the $t^2$-$x^2$ plot, where one can use linear regression.

We first show that arrival times from a flat reflector with a common data point in a medium of constant velocity lie on a hyperbola with the apex at the data point, i.e., that the event (in the image space) is hyperbolic with respect to source-receiver separation.

In fig. 14 the geometry for CDP data acquisition is shown. Besides the source-receiver combination $SG$ at separation $2x$, a second ray system is indicated corresponding to a pair at separation $2x/2$. Application of the law of cosines to the triangle $SGG'$ yields

$$v^2t^2 = 4(x^2+(h-x\sin\alpha)^2 + 2x(h-x\sin\alpha)\sin\alpha) = 4(h^2+x^2(1-\sin^2\alpha)).$$

With $t_0 = 2h/v$ we have

$$t^2 = t_0^2 + \left(2x\,\frac{\cos\alpha}{v}\right)^2 \tag{16}$$

which is a hyperbola in $x$-$t$ space (the image space).

Alternatively, (15) describes in the $x^2$-$t^2$ space a straight line with the intercept $t_0^2$ and the slope $(2\cos\alpha/v)^2$. It follows immediately that for constant velocity $v$ one has

$$v_s = v/\cos\alpha = v. \tag{17}$$

For depth-dependent velocity the points which constitute the event in a $t^2$-$x^2$ plot corresponding to CDP data over a flat discontinuity do not lie on a straight line, but still sufficiently close to one that a regression line can be drawn. As long as the dispersion of the data is small, the inverse slope of the regression line is equal to the square of $v_s$. The relation of $v_s$ thus defined to the rms-velocity is illustrated in fig. 15 (it has been assumed that $\alpha=0$).

We develop $t^2(x^2)$ into a Taylor series around some point $x_1 \neq 0$ (because data for $x=0$ rarely exist). The fact that seismic events are nearly hyperbolic means that the $t^2(x^2)$ curve is nearly straight, i.e. the first term in the series is the most significant.

$$t^2(x^2) = \sum_{n=0}^{\infty} c_n\,(x^2-x_1^2)^n = \sum_{n=0}^{k} c_n\,(x^2-x_1^2)^n + c_R(x^2-x_1^2)^{k+1} =$$

$$t_1^2 + c_1(x^2-x_1^2) + c_R(x^2-x_1^2)^2 \tag{18}$$

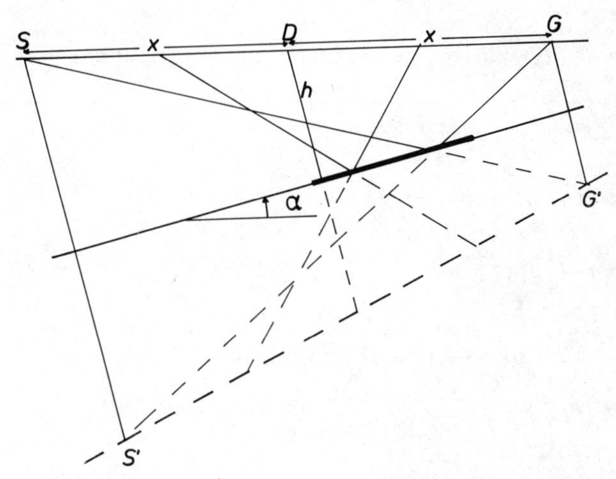


Fig. 14. Source and receiver arrangement for common-data-point acquisition. Rays are indicated for source-receiver pair with separation $2x$ and for a second pair with half the separation. Reflection points are not identical, so construction requires that the reflector is plane over the range of reflection points. $\mathcal{D}$ is the (common) data point.

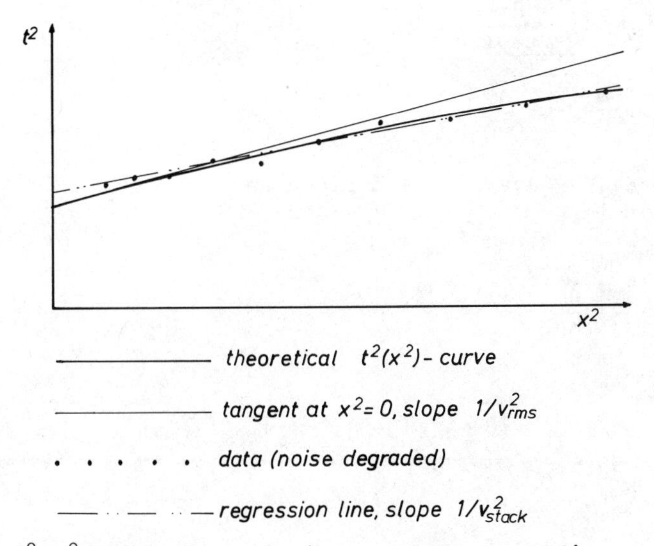


Fig. 15. $t^2$-$x^2$ plot of a seismic event corresponding to a plane discontinuity.

$$c_n = \frac{1}{n!}\,\frac{d^n(t^2)}{(d(x^2))^n}\bigg|_{x^2=x_1^2} = \frac{1}{n!}\,\mathcal{D}_n(x^2)\bigg|_{x^2=x_1^2}$$

The derivatives $\mathcal{D}_n$ can be calculated by the iteration

$$\mathcal{D}_n = \frac{1}{2x}\,\frac{d}{dx}\,\mathcal{D}_{n-1} \quad \text{with } \mathcal{D}_0 = t_1^2 \tag{19}$$

The rest term is determined as

$$c_R = \frac{1}{(k+1)!}\,\mathcal{D}_{k+1}(x^2)\bigg|_{x^2=\xi^2}, \quad x_1^2<\xi^2<x^2 .$$

The differential expressions $\mathcal{D}_n$ have to be calculated from

$$t = 2\int_0^{t_0/2} \frac{d\tau}{\cos i} \quad \text{and} \tag{20}$$

$$x = 2p\int_0^{t_0/2} v^2\,\frac{d\tau}{\cos i}, \quad p = -\frac{dt}{dx}, \tag{21}$$

where, as before, $i$ is the deviation of the ray from the vertical (fig. 16). One obtains

$$\mathcal{D}_1 = c_1 = \frac{\int_0^{t_0/2} \frac{d\tau}{\cos i}}{\int_0^{t_0/2} v^2\frac{d\tau}{\cos i}} = \frac{1}{\overline{v^2}^{\tau,i}}, \tag{22}$$

where the $(\tau,i)$- average is defined as

$$\overline{a}^{\tau,i} = \frac{\int a\,\frac{d\tau}{\cos i}}{\int\frac{d\tau}{\cos i}}, \quad \text{and}$$

$$\frac{1}{2!}\mathcal{D}_2 = c_2\,\frac{1}{4t_0^2}\,\frac{(\overline{v^2}^{\tau,i})^2 - (\overline{\frac{v^4}{\cos^2 i}}^{\tau,i}\cdot(1-\overline{\sin^2 i}^{\tau,i})}{(\overline{v^2}^{\tau,i})^2\cdot\left[(\overline{v^2}^{\tau,i})^2 + (\overline{\frac{v^4}{\cos^2 i}})^{\tau,i}\cdot\overline{\sin^2 i}^{\tau,i}\right]} \tag{23}$$

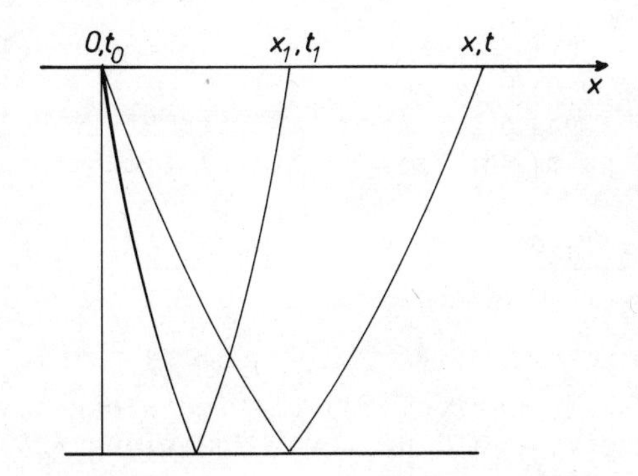


Fig. 16. System of rays used for the series development $t^2(x^2)$ around the point $x_1$.

The denominator of (23) is always positive, in the numerator the second term is close to $\overline{v^4}^{\tau,i}$, and in view of the Cauchy-Schwartz inequality $(\overline{v^2})^2 < \overline{v^4}$, $c_2 < 0$. Thus the $t^2$-$x^2$ curve is everywhere convex upwards, and the regression line has always a slope smaller than that of the tangent to the curve for $x^2=0$. The slope of this tangent is $1/(v_{rms})^2$, since for $x_1=0$ we have $i \equiv 0$ and $c_1 = 1/\overline{v^2}^{\tau}$. Thus if the data permit an estimation of the slope of the tangent for $x^2=0$, $v_{rms}$ can be obtained. Otherwise one uses the slope of the regression line, which always over-estimates $v_{rms}$.

## CONTINUOUS VELOCITY FUNCTIONS

In the previous section the velocity was assumed to be an arbitrary function of depth. In many basins one can restrict the velocity distribution one takes into consideration even further and use a closed expression for the velocity law. It is obvious that the use of closed expressions permits, in principle, closed solutions for the conversion of seismic events to discontinuities, but only the simplest types have been used so far. It can be argued that the determination of the velocity function should be treated also as a replacement problem: the parameters should be adjusted such that the horizontal coordinates are more reliably estimated than the vertical coordinates. However, in practical applications one is generally satisfied with an approximation of the actual velocity distribution by a simple power law:

$$\frac{v}{v_o} = \left(\frac{z+z_o}{z_o}\right)^{\alpha}, \tag{24}$$

where $v_o$ is the velocity at $z=0$, and $-z_o$ is the level (above ground) where the velocity would be zero.
One can express the velocity as a function of travel time along a *vertical* ray (not as a function of travel time along an arbitrary ray!):

$$\frac{v}{v_o} = \left(\frac{\tau+\tau_o}{\tau_o}\right)^{\beta}, \tag{25}$$

where now $\tau_o$ is the (negative) travel time along a vertical ray for which the velocity would be zero. Equations (24) and (25) are equivalent if

$$\beta = \frac{\alpha}{1-\alpha} \quad \left(\text{or } \alpha = \frac{\beta}{1+\beta}\right) \quad \text{and} \tag{26}$$

$$\tau_o = \frac{z_o}{v_o}\frac{1}{1-\alpha} \qquad \left(\text{or} \quad \tau_o = \frac{v_o\tau_o}{1+\beta}\right). \tag{27}$$

Equations (26) and (27) can be satisfied only if $\alpha \neq 1$ ($\beta \neq 1$).
For $\alpha=1$ one has

$$\frac{v}{v_o} = 1 + z/z_o \qquad \text{and} \tag{28}$$

$$\tau = z_o/v_o \;\; \ln(1 + z/z_o) = \tau^* \ln(1+ z/z_o) \qquad \text{or} \tag{29}$$

$$\frac{z}{z_o} = -1 + e^{\tau/\tau^*} \quad \text{with } \tau^* = \frac{z_o}{v_o}. \tag{30}$$

The coordinates $x, z$ of the reflection point corresponding to an event with $p=-dt/dx$ and the event time $t_o$ can be obtained through

$$t_o/2 = \int_0^z \frac{d\zeta}{v(\zeta)\sqrt{1-p^2v^2}} = \int_0^z \frac{d\zeta}{v_o(1+\frac{\zeta}{z_o})^{\alpha}\sqrt{1-p^2v_o^2(1+\frac{\zeta}{z_o})^{2\alpha}}} \quad \text{and} \tag{31}$$

$$x = p\int_1^z \frac{v\,d\zeta}{\sqrt{1-p^2v^2}} = p\int_1^z \frac{v_o(1+\frac{\zeta}{z_o})^{\alpha}\,d\zeta}{\sqrt{1-p^2v_o^2(1+\frac{\zeta}{z_o})^{2\alpha}}}, \tag{32}$$

where first $z$ is determined such that (31) is satisfied. Then $z$ is inserted into (32).

While no simple general statement can be made about the coordinates of the reflection points, it is easy to express the curvature of the ray as a function of depth:

The curvature is

$$\frac{1}{\rho} = \frac{d^2x/dz^2}{(1+(\frac{dx}{dz})^2)^{3/2}} .$$

But we have from (32)

$$\frac{dx}{dz} = \frac{p\,v}{\sqrt{1-p^2v^2}} , \quad 1 + (\frac{dx}{dz})^2 = \frac{1}{1-p^2v^2}$$

$$\frac{d^2x}{dz^2} = p\,\frac{dv}{dz}\,\frac{1}{(1-p^2v^2)^{3/2}} , \text{ thus}$$

$$\frac{1}{\rho} = p\,\frac{dv}{dz} = \alpha\,\frac{pv_o}{z_o}\,(1+\frac{z}{z_o})^{\alpha-1} = \alpha\,\frac{\sin i_o}{z_o}\,(1+\frac{z}{z_o})^{\alpha-1} \tag{33}$$

Velocity function (i)

An immediate consequence of (33) is that for $\alpha=1$, i.e. for

$v = v_o(1+z/z_o)$ ,

all rays are circles with radius

$$r_{ray} = \frac{z_o}{\sin i_o} \tag{34}$$

and center at

$$x_r = \frac{z_o}{tg\,i_o} , \quad z_r = -z_o \tag{35}$$

(see fig. 17). The wave fronts are spheres with radius

$$r_w = z_o \sinh(\frac{v_o t}{2z_o}) = z_o \sinh(\frac{t}{2\tau^*}) \tag{36}$$

and with the center at

$$x_w = 0 \quad \text{and} \quad z_w = z_o(\cosh(\frac{t}{2\tau^*})-1) \tag{37}$$

The dip $i$ of the reflector at the reflection point is related to the slope of the seismic event - i.e., its apparent slowness $p$ - by

$$tg\,(i/2) = e^{(\frac{t}{2\tau^*})}\,tg\,(i_o/2) \tag{38}$$

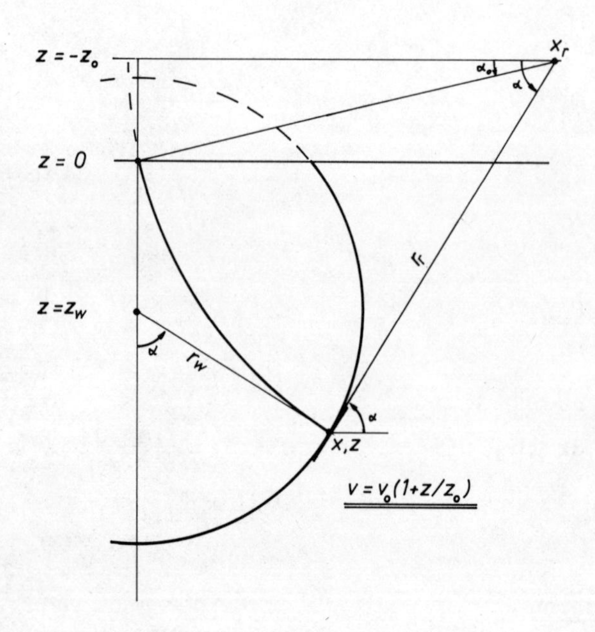


Fig. 17. In a medium where the velocity is a linear function of depth, rays are circles, wave fronts are spheres.

with $\sin i_o = pv_o$.
The coordinates of the reflection point are thus given by

$$
\begin{aligned}
x &= r_w \sin i \\
z &= z_w + r_w \cos i \\
t' &= 2\tau^* \ln(1+z/z_o).
\end{aligned}
\tag{39}
$$

The third coordinate listed in (39)($t'$) is the (two way) time to the reflector along a vertical ray. This quantity is significant because *data set migration* is often carried out in such a way that the resulting space is not the object space ($x$, $v$, $z$-space) but the $x$, $y$, $t'$-space (the technical term is "time migration"). There are several reasons for this; perhaps the most important is that a conversion to this space leaves the apparent spectrum and thus the signal shape unchanged, while a conversion to true object space also changes frequencies to wave lengths. The calculation of $t'$ thus facilitates the comparison of event migration and data set migration.

Velocity function (ii)

The velocity function discussed in the previous section is presumably the most popular velocity function, but it is not a good approximation to most velocity distributions, since the velocity gradient generally decreases significantly with depth. A better approximation can be achieved with a second function which is often referred to as "linear velocity increase with time" (vertical time!):

$$v(t') = v_o \left(\frac{\tau+\tau_o}{\tau}\right) \tag{40}$$

Obviously, this is a special case of (25) with $\beta = 1$.

The function can be converted into a function of depth with the help of (26) and (27):

$$v(z) = v_o(1+z/z_o)^{1/2} \tag{41}$$

with $z_o = v_o t_o/2$. (42)

This function has been investigated in some detail by Musgrave (1952), who designed a mechanical device that doubled as plotter and analog calculator. His findings can be summed up as follows (see fig. 18):

In a medium where the velocity is distributed according to (41), rays are right cycloids generated by a circle with radius

$$R = z_o/(2p^2v_o^2) \tag{43}$$

rolling at the plane $z = -z_o$. Wave fronts are ovals which can be generated from each other by simple scaling.

The relevant expressions for this velocity function are

$$\sin i_o = pv_o \tag{44}$$

$$i = i_o + p\frac{v_o^2}{2z_o}t \tag{45}$$

$$x_o = R(2i_o - \sin(2i_o)) \tag{46}$$

$$x = R(2i - \sin(2i)) - x_o \tag{47}$$

$$z = R(1 - \cos(2i)) - z_o \tag{48}$$

$$t' = 4z_o/v_0\left(\sqrt{1+\frac{z}{z_o}}-1\right) \tag{49}$$

Equations (36)-(49) can easily be used as a basis for programs for programmable pocket- or desk calculators. This makes the conversion of seismic events from image- to object space a convenient subject of investigations for students. A program based on the function (i) for the HP 25 was published by Michaels (1977).

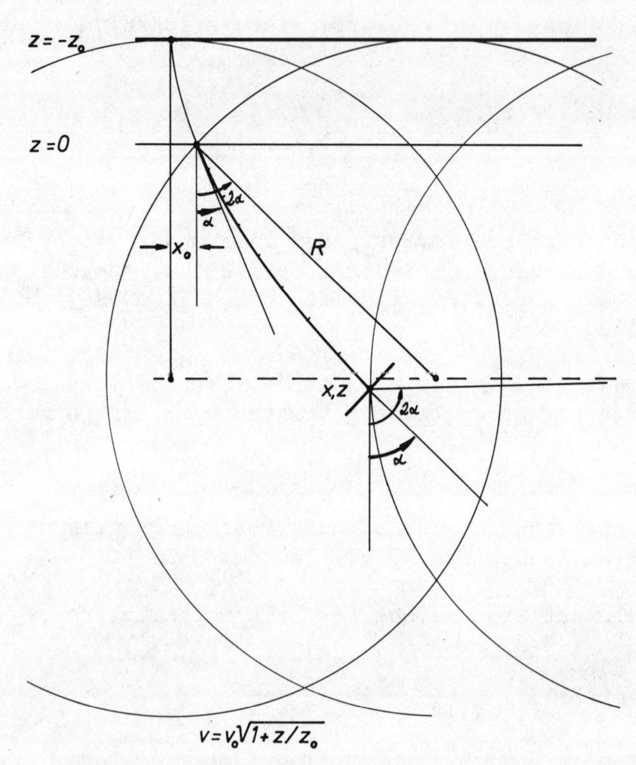


Fig. 18. In a medium where the velocity is proportional to the square root of a linear function of depth, rays are right cycloids and wave fronts ovals. The generating circle rolls on $z = -z_o$, its radius is proportional to the inverse square of the ray parameter (i.e., the slope of the event in image space). Rays belonging to different ray parameters (and wave fronts belonging to different travel times) are similar, i.e. they can be converted into each other by simple scaling.

## RAY TRACING

The discussion in the two previous chapters concerning replacement velocities and "velocity functions" contains an element of self-contradiction: we have always assumed that the velocity depends on depth only, while we discussed the positioning of steeply dipping interfaces. Generally, the presence of steeply dipping interfaces means that a ray that is vertical somewhere is not necessarily vertical everywhere.

If dip and velocity contrasts are significant, the refraction of rays at non-vertical incidence has to be taken into account. This is not difficult if the major interfaces and the velocities (or velocity functions) in the space sections between the interfaces are known. We shall discuss the corresponding methods only for source-receiver coincident data; for more general data sets, the methods are much more complicated. One establishes with any of the methods discussed so far the ray in the uppermost medium (either with a constant velocity or with a function depending on one spatial coordinate not necessarily vertical). At the first intersection of the ray with the lower boundary of the space corresponding to the current velocity (law) one calculates the time required from the surface. If this is more than half the event time, the reflection lies in the current space section, and elements of the interface can be determined with the methods discussed earlier. If not, the angle of incidence at the interface is estimated and the angle of entry into the next medium is calculated using the known velocity contrast and Snell's law. Using the velocity law applicable to this space one determines the ray, and then carries on as described before until the available time (half the event time) is used up. The reflecting element is drawn perpendicular to the ray at the endpoint of the ray. The process is repeated for individual segments of the event until the hypersurface has been converted from image space to object space.

Fig. 19 is a time section - i.e., part of the image space - showing some steeply dipping events. A migrated depth section obtained by ray tracing is shown in fig. 20. Obviously, the structural relationships between the different elements in the section - a salt dome with prominent overhangs, a rim syncline partly in the shadow of the overhang, and a fault intersecting the mother salt - are such that data set migration would have been very difficult.

It is obvious that ray tracing should be applied only in sections at right angles to strike, since for all other sections the ray is not restrained to the construction plane. In complicated structures different interfaces have different strike, thus this requirement can not always be satisfied.

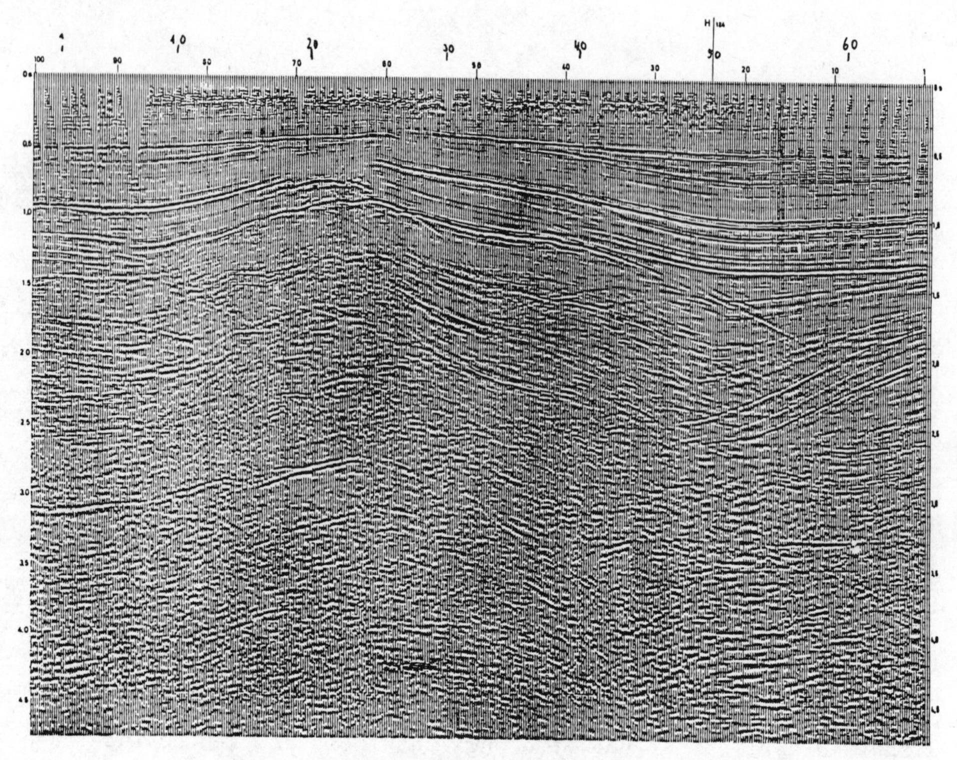

Fig. 19. Time section showing steeply dipping events over a salt dome. Numbers running from 1 to 100 starting at the northeast margin of the section are coordinates along the line, numbers running from 1 to above 60 from SW to NE identify the segments of events that have been used in the ray tracing in the next figure (courtesy of Prakla-Seismos GmbH).

Ray tracing is conceptually a simple process - particularly if it is carried out by hand or interactively. However, non-interactive computer implementations - of which the conversion from fig. 19 to fig. 20 is a typical example - lead to many difficulties that are not obvious at first sight: if the entire iteration is to run at once, the assignment of velocity functions to the different sections of space has to be carried out on the basis of information in the image space. Smooth application of Snell's law in the object space requires that for every point one and only one velocity is given, but there is no one-to-one relationship between points in object space and in image space. Moreover, the geometrical relationship of the interfaces representing the events depends on the velocities assigned to the sections between events in the image space. One result of these complications is that the object space

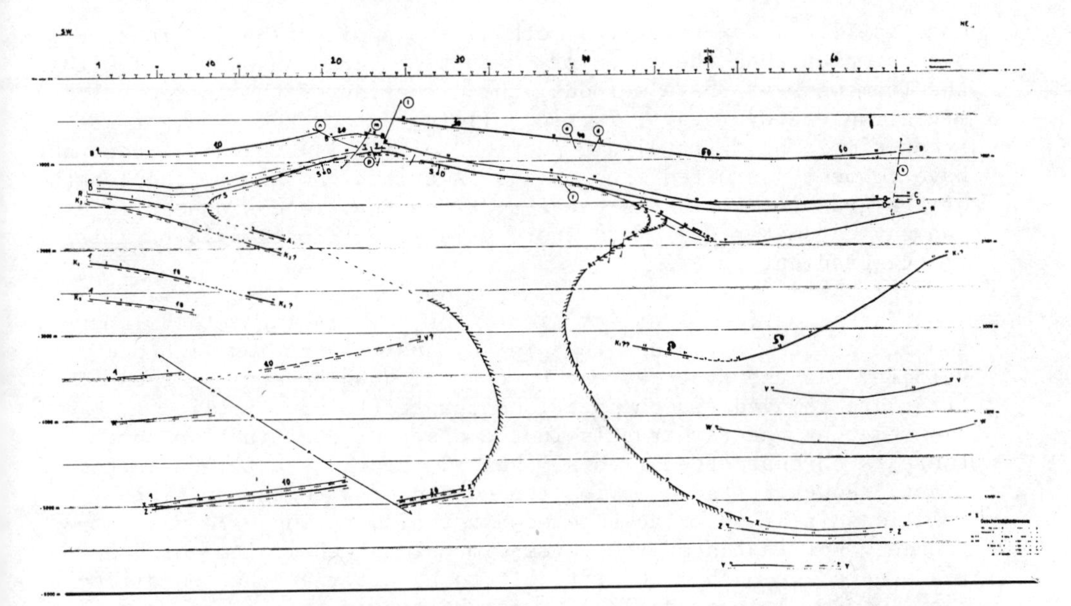

Fig. 20. Migrated depth section obtained from the time section of fig. 19 by ray tracing. The rim syncline under construction point 50 is responsible for the "bow-tie" events in the right hand side of fig. 19. Some seismic events come from the underside of the salt overhang, e.g., from construction point 14 to 18 (courtesy of Prakla-Seismos GmbH).

correspondence of lateral boundaries defined in the image space can either overlap or leave gaps. A ray entering such an area can not be continued and "gets lost." Thus the ray tracing process is not only iterative in the sense that later events can be migrated only after all earlier events have been migrated, it is also iterative in the sense that the first - tentative - assignment of velocities has to be updated for a second - and sometimes a third - migration run.

DIRECT MIGRATION OF EVENT MAPS

It was mentioned earlier that the source-receiver coincident subset of the raw data set (the zero offset data set) has the same number of coordinates as the object space. Seismic events of this subset therefore can be displayed as "isochrone maps", if the spatial coordinates span a two dimensional area. Such maps can be directly converted to depth contour maps of interfaces (i.e., object space maps) by ray geometrical migration methods. For the onestep (i.e., non-terative) methods the application is, in a way, even simpler than the customary migration based on sections: within the framework of these methods, the direction of the horizontal migration vector (e.g. $\vec{\xi}$ in fig. 11) is that of the gradient of event time. The "virtual dip section" of fig. 11 therefore does not have to be constructed from oblique sections. Moreover, the length of the gradient vector is equal to twice the trace slowness (i.e., the ray parameter $p = dt/du$ where $u$ is a coordinate axis parallel to the gradient of $t$).

The situation is not as simple for ray tracing methods. Here too one has the advantage that in any spatial section with one velocity law the true ray can be traced. Thus several sources of error are removed (in the previous chapter we have implicitly assumed that the ray travels in the plane of construction, but there is no guarantee of this, even if the ray starts out in the plane). However, the complication connected with the spatial relationship of interfaces between velocity regions are now compounded. For instance, the three-dimensional equivalent of "bow-tie events" is much more difficult to handle than the two-dimensional form. Another difficulty can be traced to the requirement of operator interference: the interpreter has to see the geometric relationship of events in order to assign velocities, he has to see the geometric relationship between interfaces to determine whether a particular ray, an interface, or a lateral velocity boundary is acceptable. There are as yet no simple methods to display several curved interfaces and the rays passing through them. For this reason three dimensional ray tracing is most difficult to apply where it is needed most: in difficult structural situations.

## THE IMAGE RAY

The criterion we have used for the necessity of ray tracing was that a ray that is vertical somewhere is not necessarily vertical everywhere (if one uses velocity functions with inclined axis, one has to replace "vertical" by "parallel to the velocity axis"). No velocity function and no replacement velocity could describe such a situation correctly.

Most current data set migration methods assume that a replacement velocity can be used. In the Sherwood-Loewenthal model (fig. 8) that forms the implicite basis of many migration algorithms, the diffractor in the object space has always the same spatial coordinates as the apex of the hyperboloid in the image space. Thus data set migration algorithms have the same shortcomings as the one-step methods for the migration of seismic events. Ray tracing is obviously impractical for data set migration, but the effects of inclined interfaces can be studied - and accounted for - with the concept of the *image ray* (Hubral 1977).

Wave fronts emanating from a point retain a spheroidal shape even after refraction at an interface, and therefore the seismic event corresponding to the diffractor in the Sherwood-Loewenthal model remains hyperboloidal even in the presence of dipping interfaces (fig. 21, fig. 22). Migration algorithms based on this hyperboloidal shape work - albeit with some degraded performance - even under these conditions, but they position the diffractor incorrectly.

The apex of the hyperboloid in the image space is that part of the event where the arrival time is stationary with respect to the spatial coordinates. In other words, it corresponds to a ray with the apparent slowness - the ray parameter - zero, i.e., to a ray arriving vertically at the surface. This is the *image ray*. Fig. 23 shows how the concept of the image ray can be used to detect positioning errors in migrated sections based on data set migration and to estimate the correct position of interface elements. The lower part shows a cross section with inclined and curved interfaces (object space) together with the source-receiver coincident rays by which the corresponding data set is generated.

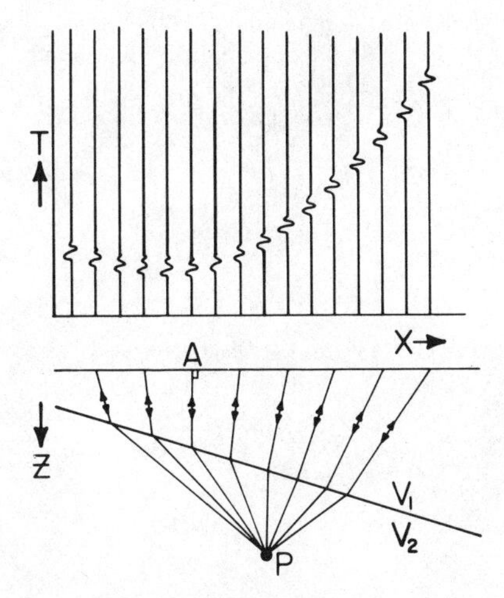


Fig. 21. Wave field from a scatterer under an inclined interface. Lower half: object space, upper half image space. The apex of the hyperbolic event has the same coordinates as the ray that arrives at the surface with vertical incidence (Western Geophysical Company).

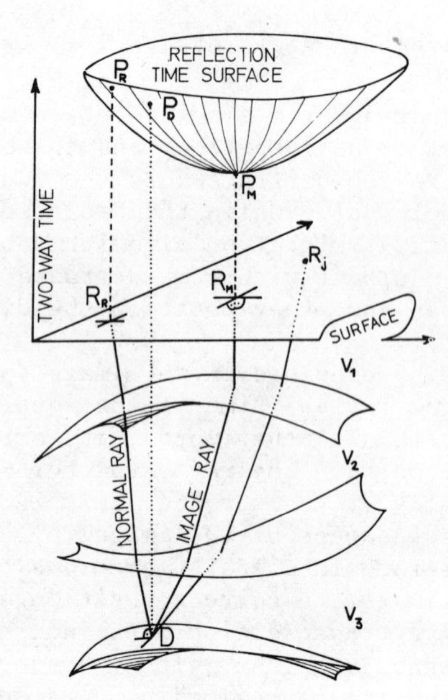


Fig. 22. Four dimensional wave field due to a scatterer below two curved interfaces. Below: object space. Above: image space. The event surface remains hyperboloidal, but the coordinates of its apex are not those of the scatterer (from Hubral 1977).

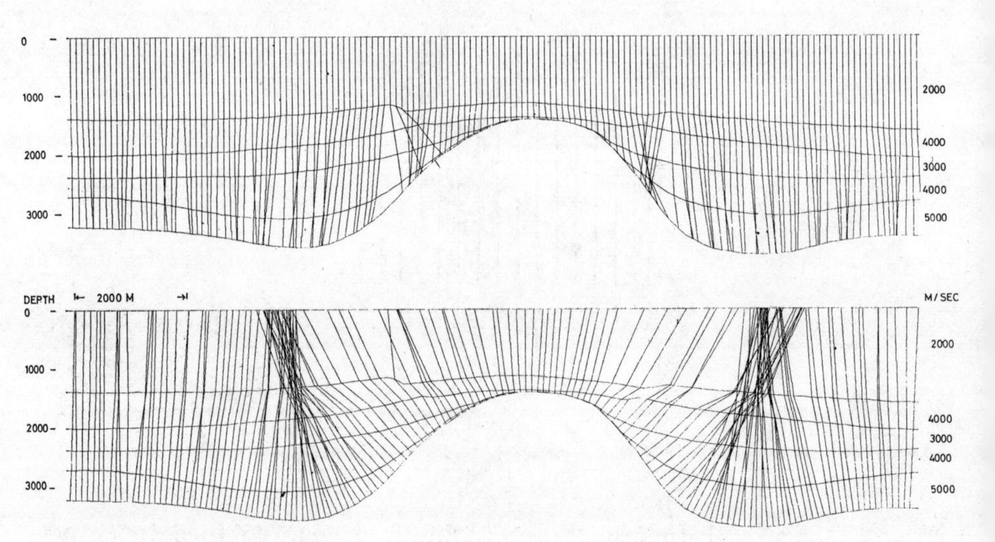


Fig. 23. Source-receiver coincident rays (below) and image rays (above) in a moderately complicated structure (from Hubral 1977).

The upper part shows the same cross section with image rays. The migration algorithm places the corresponding element of the interface vertically below the surface end of the image ray, the correct position is indicated by the intersection of the image ray with the interface.

Image ray construction is - like ray tracing - an inherently iterative process. Though there is only one image ray to any surface point, its correct position and inclination at depth is known only after the interfaces it penetrates are correctly known. Any interface below a non-horizontal interface must be checked for positioning errors - in fig. 23 that means all interfaces except the shallowest.

In applying the image ray concept one has to work in true object space, since otherwise the correct position of the intersections of image rays with interfaces cannot be determined. For this reason the corresponding methods are called *depth migration*. Again the term is slightly misleading, since any of the event migration methods leads to interfaces in the object space, and most data set migration methods could be expanded to yield results in object space. The difference of the image ray concept is not that the results are expressed in depth, but that the lateral position has been corrected in much the same way as the ray tracing methods correct lateral position.

## CONCLUSIONS

The events one can detect in the raw data set have their causes in the three dimensional subsurface. With some - more or less realistic - assumptions it is possible to derive the spatial disposition of these causes (discontinuities of impedance) from the coordinates of the events. A concept running through this paper is the clear distinction of image space (the space to which the unmigrated data belong) and object space (the real space of the subsurface, where the discontinuities exist and where we hope to find, e.g., oil and gas). Conversion of the information from the image space to the object space is thus a requirement for any structural inference, and even more so for lithological, stratigraphical, or sedimentological interpretation. This conversion can take place by data set migration (see the article of J.W. Hosken in this volume), by the methods described in this article, or in the mind of the experienced interpreter, but it cannot be omitted if one wants to avoid costly errors of judgement.

From the above one might get the impression that the migrated data set - the approximation to the object space - is better, or may be "truer", than the raw data set - the image space. While this is certainly correct in a pragmatic sense, one should never forget that the raw data set contains not only all information the migrated data set contains, but also information which was lost in the many approximations necessary in any migration process. Moreover, the raw data set depends only on the technical effort in data acquisition: repetition at a later date might yield an improved data set, but the original set will be a subset of the new data set. The migrated data set, on the other hand, depends on assumptions and approximations: ray tracing migration differs from replacement velocity processes, and migration based on isochrone maps does not necessarily yield the same result as migration in the construction plane with a later correction for "cross dip".

Migration is still a cumbersome and expensive process, so the incentive to use short cuts, or to be satisfied with *one* approximation, is strong. However, the message of the previous paragraph is clear: we can always do better, and that in two ways, by improving our computer programs and by checking migration with different degrees of approximation. One important example is the application of ray tracing *after* data set migration: the difficulties in ray tracing disappear if the velocities can be assigned to different sectors of the subsurface, i.e., in the object space. On the other hand, in ray tracing one has a greater flexibility with respect to refraction of rays on dipping interfaces, oblique velocity functions and the like.

Ray geometric migration methods for seismic events were the mainstay of structural interpretation before the advent of data set migration algorithms. Recently they have fallen somewhat into oblivion. However, the reason for discussing them again is far from simple nostalgic reminiscence: there is a need to incorporate ray geometric aspects into modern migration algorithms in much the same way as it has already happened with the image ray concept.

## REFERENCES

Al-Chalabi, M., 1973, Series approximation in velocity and travel time computations, Geophysical Prospecting, 21:783-795.

Al-Chalabi, M., 1974, An analysis of stacking, rms, average, and interval velocities over a horizontally layered ground, Geophysical Prospecting, 22: 458-475.

Hagedoorn, J.G., 1954, A process of seismic reflection interpretation, Geophysical Prospecting, 2: 85-127.

Hosken, J.W., 1980, Imaging the earth with seismic reflections, this volume.

Hubral, P., 1977, Time migration - some ray theoretical aspects, Geophysical Prospecting, 25: 738-745.

Krey, Th., 1965, Die Bedeutung der Horizontkrümmung für einige Mess- und Rechenverfahren in der angewandten Seismik, Dissertations Ludwig-Maximilians-Universität München.

Loewenthal, D., Lu, L., Roberson, R., and Sherwood, J., The wave equation applied to migration, Geophysical Prospecting, 24: 380-399.

Michaels, P., Seismic ray path migration with a pocket calculator, Geophysics, 42: 1056-1063.

Musgrave, A.W., 1952, Wave front charts and ray path plotters, Quarterly of the Colorado School of Mines, 47 (4).

Taner, M.T., Koehler, F., and Sheriff, R.E., 1979, Complex seismic trace analysis, Geophysics, 44: 1041-1063.

## IMAGING THE EARTH'S SUBSURFACE WITH SEISMIC REFLECTIONS

J.W.J. Hosken

British Petroleum Co. Ltd.
Britannic House
Moor Lane
London EC2Y 9BU

### 1. INTRODUCTION

Professor Helbig (1980) has dealt with the inversion of seismic reflection data to give a representation of the disposition of reflectors within the earth which give rise to the observed data. The data to which he referred belong to a subset of the total observed data, namely those "horizons" or "events" which the interpreter chooses to "pick", having sufficiently distinctive and continuous visual properties on the seismic sections to allow him to follow them. This may be both time-consuming and difficult particularly where horizons appear to intersect, or there may be a break in continuity. Synclinal features giving rise to "Bow-tie" patterns are often difficult to follow. Figure 1 shows the seismic response of such a feature. What the interpreter needs is a migration process which can treat all the observations in the seismic section. So today we have a plethora of signal processing methods which purport to invert all the recorded signals, be they reflections or diffractions, and to show their origins in their true positions and strengths.

The volume of computational work involved is considerable by any standards. Table 1 gives the orders of magnitude of the quantity of data to be handled in various stages and types of imaging operations, for data from typical survey dimensions.

This is the fastest-developing area of geophysical data processing today and I cannot hope to do more than to examine some of the basic concepts, outline a few of the main algorithmic methods and comment

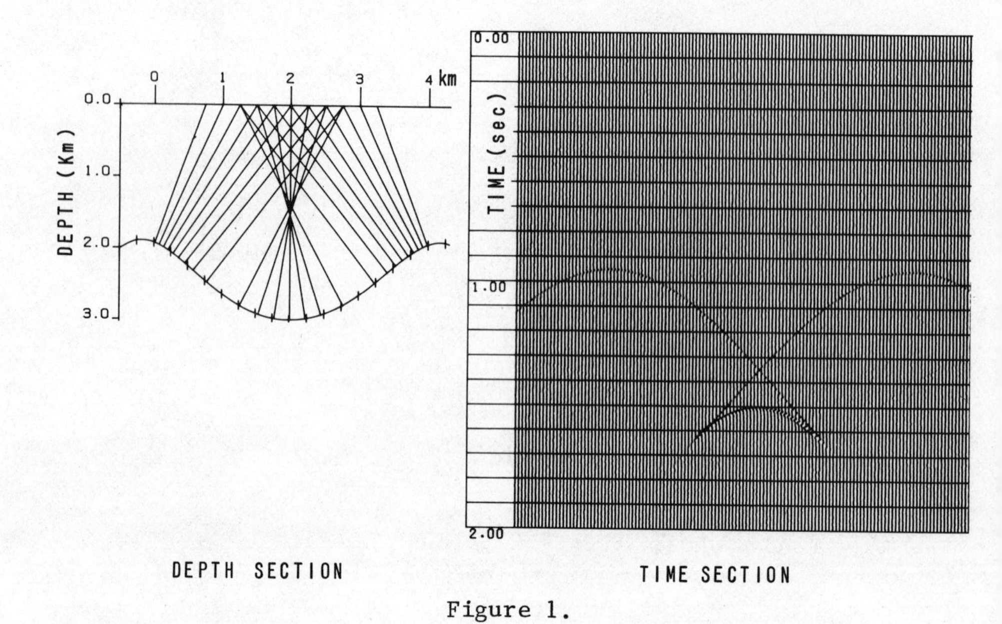


Figure 1.

TABLE 1

QUANTITY OF DATA IN SEISMIC REFLECTION SURVEYS FOR HYDROCARBONS

| | | |
|---|---|---|
| Recording density (typical 1976-1980) | 1250 samples in 5 sec. x 48 traces every 25m | $2.4 \times 10^6$ samples/km |
| Display density for interpretation (effective independent samples | ~ 500 samples in 5 sec. x 1 trace every 25m | $2 \times 10^4$ samples/km |
| | 2D Surveys | 3D Surveys |
| Total line length of survey (typical) | ~ 1,000 km | ~ 10,000 km |
| Line length requiring migration (typical) | ~ 300 km | ~ 10,000 km |
| No. of samples requiring migration (typical) | { ~ $7 \times 10^8$ recorded<br>~ $6 \times 10^6$ output | ~ $2 \times 10^{10}$ recorded<br>~ $2 \times 10^8$ output |
| Dimension of kernel $G_{ij(k)}$ of Nolet (1980) applied to model to predict data | $10^5$ (recorded data)<br>$2 \times 10^3$ (data reduced by preliminary stacking) | $10^8$ (recorded data)<br>$2 \times 10^6$ (data reduced by preliminary stacking) |

on their strengths and weaknesses as I see them. We begin with some terminology, and mention the various idealizations and approximations involved, as there must be in any practical inversion of such complex data. We then move to the simplest model for inversion, that of coincident source and receiver pairs (called "zero-offset" case, "offset" being the source-receiver interval). After a brief mention of the extension to three dimensions, we examine the inversion of non-zero-offset recorded data.

## 2. DEFINITIONS, LIMITATIONS, IDEALIZATIONS AND APPROXIMATIONS

The basic concept required in understanding the modelling and inversion of two or three dimensional wave propagation is the "wavefield". This is illustrated in Figure 2 and is a bounded region of three or four dimensional hyper-space in which time is one of the dimensions and within which the wave equation governs the behaviour of measurable quantities such as pressure or particle velocity. One boundary of this hyper-space is formed by the two or three dimensional surface in which the observations are made, again, time being one of its dimensions. This we refer to as the "image space", by analogy with geometrical optics. As we will see more clearly later, another boundary, orthogonal to the image space, is formed by the two or three dimensional surface containing the earth's discontinuities required to be mapped. Here depth is one of the dimensions. We call this the "object space". A calculation which proceeds forward in time and ends with the image space is called "forward modelling", and one which works backwards in time towards the object is "model inversion", "migration" or "imaging".

Many of the limitations and idealizations we accept on the properties of the wavefield are aimed at simplifying the wave equation.

Starting from a fairly general form of the wave equation in which the only major assumptions are the absence of external body forces and of a shear modulus of rigidity (no shear waves)

$$\frac{\partial}{\partial x}\left[\rho\frac{\partial^2\phi}{\partial t^2} - \lambda\nabla^2\phi\right] - \frac{\partial\rho}{\partial x}\frac{\partial^2\phi}{\partial t^2} = 0 \qquad (1)$$

where $\rho$ is density, $\lambda$ is the bulk modulus. $\phi$ is a displacement potential such that particle displacements

$$q_x = \frac{\partial\phi}{\partial x}\ ; \quad q_y = \frac{\partial\phi}{\partial y}\ ; \quad q_z = \frac{\partial\phi}{\partial z}\ .$$

There are two similar equations involving the derivatives with respect to y and z (see Ewing, Jardetzky and Press (1957) equation 7-10).

Equation (1) resolves to

$$\rho \frac{\partial}{\partial x}\left(\frac{\partial^2\phi}{\partial t^2}\right) - \frac{\partial\lambda}{\partial x}\,\nabla^2\phi - \lambda\frac{\partial}{\partial x}\left(\nabla^2\phi\right) = 0 \qquad (2)$$

Substituting for pressure $u = -\rho\dfrac{\partial\phi}{\partial t}$,

so that $\dfrac{\partial^2\phi}{\partial t^2} = -\dfrac{1}{\rho}\dfrac{\partial u}{\partial t}$

and $\dfrac{\partial}{\partial t}\left(\nabla^2\phi\right) = \nabla^2\left(\dfrac{\partial\phi}{\partial t}\right) = \nabla^2\left(-u/\rho\right)$,

and differentiating (2) with respect to t gives

$$\rho\frac{\partial}{\partial x}\left(-\frac{1}{\rho}\frac{\partial^2 u}{\partial t^2}\right) = \frac{\partial}{\partial x}\left(\lambda\nabla^2\left(-u/\rho\right)\right).$$

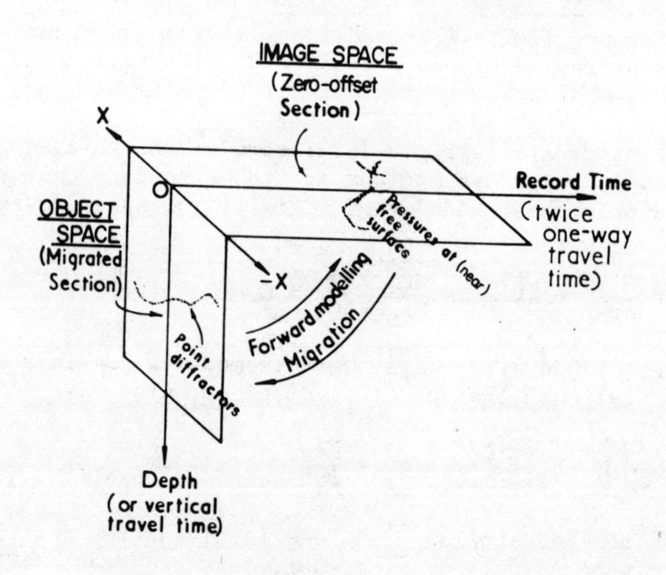


Figure 2.

Integrating in the x direction by parts,

$$\lambda \nabla^2 (u/\rho) + \int^x \frac{\partial \rho}{\partial x} \frac{1}{\rho} \frac{\partial^2 u}{\partial t^2} dx - \frac{\partial^2 u}{\partial t^2} = 0$$

$$\nabla^2 u + \rho u \nabla^2 (1/\rho) + 2\rho \nabla u \nabla (1/\rho) + \rho/\lambda \int^x \frac{\partial \rho}{\partial x} \frac{1}{\rho} \frac{\partial^2 u}{\partial t^2} dx - \rho/\lambda \frac{\partial^2 u}{\partial t^2} = 0 \quad (3)$$

Again similar equations exist which have integrals in y and z directions instead of x. (3) is sufficient however to show the magnitude of the approximation usually made, namely that $\rho$ varies slowly enough in all directions so that $\nabla(\frac{1}{\rho})$, $\nabla^2(1/\rho)$, $\frac{\partial \rho}{\partial x}$, $\frac{\partial \rho}{\partial y}$, $\frac{\partial \rho}{\partial z}$ are all small enough for

$$\nabla^2 u - \frac{1}{V^2} \frac{\partial^2 u}{\partial t^2} \simeq 0, \quad \text{where } V^2 = \lambda/\rho \quad (4)$$

It is interesting that (4) implies no restriction on the variation of $\lambda$.

To solve the wave equation exactly requires us to know V everywhere. This, in conjunction with the assumption of constant $\rho$, would imply that we knew the position and magnitude of all the reflectors, so we would not need to do the migration in the first place! Fortunately, the influence of V in the wave equation is primarily one of controlling the time-distance behaviour of travelling waves, for which purpose it is the integral or "long wavelength" properties of the spatial variations of V which matter, while in the reflection process, it is the differential or short wavelength variations of V which determine the magnitude and form of reflections. Essentially we treat V and the reflection-generating properties of the earth as independent. Nevertheless it is a general fact that when V varies significantly, we cannot properly migrate until we know those variations, and we cannot know where and how the velocity varies properly until we have migrated correctly. This suggests an iterative model-building approach which is becoming an important part of the overall procedure in severe cases of velocity variation.

Another approximation we make use of is that the source and receivers are omni-directional in their energy and response patterns. This together with assumptions made already about the medium allow us to invoke the principle of reciprocity, and lead to an extremely important simplification of the model of the experiment, namely the idea of transferring the source from the free surface down to the discontinuities required to be mapped. Apart from a shift of time origin, the upward travelling wavefield from these buried imaginary sources can be envisaged as behaving exactly like that from the true sources. This also leads to the conclusion that the signal from a continuous reflector is the same as that of a set of diffractors distributed all over its surface.

The next major assumption made in almost all migration methods is that only the upward-travelling wave energy from these buried "sources" matters. This implies that there are no multiple reflections, which include portions of travel path in the opposite direction.

Taken together these assumptions allow us to simplify the model drastically. A common name for the model we aim to invert is the "exploding reflector model". Stated in words this assumes that the wave motion recorded at the free surface can be regarded as the result of an experiment in which sources are exploded at the discontinuities in the earth at appropriate instants and with strengths proportional to the magnitude and sign of the discontinuity in acoustic impedance they mark, and in which waves only travel upwards. The instants at which the sources must be activated are those at which energy arrives at the discontinuities from the real source. This restricts the model to a single real source. As we shall see, this model becomes further simplified for zero-offset recorded data.

As a last assumption in this section, we shall simplify the treatment considerably if we confine our attention for most of this article to two-dimensional earth structures, that is one for which a direction $\mathbf{w}$ exists such that $\partial^2 u/\partial w^2$ is small enough to be neglected and so to allow $\nabla^2 u$ to be represented adequately in two orthogonal directions (x and z) only. Effectively this means that the earth can be represented everywhere by a single section, in a plane parallel with the x and z axes. w need not be horizontal nor necessarily perpendicular to the line of seismic sources and detectors, but it helps in our mental pictures if it is. Later on we shall look briefly at 3-dimensional migration when this assumption cannot be made, but for the time being we will take the wave equation to be solved as

$$\frac{1}{V^2}\frac{\partial^2 u}{\partial t^2} = \frac{\partial^2 u}{\partial x^2} + \frac{\partial^2 u}{\partial z^2}$$

## 3. MIGRATION OF ZERO-OFFSET-RECORDED DATA

When the source and receiver are coincident, the path of the energy from source to discontinuity is retraced exactly by the energy coming back from discontinuity to receiver (strictly, the ray paths can split apart if there is a second derivative of velocity in the horizontal direction in the overburden, but we will further assume that this is not a serious effect). This allows us to simplify the exploding reflector model still further: the zero-offset recorded motion at the surface can be regarded as the result of an experiment in which sources are exploded at the discontinuities in the earth at zero time and with the appropriate magnitudes as before, and with all travel times doubled by, as one of the possible devices, halving all the velocities. An important property of this model is that there is no limit to the number of source-receiver pairs which could be activated simultaneously. In other words, the data in an entire zero-offset section can be regarded as the outcome of a single experiment. This in turn implies that it is an image space of a real invertible wavefield. In the light of this model, it can be seen that the process of migration is that of working backwards through the wavefield from the image space to the object space at zero time. Forward modelling is the reverse.

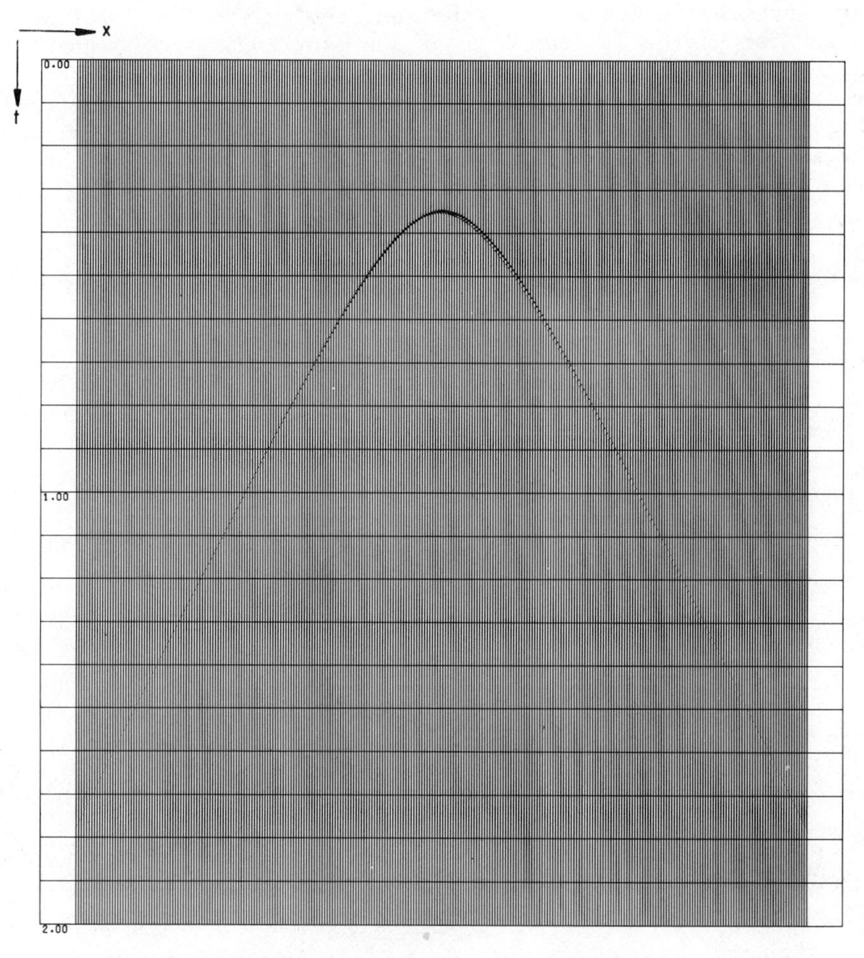


Figure 3. Zero offset section for line diffractor perpendicular to seismic line (response at surface to line source).

There are a number of different ways of looking at this process. One which is intellectually satisfying, though not computationally useful, is to see it as a deconvolution. Each point of discontinuity in the subsurface, with its associated buried "source" produces in the image space a characteristic response - a two-dimensional "impulse response". Since the total response in the image space is the linear superposition of impulse responses from all discontinuities in the object space, forward modelling can be thought of as a 2-dimensional convolution of the impulse response with the discontinuity function in the object space. Hence migration is the inverse of this convolution, i.e. a two-dimensional deconvolution (Bolondi, et al, 1978, Berkhout et al, 1979).

The forms of the impulse response and its inverse are interesting The impulse response, shown in Figure 3, has its energy concentrated along a hyperbolic space-time function, the celebrated "diffraction pattern" for a line diffractor. It does not consist of a constant-amplitude spike along this curve however, but has a "postlude" corresponding to the response of a half-differentiator (spectrally, it has a $45^{o}$ phase lead and a 3dB-per-octave increase in amplitude with frequency) and a falling amplitude with distance from the apex of the hyperbola. The inverse, see Figure 4, is like it but upside

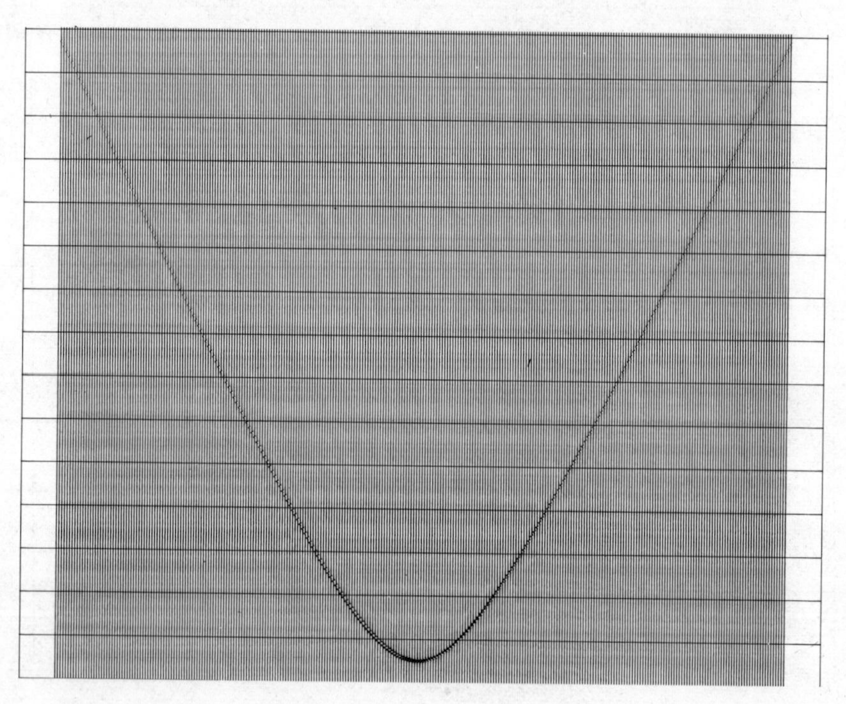

Figure 4. 'Ideal' 2-D operator to collapse a diffraction to a point.

down. Convolving one with the other is the same as turning one upside down, "flipping" it over sideways and cross-correlating with the other. When the impulse response coincides with the "flipped" inverse, the output will be very large, but it will disappear rapidly, moving away from this coincidence point.

Several concepts emerge from this: First it is obvious that any miscalculation or approximation which makes the inverse operator differ from the true one will spoil the deconvolved result, and the accuracy and definition of any practical method of migration can be assessed by inspection of the migration of an ideal diffraction pattern. Secondly, just as noise is often amplified by wavelet deconvolution in one dimension so also noise will probably produce undesirable effects through migration, and just as it is necessary to limit the bandwidth of a spectral whitening deconvolution process in order to limit the build-up of noise, so it is necessary to limit the potential resolution and "dip-bandwidth" of migration to avoid a similar trouble.

3. a) Diffraction Stacking

The main reason why two-dimensional deconvolution is not used in practical computation is that the impulse response is a time-varying function (and space-varying also if the velocity varies in a horizontal direction). This makes full deconvolution the solution of a Volterra equation of the first kind, which in general is a difficult task. In the Fourier transform domain, too, there are difficulties as we shall see. However, there is a simple approach which, though specific to the hyperbolic form of the impulse response, is a surprisingly good approximation. This consists effectively in cross-correlating the image space with a hyperbolic pattern which is matched to the expected diffraction pattern, but having a simple unit spike on the hyperbola at each intersection with a seismic trace. Computationally all that is done is to gather up the samples found along the hyperbola in the image space and sum them (in seismic data processing jargon: "stack" them) into the position of the apex mapped into the object space. This simple-minded approach, known sometimes as the "Diffraction Stack" method, is the earliest one to be generally applied in the seismic data processing industry, and with refinements which we shall outline later, it remains as one of the important processes in use today. Before we leave it, we will point out two alternative representations which help in understanding it and other methods.

We have described the Diffraction Stack in terms of the locus of input image space samples which stack into a single output object space sample. We can also show the locus of output object space samples to which a single input image space sample is sent. If $t_o$ is the travel time of a sample in object space, $t_i$ is the time of a sample in image space and X is the horizontal distance between them, the relation defining the two loci is $t_i^2 = t_o^2 + \frac{4X^2}{V^2}$ and the loci are $t_i = \left(t_o^2 + \frac{4X^2}{V^2}\right)^{\frac{1}{2}}\Big|_{t_o \text{ constant}}$ and $t_o = \left(t_i^2 - \frac{4X^2}{V^2}\right)^{\frac{1}{2}}\Big|_{t_i \text{ constant}}$

The first is a hyperbola and the second a circle, provided X is plotted in units of Vt/2. Thus the impulse response of the diffraction stack migration process is a circle of radius equal to the travel time of the impulse (see Figure 5). Indeed all migration processes have similar impulse responses.

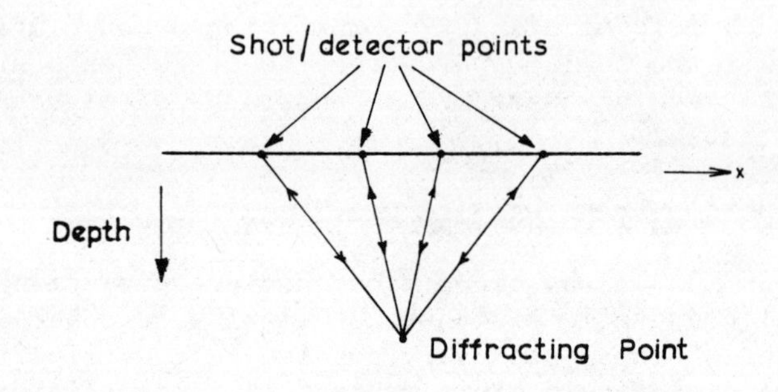


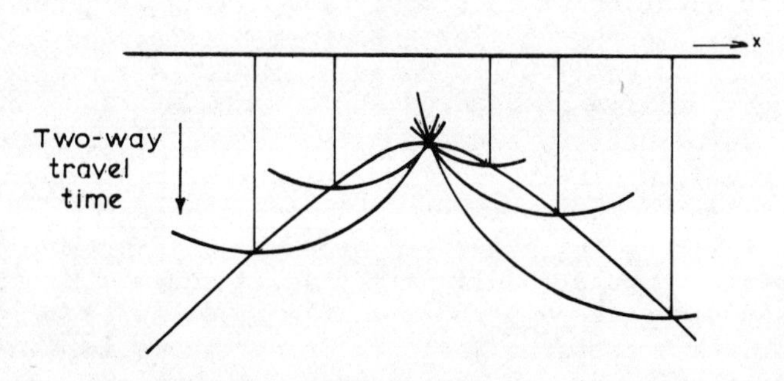


Figure 5.

This accounts for the effect of random noise on the image space section, which is to produce spurious arcs of circles ("smiles") on the migrated object space section. It also helps us in visualising how flat dipping reflectors are migrated; if the image space samples lie along a straight line, the circles they produce will abut another straight line tangential to them all and forming an envelope. This is illustrated in Figure 6. The dip of this line in the object space is the true dip of the reflector, which we will call $\alpha_D$. Then

$$\tan\alpha_D = \frac{V}{2}\frac{\partial t_0}{\partial X}\bigg|_{t_i \text{ constant}} = -\frac{2X}{Vt_0}$$

Using $\sec^2\alpha_D = 1 + \tan^2\alpha_D$ , we obtain $\cos\alpha_D = t_0/t_i$

In the image space however the dipping reflector will be tangential to the set of hyperbolae along which samples are gathered and placed at their apices. If we call the apparent dip on the zero-offset section $\alpha_T$ , then

$$\tan\alpha_T = \frac{V}{2}\frac{\partial t_i}{\partial X}\bigg|_{t_0 \text{ constant}} = \frac{2X}{Vt_i}$$

Taking into account the reversed direction of X between the two cases we can substitute

$$\tan\alpha_T = \frac{2X}{Vt_0}\cdot\frac{t_0}{t_i} = \sin\alpha_D$$

Thus migration can also be thought of as a process which rotates segments of dipping reflections from a dip of $\alpha_T$ in the image space to one of $\alpha_D = \sin^{-1}(\tan\alpha_T)$ in the object space, about a centre at their intersection with the surface of the earth. This concept will be useful when we consider migration in the frequency-wavenumber domain.

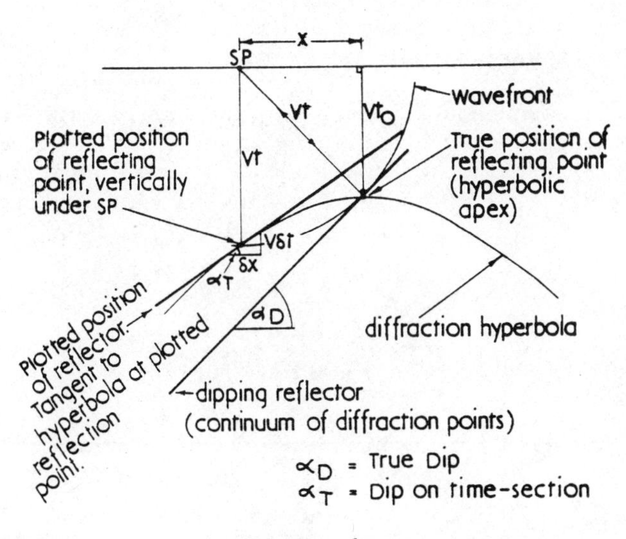


Figure 6.

3. b) Kirchhoff Integral Migration

Though this is directly related to Diffraction Stacking, we must return to the Wave Equation as its true root, and of which it is a boundary integral solution.

The theory of the boundary integral strictly requires that throughout the wavefield the velocity should be constant and it contain no sources. Provided the effects of a number of sources can be separated however in time, and that of a number of different velocities combined in an average, it is still possible to apply the ideal theory.

We begin by considering the Fourier transform of the field u by writing

$$u(x,y,z,t) = \int_{-\infty}^{\infty} U(x,y,z,\omega)\, e^{i\omega t}\, d\omega . \quad \text{Setting } k = \omega/V$$

and substituting into the wave equation we arrive at the Helmholtz equation

$$\left(\nabla^2 + k^2\right) U(x,y,z,\omega) = 0 .$$

Next we invoke Green's theorem which states that over a volume $V$ surrounded by a surface S,

$$\int_V \left(\psi_2 \nabla^2 \psi_1 - \psi_1 \nabla^2 \psi_2\right) dV = \int_S \left(\psi_2 \frac{\partial \psi_1}{\partial \underline{n}} - \psi_1 \frac{\partial \psi_2}{\partial \underline{n}}\right) dS$$

where $\psi_1$ and $\psi_2$ are arbitrary functions of x, y, z and $\omega$ which satisfy the Helmholtz equation and any boundary conditions on S, and where $\underline{n}$ is the unit outward normal vector to S.

The surface we choose lies along the earth's free surface and around a hemisphere at a great enough radius for U to be negligible over it (see Figure 7). We consider a "source" point in the object space at $x_1$, $y_1$, $z_1$ and change variables to radial vectors from this point.

$$\underline{R} = \underline{r} - \underline{r}_1 = \underline{(x,y,z)} - \underline{(x_1,y_1,z_1)}$$

The appropriate Green's functions for a free earth surface are

$$\psi_1 = U(\underline{r}) ; \qquad \psi_2 = \frac{1}{R} e^{-ikR} - \frac{1}{R'} e^{-ikR'}$$

where $R' = |\underline{r} - \underline{r}_2| = |\underline{(x,y,z)} - \underline{(x_1,y_1,-z_1)}|$

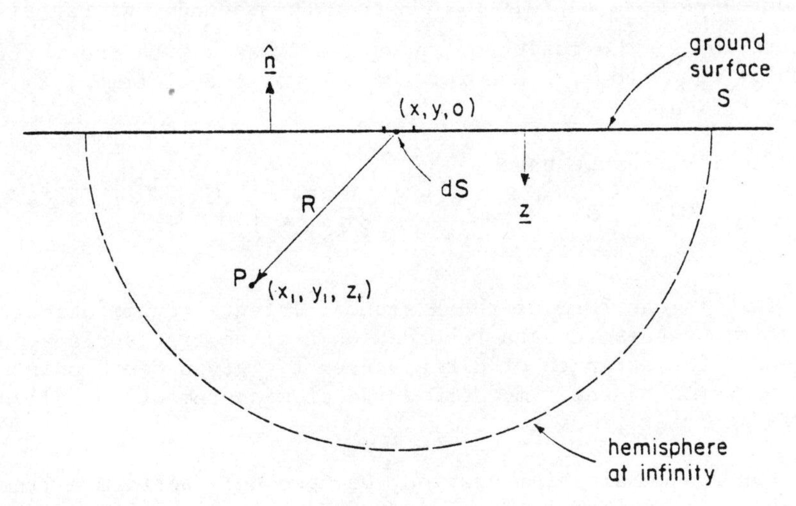


Figure 7.

the radius vector from the image of the "source" reflected in the free surface. Substituting these into the Green's equation, and performing the integrals in a manner which excludes a vanishingly small sphere around the source point, we can solve for the value of $U$ at $\underline{r_1}$ in terms of the values of U along the surface, z=0, and obtain

$$U(\underline{r_1}) = -\frac{1}{2\pi}\int_S U(\underline{r})\Big|_{z=0} \frac{\partial}{\partial \underline{n}}\left(\frac{1}{R}e^{-ikR}\right) dS$$

$$= \frac{1}{2\pi}\int_S \left[\frac{Ue^{-ikR}}{R^2}\cdot\frac{\partial R}{\partial \underline{n}} + \frac{ikUe^{-ikR}}{R}\frac{\partial R}{\partial \underline{n}}\right] dS$$

Transforming back to the time domain and advancing the time earlier to t=0, gives the migration output

$$u(x_1, y_1, z_1, t=0) = \frac{1}{2\pi}\int_S \left\{\frac{1}{R^2}\frac{\partial R}{\partial \underline{n}}[P]_s - \frac{1}{VR}\cdot\frac{\partial R}{\partial \underline{n}}\left[\frac{\partial P}{\partial t}\right]_s\right\} dS$$

where $[P]_s = u(x, y, z=0, t=R/V)$ , $\left[\frac{\partial P}{\partial t}\right]_s = \frac{\partial}{\partial t}\{u(x, y, 0, R/V)\}$

and $V = \frac{dR}{dt}$ , the velocity.

So long as $z_1$ is sufficiently deep for R always to be at least a few wavelengths, the first term in the integral may be ignored as it decreases more rapidly with R than the second. We can also substitute for $\frac{\partial R}{\partial n}$ the cosine of the angle R makes with the vertical, which is equal to $z_1/R$, and for the one-way travel times, $t_o = z_1/V$ and $t_i = R/V$.

Finally then we have

$$u(x_1, y_1, Vt_o) \simeq -\frac{1}{2\pi}\frac{t_o}{V^2}\int_S \frac{1}{t_i^2}\left[\frac{\partial P}{\partial t}\right]_{S,t_i} dS$$

This is the form of the Kirchhoff Integral applicable to 3-D migration of seismic data recorded on a plane free surface, and expresses the strength of disturbance at a given depth point at zero time in terms of the time derivative of measurements at all surface points at later times $t_i$.

For 2-dimensional migration, we need only perform a line integral instead of a surface integral. To find this line integral we perform the surface integral over the y direction first under the assumption that $\left[\frac{\partial P}{\partial t}\right]_S$ is independent of y. Noting that $y = V(t_i^2 - t_x^2)^{1/2}$, where $t_x = \{(x-x_1)^2 + z_1^2\}^{1/2}/V$ , we see that the integral over y becomes an integral over t. This is not a simple integration and we will not show the steps, but simply quote the outcome, which is

$$u(x_1, y_1=0, z_1) \simeq \frac{1}{\pi}\int_{-\infty}^{\infty} \frac{t_o}{V^2 t_x^{3/2}}\left[f(t_x) * P(x,0,0,t_x)\right] dx$$

where * expresses convolution, and

$$f(t_x) = (2|t_x|)^{-1/2}\delta(t) - (2|t_x|)^{-3/2}(1 - H(t))$$

$\delta(t)$ being the Dirac impulse and H(t) the Heaviside unit step $\int_0^t \delta(t)\,dt$

$f(t_x)$ is describable as a maximum-phase half-differentiator filter operator, giving $45^\circ$ phase lag and a 3dB-per-octave rising amplitude with frequency.

The above approximate equation is close enough again for $z_1$ more than a few wavelengths, and expresses the 2-D Kirchhoff Integral for migrating data recorded on a plane surface along a line at right angles to the strike direction.

The relation between the Diffraction Stack migration method and the Kirchhoff Integral solution of the Wave Equation now becomes clear. The Diffraction Stack is a summation approximating to the integral

$$\int_{-\infty}^{\infty} P(x, 0, 0, t_x)\, dx$$

since $t_x$ simply follows the image space locus $t_1$ given in the previous section. It departs from the full Kirchhoff Integral in omitting the half-differentiator filter and the variation in amplitude with $t_o$ and $t_x$. Neither of these have very seriously damaging effects, but in some circumstances they are noticeable. For instance the lack of the amplitude factor makes the response to steeply dipping events stronger than it should be, and exaggerates noise "smiles". The absence of the half-differentiator attenuates high frequencies noticeably, and though subsequent deconvolution can easily correct this, it will inject a noticeable $90^o$ phase-lead error.

More serious, on occasions, are the errors introduced by the replacement in practice of the continuous infinite integral by a finite discrete sum of samples from a set of traces. This gives rise to two sets of errors, one due to truncation and the other due to discretization. The principal effect of truncation is to limit the magnitude of dip which can be migrated, fairly abruptly, to $\tan^{-1}(X_{max}/z)$ where $X_{max}$ is the distance of the farthest input trace from the migrated output trace and z the depth at the output trace position. Truncation also injects small noise pulses running approximately parallel with and shallower than strong continuous reflections, and occurring at the extremities of the circular arcs swept out by the impulse response. The resolution of discontinuities is also damaged by excessive truncation, as the imaged diffractor point spreads out, exactly as in optical imaging when a lens aperture is restricted.

Discretization errors, though their effects remain invisible in most practical cases so long as the earth's surface is adequately sampled by the source-receiver pairs, can produce serious noise wherever the migration of steep dips is being attempted with sparsely sampled data such that there is a danger of wavenumber aliassing occurring. It takes the form of bands of high frequency noise running around, mostly above, the strong dipping reflections, though in extreme cases, it can even occur above horizontal reflections. It occurs whenever on an output trace, the samples from the same reflection on successive input traces are stacked in at intervals which exceed the reciprocal of the cut-off frequency of the high-cut filter in the system. Figure 8 shows the danger. If it is found to occur, more severe high-cut filtering of the input traces is necessary.

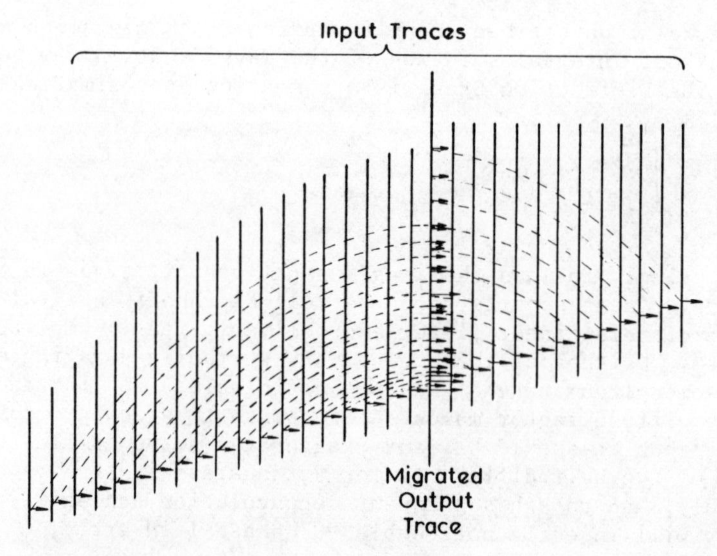


Figure 8. Diffraction stack of single dipping reflection.

It is possible to make the method take account of variations in the velocity V(x, z). The velocity only affects the result in two ways: as a scalar multiplier of amplitudes, which is not important in practice, and as a factor in timing the oblique travel path $t_i$, which is crucial.

The timing may be calculated to any required degree of accuracy, given the velocity model V(x, z), but, depending on the degree of sophistication employed, this can add greatly to the computational cost. Hitherto, however, the process has been predominantly "input-output-bound" rather than "compute-bound", that is the time of data transfer has exceeded the arithmetic time in the computer, so there is no cost penalty for a certain amount of extra computational work. The method also lends itself to a recent development called "datuming" (Berryhill, 1978) where the computation proceeds in stages from one major interface to another, the velocity being taken as constant between the interfaces. It is necessary for the user to define these interfaces to the process, which requires a preliminary migration to have been done beforehand.

The strength of the Kirchhoff Integral method lies in it being understandable to the user, who can readily appreciate the effects of the parameters under his control, such as the "aperture" $X_{max}$.

Its weakness is that it is somewhat more expensive to run than other methods because of the greater amount of data requiring to be re-read into computer memory. Another weakness, strangely, is that it can migrate accurately, maintaining the amplitudes of steeply dipping events. In the presence of noise, this is often not the best thing to do, as noise "smiles" can seriously interfere with the interpretation in regions of weaker signals.

### 3. c) The Finite Difference Method

This is by far the most popular approach to migration today, which is at first sight somewhat surprising as it works its way from the image space to the object space in a series of intermediate steps, instead of the single step of the Kirchhoff Integral. (In engineering applications, the trend seems to be away from finite difference calculations towards boundary integral methods). However the small step-at-a-time approach can prove to be an advantage. In theory at least it should allow the calculation to take account of detailed variations in velocity through the x, z space, though as we shall show, to do so involves making the data-handling more complex. Otherwise when velocity is taken as a function of z only, unless the step size is made exceedingly small, the quantity of data needing to be transferred in and out of computer memory can be less than the Kirchhoff Integral method requires. Though the amount of arithmetic to be done is far greater, this can be handled efficiently in modern array processors, so that the method is potentially cheaper. In outline, the method divides the wavefield hyperspace into a grid of space-time elements. The wave equation is approximated by finite difference relationships between adjacent elements, which are solved in a regular order, working away from the image space and finishing when the whole required object space has been calculated. The details of the calculation vary considerably from one implementation to another. A number of compromises between accuracy and speed of computation are open to the programmer who must settle on the order of approximation to be accepted for each derivative, and the precise point within each element at which it is to be estimated (the centre is not necessarily the best).

One very common trick is to exploit the rapid dispersion and dissipation of rapidly travelling energy in a finite difference grid in order to eliminate downward travelling waves, by using a moving frame of reference which keeps upward travelling waves approximately stationary. (See Claerbout, 1976, p. 208 et seq. for a full treatment).

By way of example we will follow through the derivation of a particular finite difference approximation and indicate the algorithm which would be derived from it. This one, in common with most in practical use, performs "downward continuation", that is the order of computation is organised in major steps downwards in z and computes the wavefield as it would be recorded at successively increasing depths. An alternative scheme would be to take the major steps in t, and show the wavefield as a sequence of snapshots in time, a procedure sometimes called "wave-marching".

We start from the Helmholtz equation in its 2-dimensional form

$$\frac{\partial^2 U}{\partial x^2} + \frac{\partial^2 U}{\partial z^2} = -\frac{\omega^2}{V^2} U \qquad \text{where } U = \int_{-\infty}^{\infty} u \cdot e^{-i\omega t} \, dt$$

The solution may be sought in the form

$$U = A \exp\left\{ z \left(\frac{\partial^2}{\partial z^2}\right)^{\frac{1}{2}} \right\} + B \exp\left\{ -z \left(\frac{\partial^2}{\partial z^2}\right)^{\frac{1}{2}} \right\}$$

where $\left(\partial^2/\partial z^2\right)^{\frac{1}{2}} = \partial/\partial z$ which transforms to $ik_z$ in the vertical wavenumber domain, so that the two terms can be seen to represent upward and downward travelling waves. To solve for the upward travelling waves we take the first term only:

$$\frac{\partial U}{\partial z} = \frac{i\omega}{V} \left(1 + \frac{V^2}{\omega^2} \frac{\partial^2}{\partial x^2}\right)^{\frac{1}{2}} U$$

This equation is a form of the exact one-way wave equation as may be checked by differentiating with respect to z and assuming V is independent of z, but is only valid for constant V. For variable V it is necessary to approximate the square root. A useful approximation is the "continued fraction":

$$(1 + a)^{\frac{1}{2}} = 1 + \cfrac{a}{2 + \cfrac{a}{2 + \cfrac{a}{\cdots\cdots}}}$$

Truncation of this expansion gives rise to approximations to the wave equation which produce errors in the effective velocity of propagation as a function of the direction. For instance the velocity is in error by more than about 1% for propagation directions more than 15$^\circ$ from the vertical if the series is truncated to $1 + a/2$. This produces what is known as the "15$^\circ$ algorithm". The angle for this error rises to about 45$^\circ$ if the series is truncated to

$$1 + \frac{a}{2 + a/2} = 1 + \frac{2a}{4 + a}$$

Substituting with $a = \frac{V^2}{\omega^2} \frac{\partial^2}{\partial x^2}$ gives

$$\frac{\partial U}{\partial z} = \frac{i\omega}{V} \left[ 1 + \frac{2 \cdot V^2/\omega^2 \cdot \partial^2/\partial x^2}{4 + V^2/\omega^2 \cdot \partial^2/\partial x^2} \right] U$$

Now we transform to a frame of coordinates which moves upwards with a velocity which is in the region of the local velocities, say $\bar{V}$, by means of the substitution:

$$U = U' \exp\left(\frac{i\omega z}{\bar{V}}\right)$$

Hence

$$\frac{\partial U'}{\partial z} = i\omega\left(\frac{1}{V} - \frac{1}{\bar{V}}\right)U' + \left[\frac{\frac{2iV}{\omega}\,\partial^2/\partial x^2}{4 + \frac{V^2}{\omega^2}\frac{\partial^2}{\partial x^2}}\right]U'$$

This equation may be split into two equations

$$\frac{\partial U'}{\partial z} = i\omega\left(\frac{1}{V} - \frac{1}{\bar{V}}\right)U' \quad \text{and} \quad \frac{\partial U'}{\partial z} = \frac{2iV}{\omega}\left(\frac{\partial^2/\partial x^2}{4 + \frac{V^2}{\omega^2}\frac{\partial^2}{\partial x^2}}\right)U'$$

which if solved separately and applied in succession to the same depth step gives rise to little error.

The first can be solved simply as

$$U'(x, z+\Delta z, \omega) = U'(x, z, \omega)\exp\left[i\omega\int_z^{z+\Delta z}\left(\frac{1}{V} - \frac{1}{\bar{V}}\right)dz'\right]$$

and in the time domain as

$$u'(x, z+\Delta z, t) = u'\left(x, z, t + \int_z^{z+\Delta z}\left(\frac{1}{V} - \frac{1}{\bar{V}}\right)dz'\right)$$

The second equation can be written

$$\left[-(i\omega)^2 + \frac{V^2}{4}\frac{\partial^2}{\partial x^2}\right]\frac{\partial U'}{\partial z} = \frac{i\omega V}{2}\frac{\partial^2 U'}{\partial x^2}$$

which transforms into time as

$$\frac{\partial^3 u'}{\partial z\,\partial t^2} - \frac{V^2}{4}\frac{\partial^3 u'}{\partial x^2\,\partial z} + \frac{V}{2}\frac{\partial^3 u'}{\partial x^2\,\partial t} = 0$$

This is the "$45^{\circ}$" approximation to the wave equation cast in the moving frame of coordinates.

To find the difference equation we discretize the wavefield into grid points (k, n, j) and sample u' so that

$$u'^{\,j}_{k,n} = u'(x = k\Delta x,\ z = n\Delta z,\ t = j\Delta t)$$

where $\Delta x$, $\Delta z$ and $\Delta t$ are the grid intervals.

We now introduce finite difference operators

$$D_x\, u^j_{k,n} = \left( u^j_{k+1,n} - u^j_{k,n} \right) \frac{1}{\Delta x} \simeq \frac{\partial u}{\partial x}, \text{ etc.}$$

$$D_{xx}\, u^j_{k,n} = \left( u^j_{k+1,n} - 2u^j_{k,n} + u^j_{k-1,n} \right) \frac{1}{\Delta x^2} \simeq \frac{\partial^2 u}{\partial x^2}, \text{ etc.}$$

A closer approximation to the second derivative is desirable particularly in the x direction when dips may be steep, such that a considerable phase shift occurs between successive elements. This is sometimes obtained by taking

$$\frac{\partial^2 u}{\partial x^2} \simeq \frac{D_{xx}}{1 + \alpha D_{xx}} \qquad \text{where } 0 \leqslant \alpha \leqslant 1/4$$

$\alpha = 0.13$ is a suitable value. A similar refinement for $\frac{\partial^2 u}{\partial t^2}$ is hard to make but is not so necessary as the data are usually more densely sampled in time than is necessary.

Stability is improved and grid dispersion reduced if the first derivative in t is located slightly deeper than the estimated derivative in z. This is done by use of an "averaging operator" of the form:

$$A\, u^j_{k,n} = \frac{1}{2}\left[ (1-\theta)\, u^j_{k,n} + \theta\, u^j_{k,n+1} + (1-\theta)\, u^{j+1}_{k,n} + \theta\, u^{j+1}_{k,n+1} \right]$$

where $\theta$ is usually between 0.50 and 0.52.

Then the difference equation based on the 45° approximation to the wave equation becomes:

$$\left[ (1+\alpha D_{xx}) \cdot D_{tt} \cdot D_z - \frac{V^2}{4} D_{xx} \cdot D_z + \frac{V}{2} A \cdot D_{xx} \cdot D_t \right] u^j_{k,n} = 0$$

Substituting for the difference and averaging operators, and sorting terms produces an equation of the form:

$$[Q]\, u^{j-1}_{k,n+1} = [R] \cdot [u]$$

where $[u]$ contains later and shallower values already calculated. The $[Q]$ matrix is inverted (this is readily done, for the Dxx operator implies a simple tridiagonal matrix) and $u^{j-1}_{k,n+1}$ solved for all k. j is decremented for the next, intermediate, loop and on arriving at zero time (j=0) the outer loop with n incremented is begun after applying the time shifts given above and outputting the results in the vector u' (x, $n\Delta z$, 0)

It should be noted that here, as for all algorithms inverting the zero-offset data model, the values of V used must be one half of the true velocities in the earth, in order that t may represent two-way time as in the recorded section.

The significance of the two steps of solving the two equations at each depth step, one the finite difference $45^o$ equation and the other applying a simple shift to the data is as follows: The use of local velocities V in the $45^o$ equation implies that the coordinate frame was taken to have moved down at that local velocity, which being variable in x means that it would have become distorted. To straighten it and assign a fixed depth to the current downward continuation output, it is necessary to correct the data by means of appropriate time shifts to the frame considered as moving at velocity $\bar{V}$. The shift is such as to take account of refraction of the image ray, (Hubral, 1977). In practice the shift is not very readily applied, because for ease of applying the difference equation to vectors in the x direction, the data are held in multiplexed form, and are not easily "distorted" in any other direction.

Nevertheless, the strength of the finite difference approach in taking direct account of local variations in $V_{k,n}$, without a long ray-tracing or integration, is apparent. Its weakness lies in the fact that its response to dipping events is neither apparent to, nor generally controllable by, the user. It depends on the choice of parameters such as $\theta$ as well as on grid intervals $\Delta x$, $\Delta z$, $\Delta t$ in ways which are far from obvious. Its performance with near-horizontal wavefronts is exemplary, but at steeper dips it attenuates, and can introduce dispersion of, higher frequency energy, and may even send energy in the wrong direction. The appellation "$15^o$" or "$45^o$" to an algorithm is no guarantee that it will perform satisfactorily up to that angle under every circumstance.

Figure 9 shows synthetic test data corresponding to a "fan" of reflectors with dips increasing from $0^o$ to $50^o$. Migrating these data gives the results shown in Figure 10 for a number of $\theta$ values. These figures are from Hood, 1978.

3. d) Frequency-Wavenumber Domain Migration

This is potentially the fastest algorithm to compute as it involves least data transfer and possibly somewhat less arithmetic than finite differences. It does not intrinsically require any limitation in dip response, unless this is called for by the user, as there is no approximation in evaluating derivatives, unlike the finite difference method. It is, however, very limited in its ability to

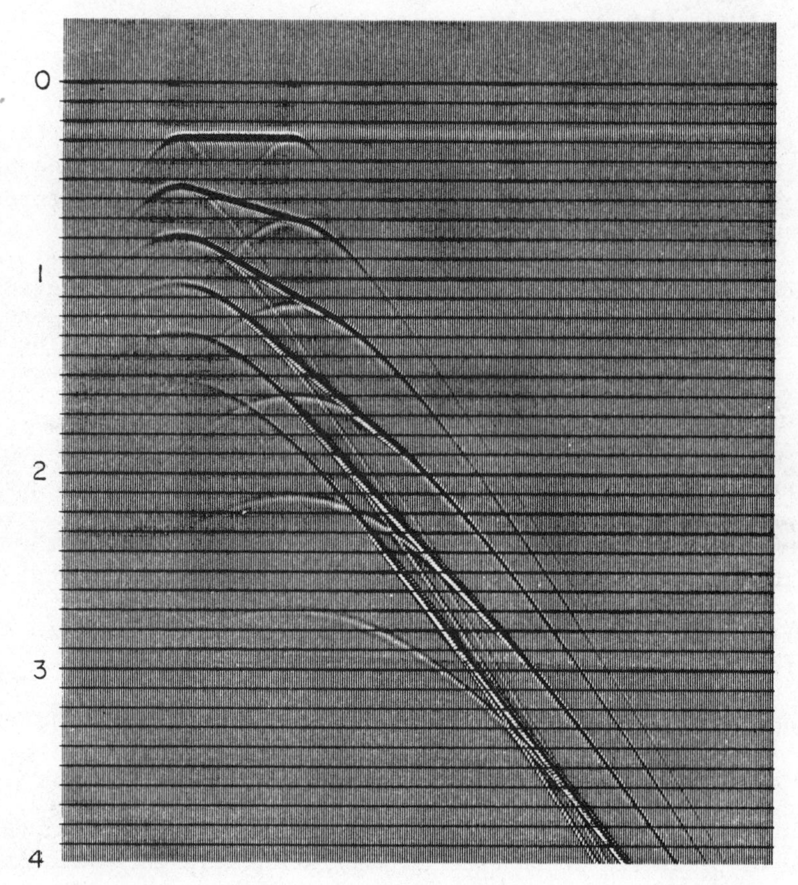


Figure 9. Dipping planes 0-50$^{o}$; v = 3048 m/sec, $\Delta x$ = 30.5 m.

respond to variations in velocity. Much research, however, is going on into improving this approach, and it may well be that before long we shall see Fourier transform type migrations without this restriction.

There are a number of possible methods which start by Fourier transforming the data - some merely transform time to frequency and not distance to wavenumber, some are modelled on a finite difference approach in the frequency domain, others on the integral method. We can only take a brief look at one method, the frequency-wavenumber method of R.H. Stolt. (1978)

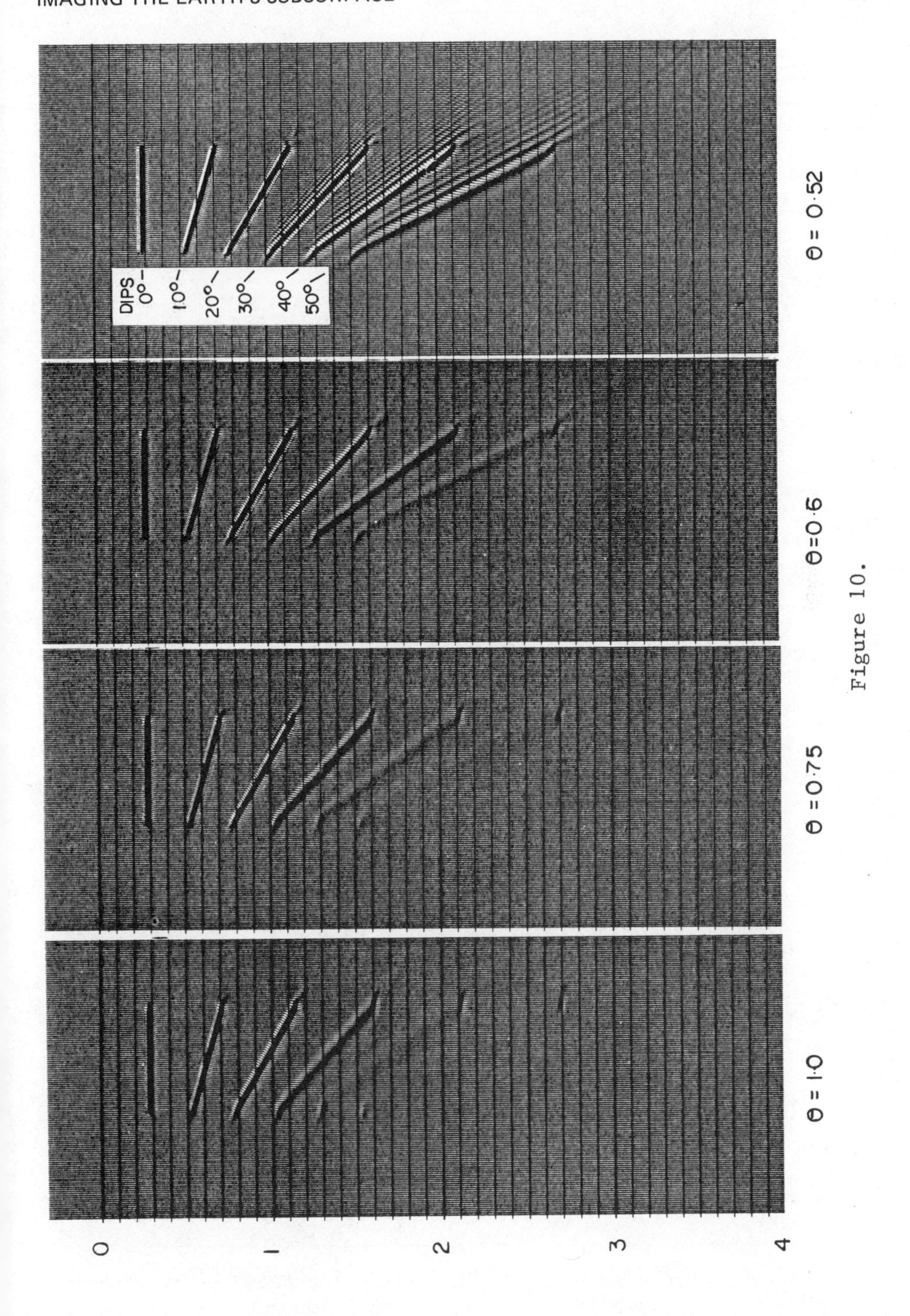


Figure 10.

We start by Fourier transforming the data in the image space in two dimensions:

$$U(k_x, z=0, \omega) = \iint u(x, z=0, t) \exp\{-i(k_x x + \omega t)\} \, dx \, dt$$

and we call to mind the one-way form of the wave equation:

$$\frac{\partial U(x,z,\omega)}{\partial z} = \frac{i\omega}{V}\left(1 + \frac{V^2}{\omega^2}\frac{\partial^2}{\partial x^2}\right)^{\frac{1}{2}} U(x,z,\omega)$$

to derive

$$\frac{\partial U(k_x, z=0, \omega)}{\partial z} = \frac{i\omega}{V}\left(1 - \frac{V^2}{\omega^2}k_x^2\right)^{\frac{1}{2}} U(k_x, z=0, \omega).$$

Now the 2-dimensional transform of the required object space is

$$U(k_x, k_z, t=0) = \iint u(x, z, t=0) \exp\{-i(k_x x + k_z z)\} \, dx \, dz$$

for which

$$\frac{\partial U(k_x, k_z, t=0)}{\partial z} = ik_z \, U(k_x, k_z, t=0).$$

We can make the two obey the same wave equation by mapping from the $k_x, \omega$ plane into the $k_x$, $k_z$ plane using the relation

$$k_z = \left(\frac{\omega^2}{V^2} - k_x^2\right)^{\frac{1}{2}}.$$

It is then permissible to transform back into the object space:

$$u(x, z, t=0) = \frac{1}{4\pi^2}\iint U(k_x, z=0, \omega) \exp\left\{ik_x x + i\left(\frac{\omega^2}{V^2} - k_x^2\right)^{\frac{1}{2}} z\right\} dk_x \, d\omega$$

This transform can be effected by substituting $k_z$ from the relation above

$$u(x, z, t=0) = \frac{1}{4\pi^2}\iint U\left\{k_x, z=0, k_z V\left(1 + \frac{k_x^2}{k_z^2}\right)^{\frac{1}{2}}\right\} \frac{V}{\left(1 + \frac{k_x^2}{k_z^2}\right)^{\frac{1}{2}}} \exp\{ik_x x + ik_z z\} \, dk_x \, d\omega$$

The form of the mapping which takes place between the forward and inverse transforms is best thought of as an amplitude scaling and a positional shift. The positional shift is illustrated in

Figure 11. A point at $(k_x, \omega)$ is moved from the corresponding wavenumber space $(k_x, \omega/V)$ to a point at $(k_x, \sqrt{\omega^2/V^2 - k_x^2})$ which corresponds to dropping it onto a circle centred at $k_x = \omega = 0$ and of radius $\omega/V$. Its amplitude is reduced by the cosine of the angle the radius to its new position makes with the $\omega$-axis. The sine of this angle is the same as the tangent of the angle between the radius to the old position and the $\omega$-axis. When we remember that these angles are those made by energy aligning in the wavenumber domain from alignments in the x,z domain corresponding to the same angles of dip, we see that this rotation is exactly the amount required for the migration of segments of dipping reflectors.

When V is a function of z, Stolt uses a special transformation which distorts x, t space into x, d space where d is a range distance, so that the first transformation is into two wavenumber dimensions $k_x$, $k_d$ and the procedure carries on as for constant V. Unfortunately this is only an approximate method, especially if V is really a function of x also. The usefulness of the method at present is, I feel, to provide a cheap first migration to show all the migrated features of any dip up to $90^o$, and to give an approximate velocity model to start accurate migration with other methods.

## 4. MIGRATION OF 3-D DATA

There is no difficulty in extending the concepts of migration of zero-offset data to another space dimension. If we relax the constraint that the earth must contain only discontinuities which are describable by a single 2-dimensional section, then we must allow the object space and the image space to expand to three dimensions. Data must be gathered over the surface of the earth in a two-dimensional pattern of samples which are dense enough to avoid wavenumber aliassing problems, and extensive enough to "catch" signals from as wide a subsurface area as is interesting. As we saw in Table 1 the quantity of data from such a survey is typically an order of magnitude greater than that from a conventional set of lines shot over the same area. That in itself shows that the main problem associated with processing is the practical one of organising so much more data in and out of the computer. If the computer were big enough to store all the data and have it all immediately accessible, there would only have to be the extra complication of the programs and the extra time to run them implied in the extra dimensionality. In the simple case of velocity being independent of x and y, then the impulse response of a single diffractor point would appear in image space as a hyperbola of revolution. The impulse response of the Diffraction Stack migration method would become the inverted cap of a sphere. The Kirchhoff Integral method becomes the areal summation of a weighted set of derivatives, as explained in section 3b. The Finite Difference and Wavenumber Transform methods require that we take into account the derivatives and wavenumbers respectively in

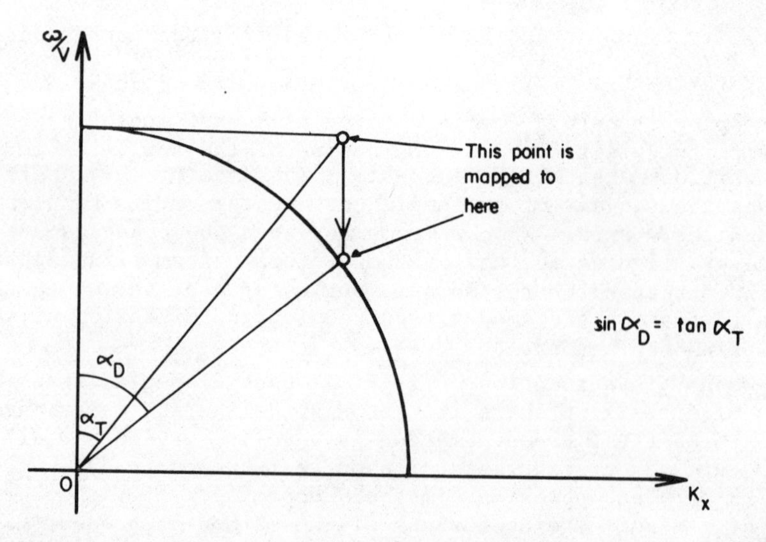


Figure 11.

the horizontal direction y. Some computer systems have large enough memories for all these operations to take place without massive input and output of intermediate data, but for economy it is at present found more efficient to organise the processing in a manner such that smaller computers can handle the work, but take longer because they require more intermediate input/output transfers.

One of the most efficient procedures is to split the 3-D migration into a series of 2-D migrations. If the data were recorded on a rectangular grid, they are first sorted into lines parallel to one side of the grid, and these lines are migrated in a standard 2-D process. The results are then re-sorted into lines normal to the first, and are migrated again by 2-D processing. Figure 12 shows the basic idea of moving energy from the point x, y, t to the apex (0, 0, $T_o$) in two stages. It can be shown that this gives an accurate final result, provided that the velocity is constant in both depth and horizontal position. For the common practical case of velocity varying in every way, the results are inaccurate, but often by not more than would be the case when using velocities which are in error by amounts commensurate with their usual uncertainty.

## 5. MIGRATION OF NON-ZERO-OFFSET DATA

Provided all interfaces are essentially planar, the process of Common-Mid-Point stacking of seismic data recorded at a variety of offsets, and corrected in time by Normal Moveout, proves to be remarkably robust, and to provide a close approximation to a true

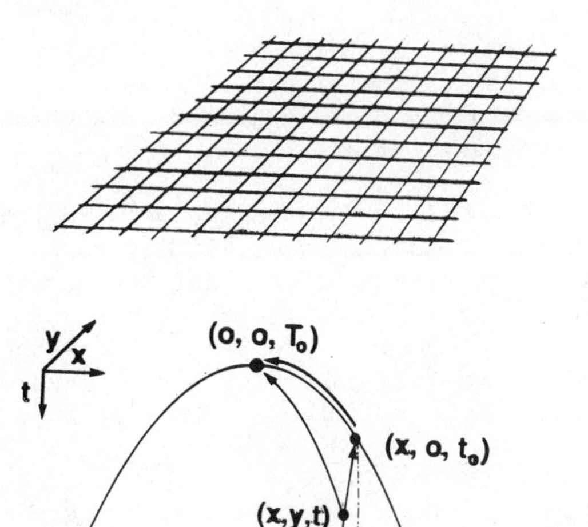


Figure 12.

zero-offset recorded data section. At first sight this is surprising since when a reflector is dipping, the point at which the reflected ray impinges on it departs more and more from the zero-offset reflection point as the source and detector diverge on either side of the midpoint. In fact, it moves up-dip by a distance $\frac{h^2}{z}\sin\theta$ where h is half the offset, $\theta$ is the dip and z the reflector depth below the midpoint. In spite of this, the travel time-vs-offset curve remains hyperbolic, though with a curvature corresponding to a higher velocity that the expected R.M.S. value for the overburden, and for plane continuous reflectors no damage is caused to the stack provided this higher velocity hyperbola is used for correction first.

But for curved and discontinuous interfaces, the stacked section will differ from a zero-offset recorded section, as the moveout time-offset curve will be non-hyperbolic and the displacement of the reflector point as a function of offset will smear the detail of the reflector. Further difficulties occur where reflections with different dips apparently intersect (as in the case of synclinal "bow-ties"). These events require different moveout hyperbolae to stack them properly, and where they coincide on the section there has to be a compromise and some mis-stacking will take place. All of this means that in regions of rapidly varying velocity and dip or of faulting or other truncation of beds, Common Midpoint Stacking destroys high resolution data and no amount of sophisticated processing after stack can recover it.

This again is a field of intensive research and development, and anything said about it today is very likely to be out-of-date tomorrow. In the current state of the art, three different approaches are on offer. To relate these, it is desirable to comprehend how the additional "dimension" of offset alters the model, the inversion of which is migration. The zero-offset-data "exploding reflector" model, which has served so well up to now, must be discarded.

The image space now contains data with independent source and detector positions. We will confine our attention to the 2-D migration case where those positions are co-linear (3-D before-stack migration is still an unthinkable problem). We call the horizontal co-ordinate of the midpoint between source and receiver y, and half offset h, and can write for the source position (y-h) and the detector position (y+h). Figure 13 shows the geometry. Considering the travel time t from source to detector via a diffractor point at depth z below the origin of coordinates,

$$Vt = \left\{z^2 + (y+h)^2\right\}^{\frac{1}{2}} + \left\{z^2 + (y-h)^2\right\}^{\frac{1}{2}} ,$$

the surface of constant t defines the hyper-surface of equal phase referred to by Helbig (1980). An isometric projection of it is given in Figure 14. It is not a hyperbola of revolution in y,h,z space, but rather nearer to a pyramid with the corners rounded off. Either section through the centre with h=0 or y=0 is hyperbolic, but section parallel to these are flattened in the central region. One conclusio that is immediately clear is that CMP stacking with hyperbolic moveout correction cannot stack diffraction events properly except near their apices. (A continuous line diffractor, on the other hand, such as would lie along the surface of a plane dipping reflector, traces out a prism with hyperbolic section, so this is properly stackable).

One of the methods of tackling the problem is effectively to apply corrections to the time of the data so as to convert the pyramidal surface into a hyperbola of revolution, any section through which is hyperbolic. This is done in practice by dividing the data into sections along lines of constant h and submitting them to a process which moves events later in time by an amount which is a function of their dip as well as the offset of the section. This process is marketed under the name of "Devilish" (Judson, et al, 1978). The resulting sections can then be regathered into common mid-point (constant y) sets of traces, stacked after NMO correction and the final stacked section migrated as a zero-offset section.

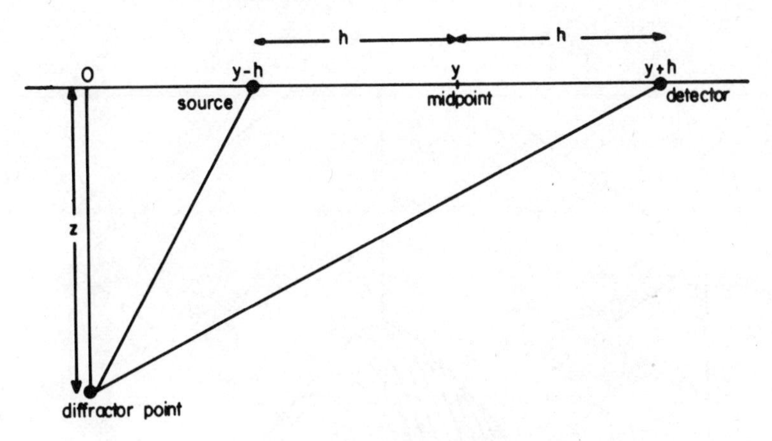


Figure 13.

Another method is to start with the common offset (constant h) sections and to attempt to migrate them, that is to collapse the flat-topped responses from the diffractor points to points at their centres. Again the results are regathered into CMP (constant y) sets of traces and conventionally corrected for NMO and stacked, with no further migration. There is a fundamental objection to this approach, however, in that it attempts to treat a common-offset section as an unique image space of a special wavefield which incorporates the effect of the offset h dimension, but from which the normal moveout correction is separated. In theory, this separation cannot be done. In practice, the flat-topped nature of the diffraction response curves, which define the impulse response of the inverse process as being elliptical and flat in the centre, means that images tend to be smeared in the horizontal direction, particularly those from the far-offset sections.

The third method available, the most elegant but also the most expensive, is describable as "Source and Detector Downward Continuation". It has been explained by Schultz and Sherwood (1980). This depends on the fact that the response to the earth to a single shot is a true invertible wavefield. The data are first sorted into a set of common-source-gathers. Each gather in turn is downward continued through a depth step to produce an estimate of what would have been recorded with a set of detectors placed at the new depth in the earth. The results for all gathers are then re-sorted into common-detector-gathers. The principle of reciprocity is invoked to treat these as if they were common source gathers with the source at the new deeper level and the detectors still at the previous level. Again the gathers are subjected to downward continuation to compute the signals which would have been received had all sources and detectors been at the new lower level. This two-stage process

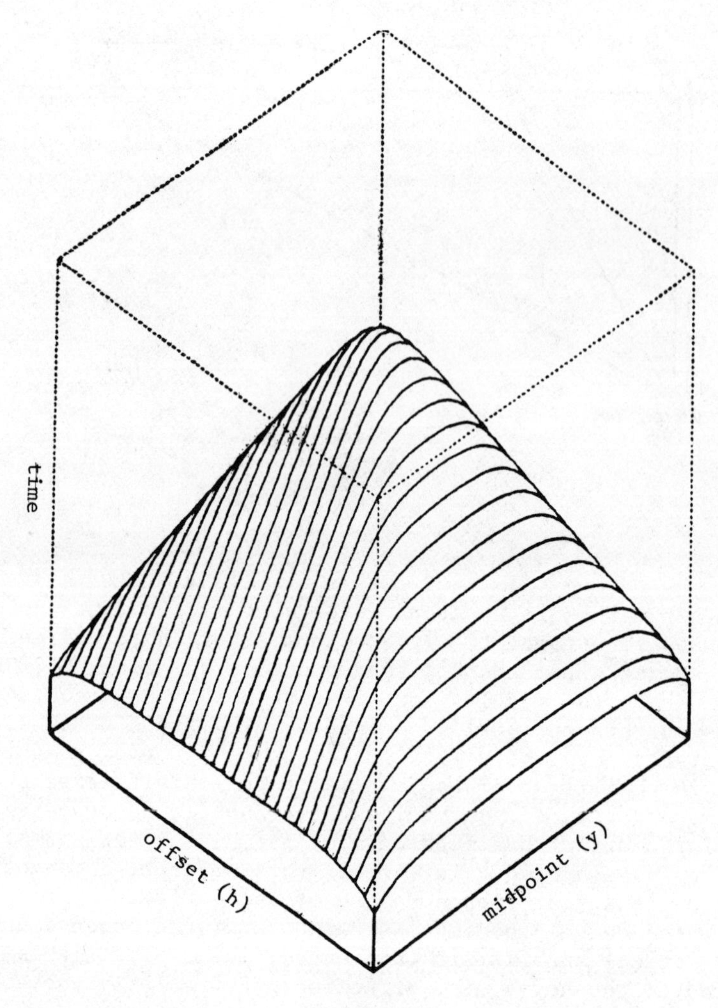


Figure 14.

is repeated as the level is lowered progressively through the object space. It is to be expected (and it is borne out in practice) that the energy migrates towards traces with smaller offsets. The appropriate portion of the zero-offset trace is output to the object space corresponding to the depth step which has just been made, after the migration of each common detector gather. (If no zero-offset traces were recorded originally they must be inserted in the original data as dummy traces). The process accomplishes implicitly the operations of normal moveout correction and stacking. Consequently for that portion of the section which has been completely migrated it denies the opportunity to carry out normal moveout velocity analysis, though such analysis can be made of the partially migrated data at intermediate stages. As it is a process which is more sensitive to the velocity model defined to it than is zero-offset migration, the user should have a good knowledge of that velocity model before embarking on the approach. On the other hand, if he is prepared to afford a number or runs, it is an excellent means of refining the velocity model.

## 6. CONCLUSION

We have attempted to review the principal processing techniques currently used in the seismic data processing industry for imaging the subsurface by means of reflection data recorded on the surface, and to make clear the limitations and assumptions on which these methods are based. The seriousness of these limitations is always a matter of judgment of the geophysicist, bearing in mind the nature of the geological features he is trying to image, the quality of the data he has to invert and the computational cost he is prepared to pay. Much research and development work is still going into improving the benefit/cost ratio, and no doubt this will bring forth more efficient and more accurate processes as time goes on, but the need for the user to be aware of and to weigh the importance of the assumptions and approximations that every practical inversion procedure must entail will always remain. The effects of choice of parameters either under his control or hidden within the computer program must also be kept clearly in view.

## 7. ACKNOWLEDGMENTS

The author thanks the Chairman and Board of Directors of The British Petroleum Company for permission to publish this paper, and gratefully acknowledges the help and advice given by many of his colleagues in the Geophysics Research Branch, principally Messrs. S.M. Deręgowski, M.G. Devey, P. Hood and P.N.S. O'Brien.

## 8. REFERENCES

Berkhout, A.J. and Van Wulfften Palthe, D.W., 1979
Migration in terms of spatial deconvolution. Geophysical Prospecting 27 (1), 261

Berryhill, J.R. 1978
Wave equation datuming. Paper presented at the 48th Annual Meeting of the SEG, San Francisco.

Bolondi, G., Rocca, F., and Savelli, S., 1978
A frequency domain approach to two dimensional migration. Geophysical Prospecting 26 (4), 750

Claerbout, J.F., 1976
Fundamentals of geophysical data processing with applications to petroleum prospecting. McGraw Hill Book Co., New York

Ewing, W.M., Jardetzky, W.S. and Press, F. 1957
Elastic waves in layered media. McGraw Hill Book Co. Inc. New York

Helbig, K., 1980
Ray geometric migration in seismic prospecting. This volume, p. 141.

Hood P., 1978
Finite difference and wave number migration. Geophysical Prospecting 26 (4), 773

Hubral, P., 1977
Time migration - some ray theoretical aspects. Geophysical Prospecting 25, 738

Judson, D.R., Schultz, P.S. and Sherwood J.W.C., 1978
Equalising the stacking velocities via DEVILISH. Paper presented at the 48th annual meeting of the SEG, San Francisco.

Nolet, G., 1980
Linearized inversion of (teleseismic) data. This volume, p. 9.

Schultz, P.S. and Sherwood, J.W.C., 1980
Depth migration before stack. Geophysics 45 (3), 376

Stolt, R.H., 1978
Migration by Fourier transform. Geophysics 43 (1), 23

GRAVITY AND MAGNETIC FIELDS

# SOME ASPECTS OF THE INTERPRETATION OF GRAVITY DATA FOR THE STUDY OF REGIONAL AND LOCAL STRUCTURES

Antonio Rapolla

Institute of Geology and Geophysics

University of Naples, Naples, Italy

## INTRODUCTION

During last years a tremendous improvement has been made in all the fields of geophysical prospecting. While other geophysical techniques have strongly gained efficiency also by the advent of a new generation of equipments and of field techniques, in the gravity prospecting field, this improvement has only been caused by the advent of computers which are more and more used in all the phases of treatment of data, i.e. data reduction, data analysis, data interpretation. Basic concepts in gravity prospecting techniques and instrumentation have in fact practically remained inaltered. The precision and the detailing that have been gained in gravity data interpretation have enormously increased the benefit to cost ratio for this methodology in respect to other geophysical techniques.

This paper shall briefly discuss some of the several problems arising from the computer quantitative interpretation of gravity data in terms of geological structures, giving few case histories taken from the author experience and from literature. Although it is focused on the interpretation of gravity data, it will pay attention also on certain aspects of data reduction and processing, which the interpreter should take into account while handling the data.

## COMPUTER INTERPRETATION OF GRAVITY ANOMALIES

As in other geophysical fields we shall distinguish the direct problem of calculating the gravitational field of a given body of known geometry and density, from the inverse problem of evaluating the source of a given anomaly, namely its geometry and density. While the former problem has a single solution the latter has not, as the same gravitational field may be caused by different masses distributions. The latter is the problem which the interpreter is generally asked to solve. Depending on the number and quality of available information about the source, given from other references, the interpreter can considerably reduce the number of possible alternative models. It should be noticed that the observed data have a limited information content and that, at the very end, the most important role in evaluating a given anomaly source model is played by the experience and the "feeling" of the interpreter.

Even if references are given in the literature of complex and sophysticated interpretation of a gravity anomalies in terms of three dimensional multilayered Earth, the most common approach is to fit each of the anomalies composing a gravity field by a simple two-dimensional arbitrary shaped model and afterwards add all results to compose a complex geological structure.

Practically, the interpreter calculates the gravity effect of a first guess model and on the basis of the comparison between the calculated and the observed anomaly, the model is corrected to better fit the observed data.

A well suited computer algorithm for the evaluation of the gravity effect of a two dimensional model of arbitrary polygonal cross section with vertices $(x_j, z_j)$ and density contrast $\rho$ is given by Talwany et al. (1959).

Assuming the coordinate $x_j$ as known, the inverse problem may be defined as the evaluation of a dependent variable $z=z(x)$ on the basis of the experimental data of a non-linear function (g)

$$g = F\left[x,\ z(x)\right]$$

the evaluation of $z(x)$ means the evaluation of a finite number of its values $(z_j)$ so that a numerical representation of this function may be obtained.

A proper solution of this problem suitable for a computer processing can be obtained by an analitical procedure reducing the function (g) to a linear form by means of the Taylor expansion relative to the variable $z$ and limited to the terms of the $1^{st}$ order. Subsequently the resulting linear equations are solved by an iterative procedure by applying a least-square method. A modified form of the least-square technique as proposed by Marquardt (1963) appears to be very suitable for this problem, in that it improves the convergency of the iterative procedure.

Marquardt technique leads to the following system (in matrix notation):

$$(A^T A + \lambda I)\mathbf{z} = A^T \mathbf{g}$$

This approach differs from the normal least-square technique only for the term $(\lambda I)$, where I is the unitary matrix and $\lambda$ is an arbitrary constant which is added to the terms of the matrix principal diagonal. Such constant reduces the fluctuation of the $\Delta z$ elements forming the $\mathbf{z}$ vector and therefore acts as a damping coefficient. The value to be assigned to $\lambda$ depends on the magnitude of the A diagonal terms. Therefore, it is worthwhile to normalize A by reducing such terms to unity and to go on with the iterative procedure by assigning different values to $\lambda$.

Practically, by this method the geological body is schematically but often with a sufficient precision, described by a closed polygonal having vertices $x_j$, $z_j$ (Fig. 1). Without entering the relative formulas for which we refer to the paper by Talwany et al. (1959) let us say that it is quite easy to calculate by means of a computer the gravity effect of that body on a given point of the Earth surface.

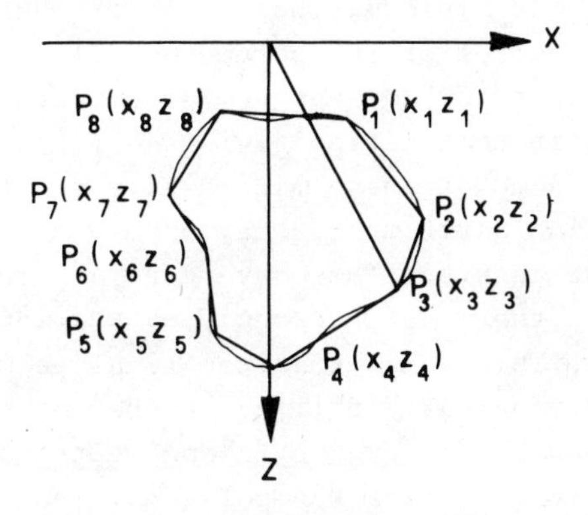


Fig. 1. A two-dimensional body of arbitrary shape can be schematically represented by a polygonal with vertices $P_j(x_j, z_j)$.

The gravity anomaly may be defined carrying out the computation on a certain number of points along a profile. If, as often is the case, the anomaly to be interpreted represent the effect of a varying morphology of, for example, a basement, and therefore we are not in presence of a closed polygonal, the interpreter must fictitiously close the body far away from the considered area. The computed anomaly is then compared with the observed one and the differences

$$\left(g_{calc.}\right)_i - \left(g_{obs.}\right)_i = v_i$$

are computed for each point $x_i$.

On the basis of the differences $v_i$, it is possible, as shown for example by Corbato (1965), to change accordingly the depth $z_j$ of the polygonal vertices and then reenter the gravity effect computation for a new fitting with the observed data (Fig. 2). The process is generally automatically repeated until an acceptable fitting is obtained (Fig. 3).

When the rms reaches a value less than a given one, the computation automatically ends. It should be noticed that by using automatically the rms as the parameter for the measure of the fit we do not account for a differentation be made among different part of the profile, for which, for example, we may ask for a different degree of fitting.

The model obtained at the end of the iterative process may represent only a relative optimum. In fact,as shown by Al Chalabi (1970), the function describing the differencies between computed and observed data, and for which we are seeking for a minimization, presents more than one minimum. The number of such local minima increases with the number of the parameters of the model, which stands for saying that for a large number of parameters there is a large number of their combination giving a fitting between calculated and observed data. Some of these minima give a better fit then other do. The corresponding Earth models may or may not represent geologically significant results. The convergency of the procedure toward one or the other of the minima strongly depends on the choosing of the starting model. This means that the final result is in any case dependent on the geological "feeling" of the interpreter.

The fitting can be improved by changing the density contrast or the location of the vertices of the polygonal model along the x direction, or even by adding new vertices. In order to achieve this aim fully automatic computer techniques for matching the gravity

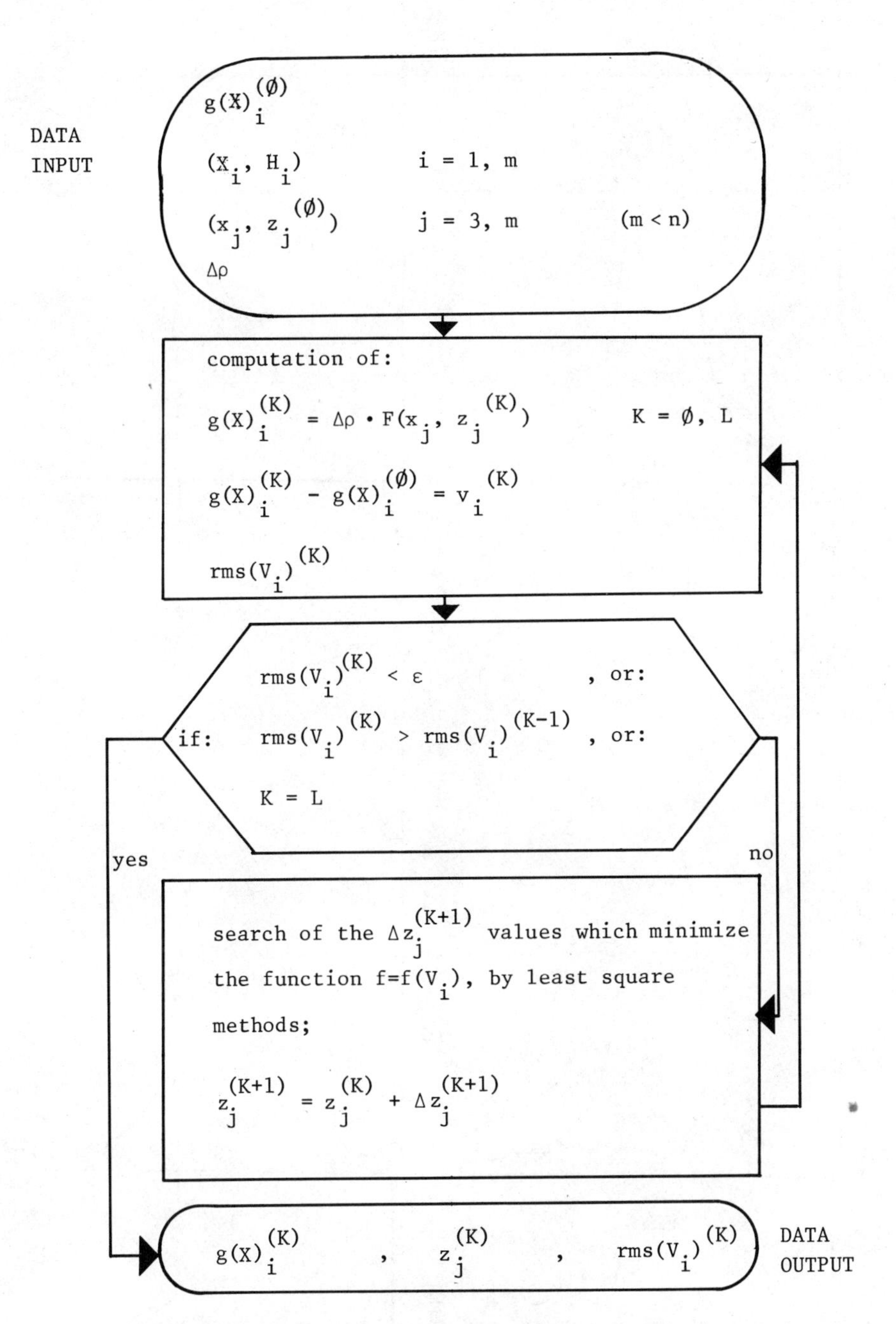


Fig. 2. Schematic flow chart of a gravity anomaly iterative best fitting computer programme.

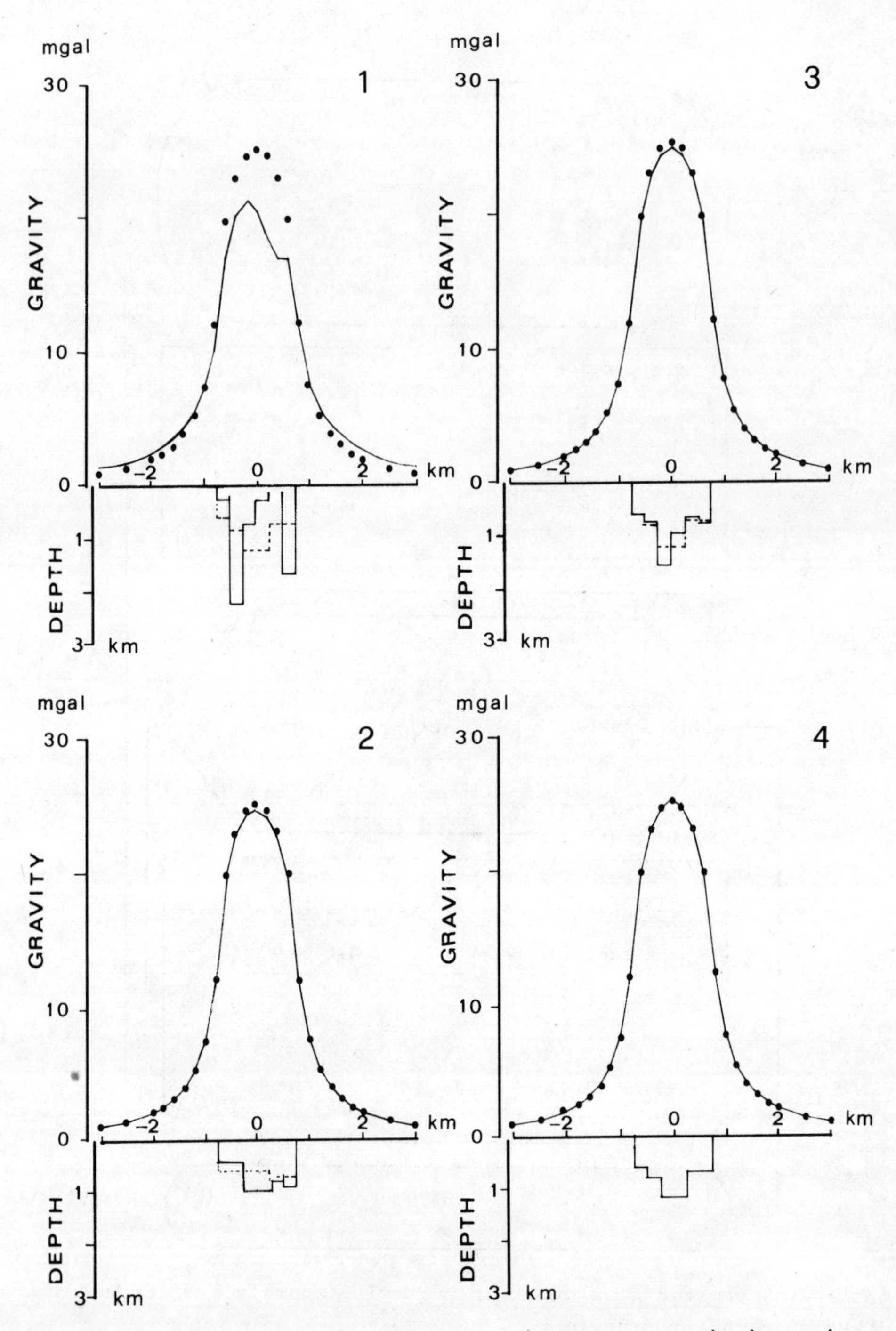


Fig. 3. The figure reports four steps in the automatic iterative procedure for the evaluation of the best fitting model.

anomalies are more and more being substituted by interactive programming where the interpreter can exert during the computation a sort of influence upon the geometry of the model, addressing the result toward a solution which is both gravity and geologically compatible. The responsibility of the interpreter is increased which, provided he has a sound based knowledge of the area under study, can strongly help to reach a fast interpretation, or, better, to make a geologically reliable selection among alternatives with equal mathematical validity. This has been recently rendered possible by the use of videographic devices.

## THE REFERENCE SURFACE

It may not be trivial to remember that a gravity anomaly in a given point of the Earth surface is defined as (disregarding the Bouguer and topographic correction):

$$A = g_{obs.} - \gamma_o + (\Delta_\gamma \cdot h)$$

where: $g_{obs.}$ is the observed gravity value at a point of elevation h above the reference ellipsoid

$\gamma_o$ is the normal gravity on the reference ellipsoid at the given latitude

$\Delta_\gamma$ is the normal vertical gradient of gravity (free-air)

As $\Delta_\gamma$ is defined and computed as the vertical gradient of the normal gravity field and not as the vertical gradient of the actual gravity field,which is generally unknown, the above formula should be seen as:

$$A = g_{obs.} - (\gamma_o - \Delta\gamma \cdot h) = g_{obs.} - \gamma_h$$

That is,the anomaly is given by the difference between the observed value of the gravity field at the actual surface of the Earth and the height-corrected normal gravity value. It is worthwhile to recall that, strictly speaking, by the so called Faye reduction, the observed value should be corrected by the actual gravity gradient in order to transfer it down to the ellipsoid and then substracted of the normal gravity (Sazhine and Grushinky, 1971). In order to achieve this, the actual gravity gradient should be known or measured, as it rarely is the case (Morelli, 1968).

Therefore, after the free-air reduction is made, the anomaly field is not referred to the ellipsoid (or other) reference plane but instead it is referred to the actual physical surface of the Earth. Consequently, the gravity effect of a given body, to be compared with the observed anomaly, should be computed on the actual surface of the Earth and not on the plane H=0. Obviously, this fact is of practical importance only where the topography is strongly uneven as compared to the depth of the anomaly source, as it occurs for example in volcanic areas. Three examples are given in Figs. 4, 5 and 6. The first refers to a theoretical model the second to Mt. Vesuvius, Italy (Carrara et al., 1974) and the other to the Puna district, Hawaii (data from Broyles et al., 1979). The practical equivalence of free-air and Faye correction in flat areas has been the cause of the confusion of the two terms (free-

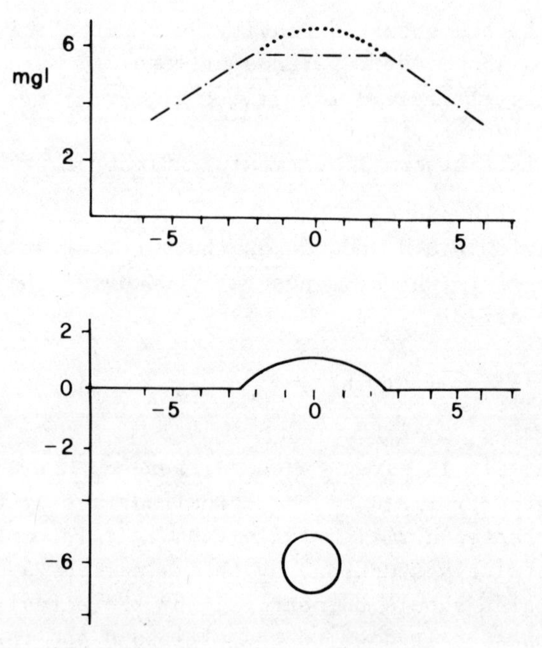


Fig. 4. The gravity effect of a two-dimensional body of circular section (diameter=2km, density=1gr/cm$^3$) on the topographic surface is shown by lines and dots. The gravity effect which would have been measured along the plane H=0 is reported by dots.

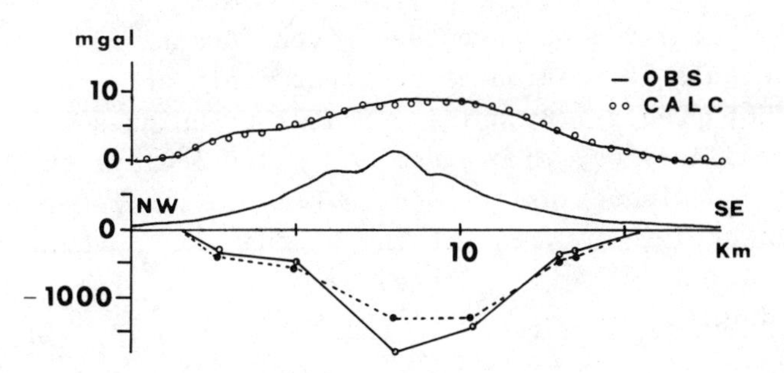


Fig. 5. Gravity profile along Mt. Vesuvius, Italy. The two models (● and o) represent both best fitting models, the first considering the observed data as reduced to the plane H=0; the second - more correctly - considering the observed data as reduced to the actual topographic surface.

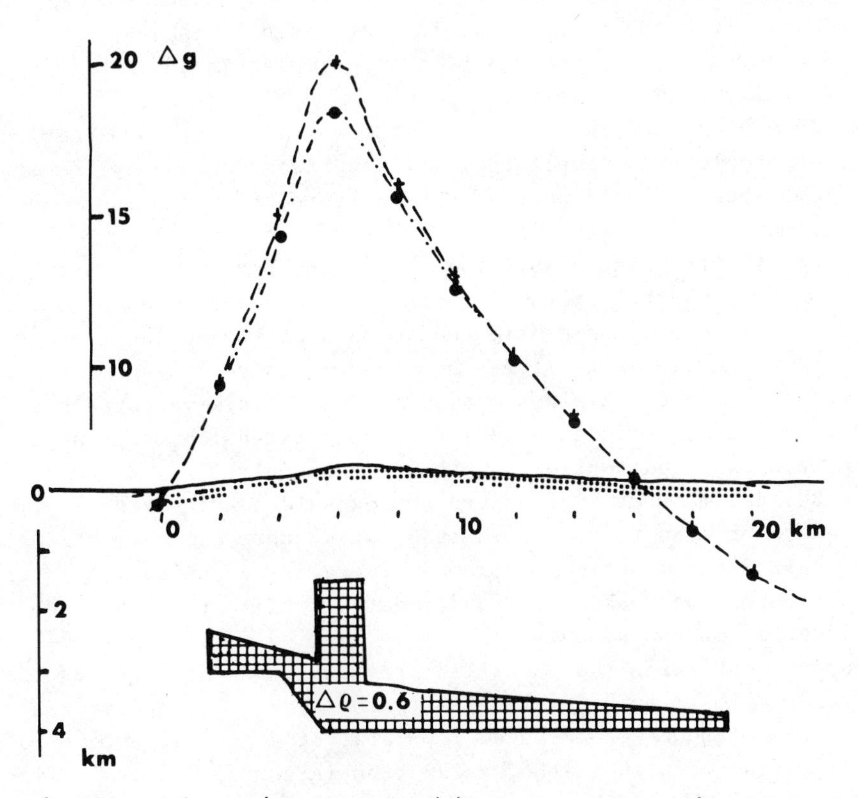


Fig. 6. Observed gravity anomaly (+) along a NS profile, Puna district, Hawaii and a best fitting model as computed by Broyles et al., 1979. The actual gravity effect of this model on the topographic surface is reported (●).

air and Faye). Formulas have been given (Grant and West, 1974) to compute this "elevation effect" but their use is often troublesome and not precise. By using computer techniques we have adopted a simple and easy way to overcome this difficulty, that is to refer the coordinates of the source body ($x_j$, $z_j$) to a coordinates system centered in ($X_i$, $H_i$), being $H_i$ the height of the gravity station, instead of ($X_i$, 0) as erroneously is often done.

## THE RESIDUATION OF THE ANOMALIES

The gravity anomalies which are to be interpreted have generally been previously residuated from a gravity Bouguer map where the gravitational effects of a number of anomalous bodies add each other, which often results in a very complicated pattern. Unfortunately only in few cases the gravity effect to be interpreted shows up clearly from the background. In such cases it is very easy to residuate the anomaly and the best and simplest method to be used is generally hand smoothing.

In most cases, residuation of the anomalies by more complex methods should be accomplished. This represents probably the most difficult and delicate part of the interpretation process being one of the main error source in the whole process.

All the techniques so far utilized are based on the assumption that being true that the deeper the occurrence of an anomalous body the more extensive and gentle the anomaly at the surface, it is also true the viceversa, i.e. the more extensive the anomaly, the deeper the source. This assumption may not be always correct as broad bodies may produce anomalies having extension much larger than what expected considering their depth only.

Besides methods like upward or downward continuation or first or second or even third derivatives, which were commonly utilized for their discrimination power when a precise quantitative interpretation was not asked to gravity data, until a few years ago the residuation of the anomaly was mainly carried out by numerical procedures identifying the so called "regional" as some kind of local average. Unless for special cases (when for example the area over which the regional is computed is by far of a greater extent compared to the area where the residuation is carried out) all averaging methods, whatever complex they are, do not generally give acceptable results, mainly because they lack to take into account geological or other type of information about the structure of the gravity field.

Filtering procedures have proved to be a simple and effective means for analyzing gravity data (as well as other geophysical data). Several authors (Bozzi Zadro et al., 1968; Ku et al., 1971; Bath, 1974; Corrado and Rapolla, 1979) have shown that the results of the application of filtering procedures to gravity data are undoubtly preferable to other procedures of residuation of the field as they let separate the anomalies in a way as objective and effective as possible. A sensible decrease of the ambiguity inherent in this type of procedure - as well as in others - comes from the fact that the results can be conditioned by the data on the different structures, available from seismic refraction surveys, from deep drillings and from geological considerations.

The analysis procedure which we follow for the residuation of the anomalies is based on the use of bidimensional filters as described by Zurfluech (1967) and Lavin and Devane (1970). As an example we quote the analysis of the gravity field of Italy (Corrado and Rapolla, 1979; Pinna and Rapolla, 1979). Assuming that the Bouguer field (Fig. 7) was essentially due to the sum of the effects of the morphology of three main discontinuities, i.e.: a) top of the upper mantle, b) top of the "crystalline" layer, c) top of the rigid sedimentary sequence, the analysis procedure was based on the consideration that, given the different average depth of such discontinuities, this would result in a different extension of their anomalies and hence the application of bidimensional filtering procedures could let single out the different residual fields. The choice of the cut-off frequencies of the filters was made on the basis of a previous spectral analysis and of geological criteria. Fig. 8 shows typical examples of the obtained amplitude spectra. The main components characterizing the spectra are: one with a wavelenght longer than 215 km; a second with a wavelenght lower than 100 km. Other components having wavelenghts lower than 30 km were neglected because of their low energy and the low significance of the higher frequency components due to the relatively large grid spacing.

It should be noticed here the division of the field into a series of components and hence their quantitative interpretation implies the choice of a inherent reference crustal model, that is the model for which the anomalies are null. This means that the anomalies present in the field represent the effects of the varying morphology of the different discontinuity and/or of the horizontal

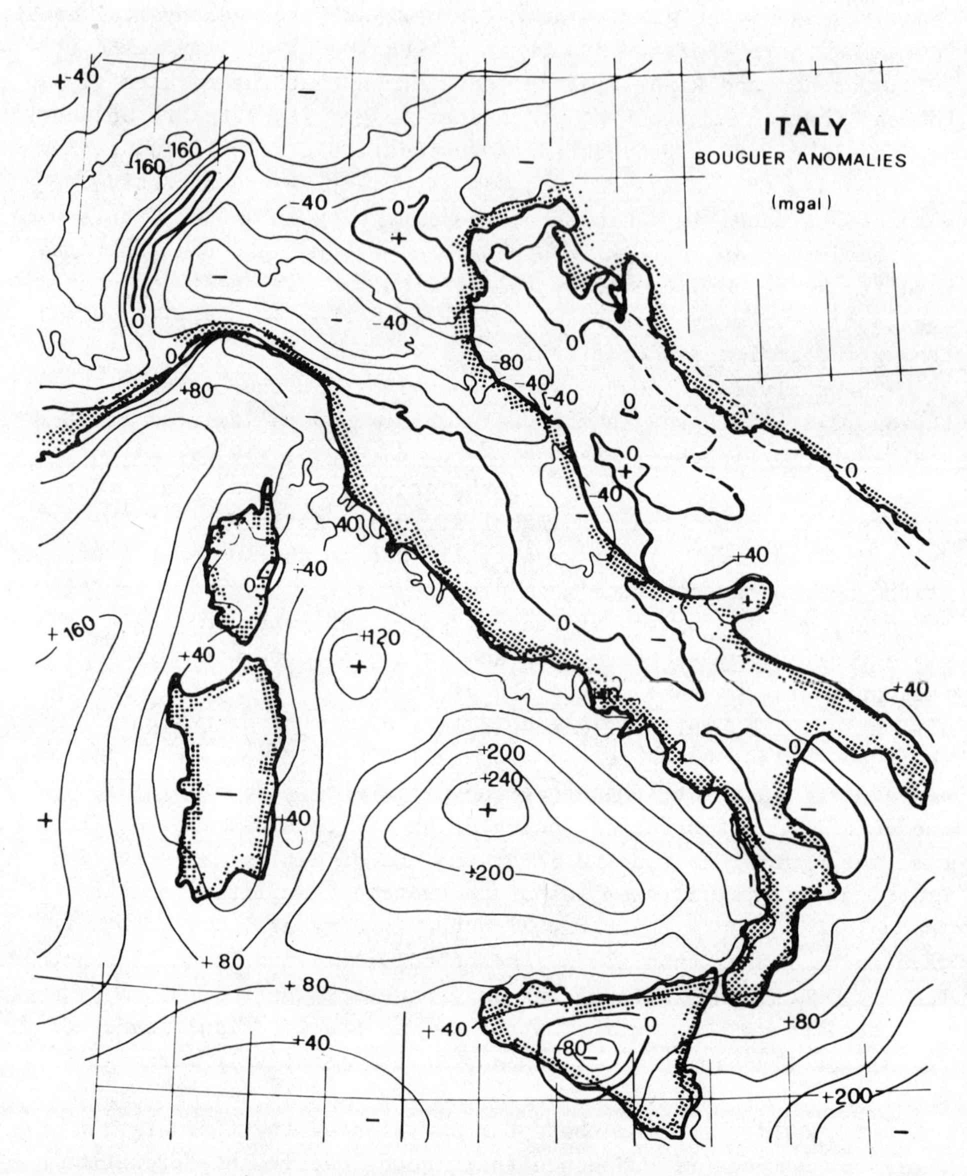


Fig. 7. Bouguer anomaly map of Italy.

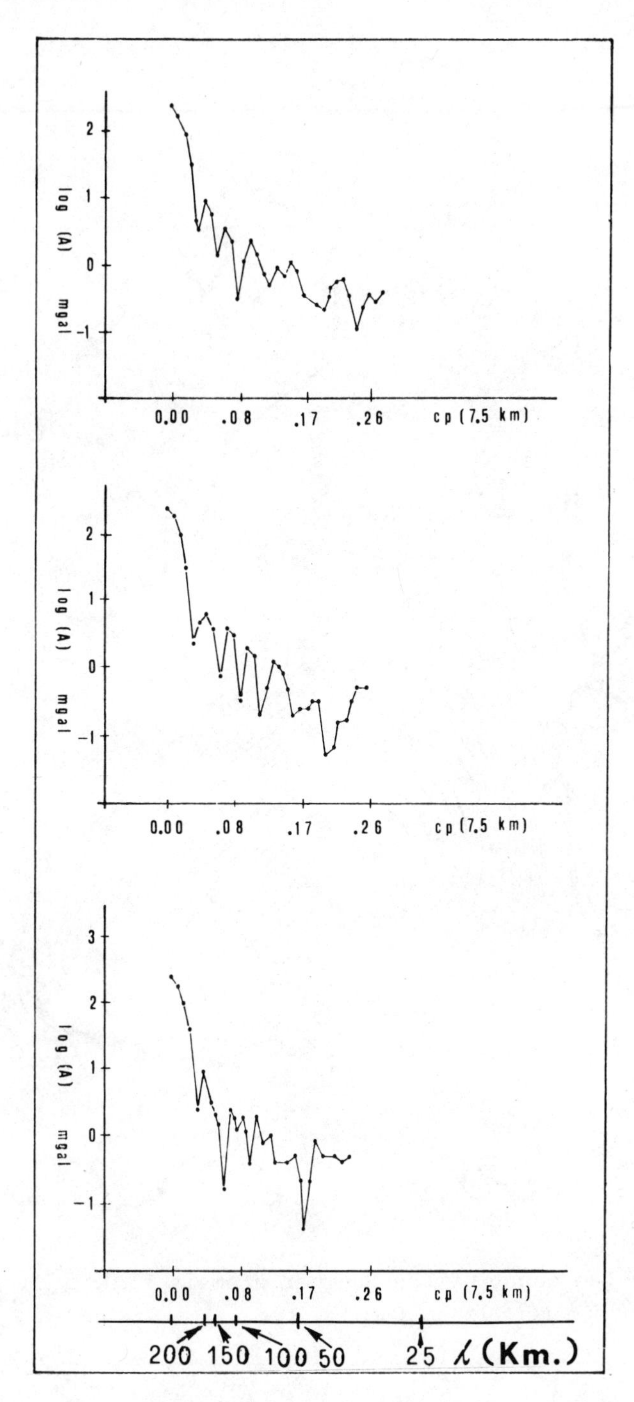


Fig. 8. Amplitude spectra of the gravity Bouguer anomaly of Italy along three SE-NW profiles.

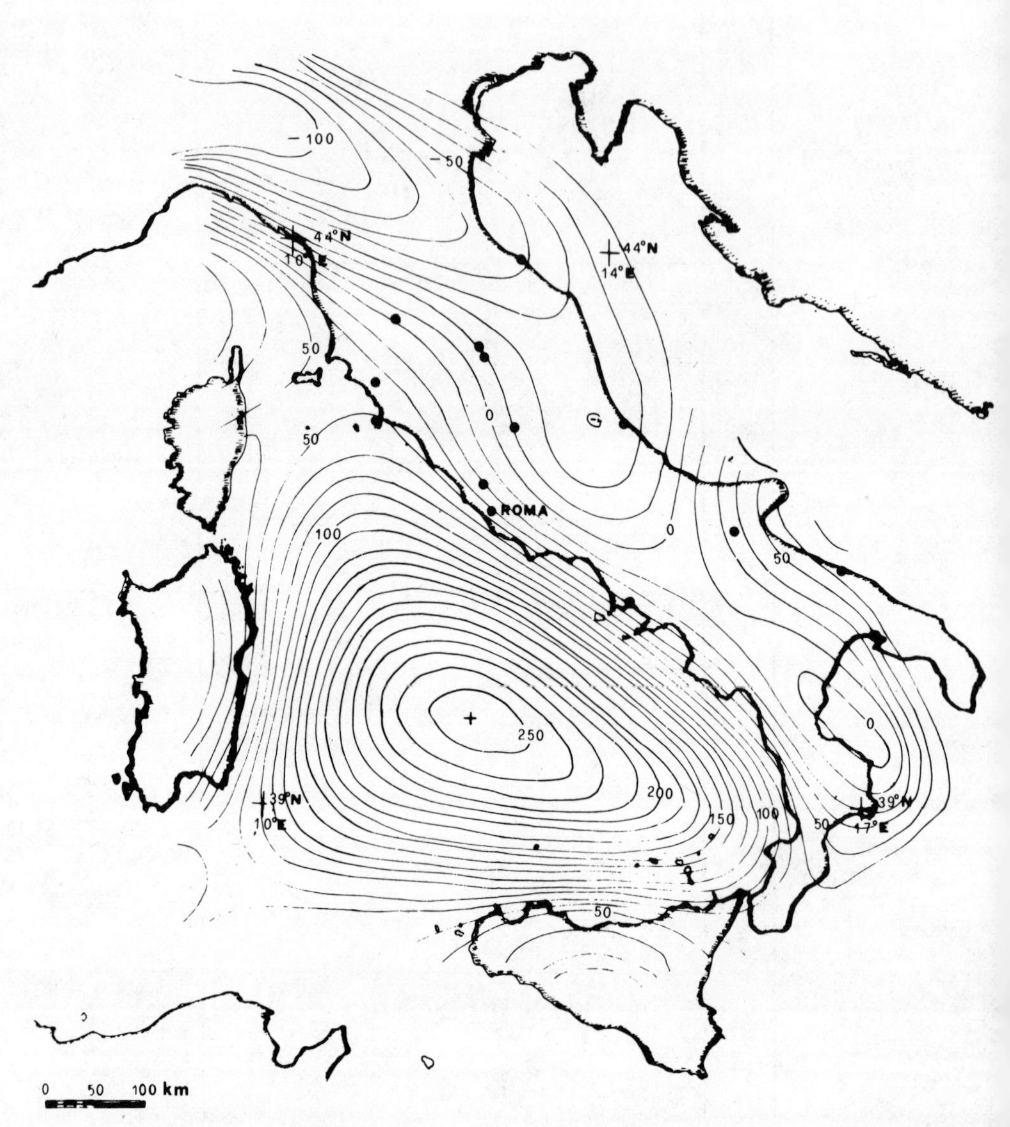


Fig. 9. The component of the Bouguer anomaly field of Italy having wavelength greater than 215 km (mgal).

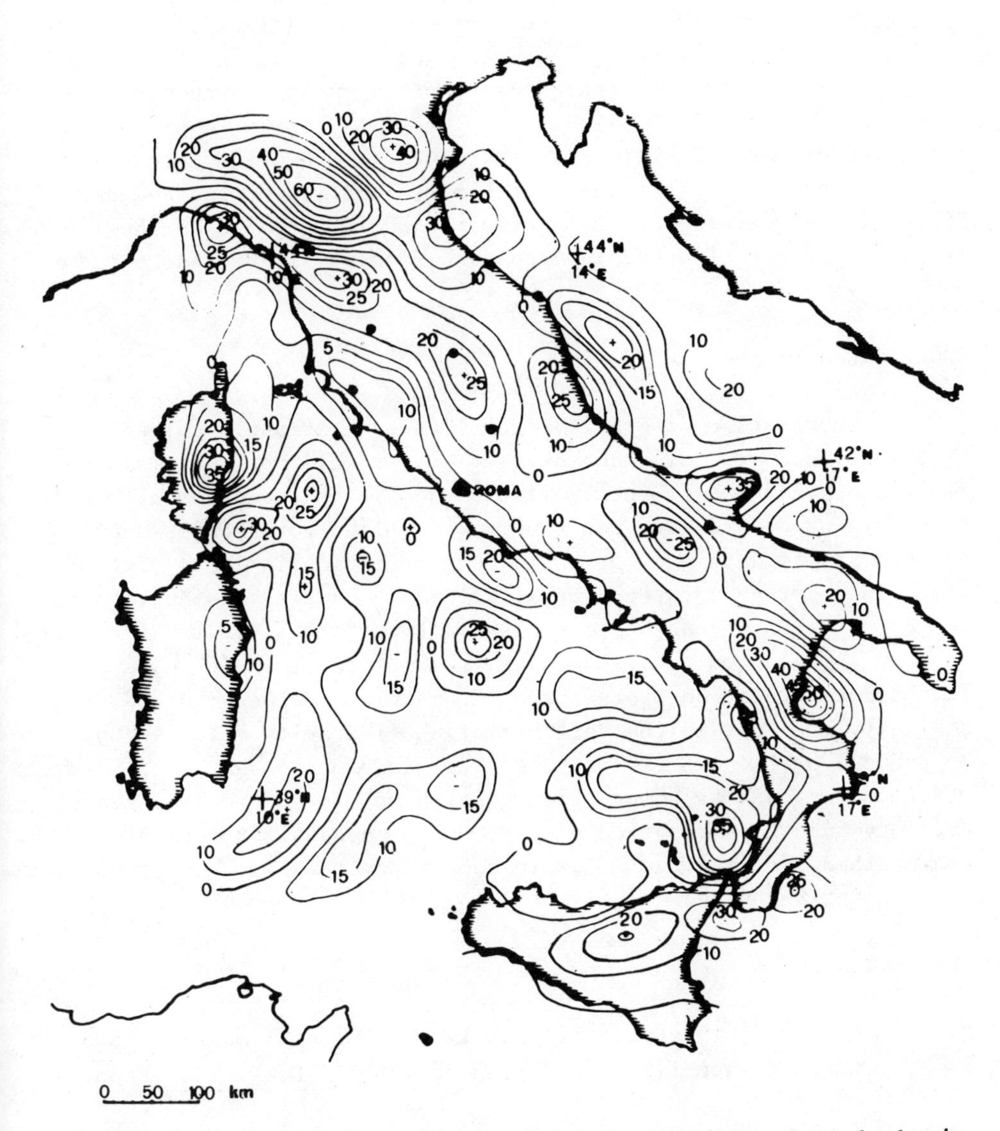


Fig. 10. The component of the Bouguer anomaly field of Italy having wavelength comprised between 215 km and 100 km (mgal).

variation of the density relatively to the reference "normal" crust. Obviously, the use of a given reference crustal model has the only aim of defining a "normal" section which is assumed to be the most geologically significant for the area (Müeller, Talwani, 1971). In the reference crustal section choosen for this example, the upper mantle ($\rho$=3.3 gr/cm$^3$) lays at a depth of 27 km while the crystalline layer (average density: $\rho$=2.9 gr/cm$^3$) has its top at 10 km and is overlayed by a rigid sedimentary sequence ($\rho$=2.6 gr/cm$^3$) reaching the Earth surface. The results of the residuation of the Bouguer field of Italy are reported in Figs. 9 and 10. Details of the interpretation are reported in Pinna and Rapolla, 1979, and Corrado and Rapolla, 1979.

## CONCLUSIONS

Interpretation of gravity anomalies have strongly gained in efficiency and precision during last years. This has been mainly, if not totally, due to the availability of high-speed computers. Computers procedures are now routinely applied for the residuation of the gravity data and for their interpretation in terms of two-dimensional or even three-dimensional Earth models. Residuation of the anomaly, which can be accomplished by numerical filtering procedure in a way often very effective, remains the most delicate phase in the interpretation process and the main source of errors. The ambiguity inherent in the interpretation of gravity data can only be reduced by the availability of other information about the source which can be used as computation constrains. In such cases, gravity data may furnish a detailed picture of the crustal structures at a cost to benefit ratio very low in comparison to all other geophysical prospecting techniques.

## REFERENCES


Al-Chalabi, 1970, Interpretation of two-dimensional magnetic profiles by non-linear optimisation, Boll. Geof. Teor. Appl., 12, 45-46:3-20.

Bath, M., 1974, Spectral analysis in geophysics. Elsevier, 563.

Bozzi Zadro, M.,and Caputo, M., 1968, Filtri multidimensionali e loro applicazioni geofisiche, Ann. Geof., 3:317-336.

Broyles, M.L., Suyenaga, W.,and Furumoto, A.S., 1979, Structure of the lower east rift zone of Kilauea volcano, Hawaii, from seismic and gravity data, Journ. of Volc. and Geoth. Res., 5:317-336.

Carrara, E., Iacobucci, F., Pinna, E., and Rapolla, A., 1974, Interpretation of gravity and magnetic anomalies near Naples, Italy, using computer techniques, Bull. Volc., 38:4.

Corbato, C.E., 1965, A least squares procedure for gravity interpretation, Geoph., v.XXX, 2:228-233.

Corrado, G., and Rapolla, A., 1979, The gravity field of Italy: analysis of its spectral components and definition of a tridimensional crustal model for central-southern Italy, Geodyn.Sp.Project, Publ.n.308.

Grant, F.S., and West, G.S., 1965, Interpretation theory in applied geophysics, McGraw-Hill Inc., 584.

Ku, C.C., Telford, W.M., Lim, S.H., 1971, The use of linear filtering in gravity problems, Geoph., 35-36:1174-1203.

Lavin, P.M., and Devane, J.F., 1970, Direct design of two-dimensional digital wavenumber filters, Geoph., 35:1073-1078.

Marquardt, D.W., 1963, An algorithm for least-squares estimation of non-linear parameters, J. Soc. Ind. App. Mat., 11, 2:431-441.

Morelli, C., 1968,"Gravimetria," Del Bianco Ed., Udine, pp.575.

Müeller, S., Talwany, M., 1971, A crustal section across the eastern Alps based on gravity and seismic refraction data, Pageoph., 85, 11:226-239.

Pinna, E., and Rapolla, A., 1978, Strutture crostali nell'Italia meridionale da dati gravimetrici, in:"Contributi preliminari alla sorveglianza e rischio vulcanico (Etna-Eolie)", G.Luongo, ed., Geodynamic Special Project, CNR, Publ. n.235, Naples.

Sazhine, N., and Grushinsky, N., 1971, Gravity prospecting, MIR publ., Moscow, pp.491 (trans. from Russian by A.K. Chatterejee).

Talwany, M., Worzel, J.L., Landisman, M., 1959, Rapid gravity computations for two-dimensional bodies with application to the Mendocino submarine fracture zone, J. Geoph. Res., 64, 1:49-59.

Zurfluech, E.G., 1967, Application of two-dimensional linear wavelenght filtering, Geoph., 32, 6:1015-1035.

# INTERPRETATION OF LONG WAVELENGTH MAGNETIC ANOMALIES

Paolo Gasparini (°), Marta S.M. Mantovani (°°), Wladimir Shukowsky (°°)

(°) Istituto di Geologia e Geofisica, Università di Napoli, Italy - Osservatorio Vesuviano, Napoli, Italy.

(°°) Instituto Astronomico e Geofisico, Universidade de Sao Paulo, Brazil.

## INTRODUCTION

The Earth's magnetic field (emf) measured at or above the Earth's surface has two main components of internal origin: a primary (or main) field which is originated in the outer core and appears as large scale features at the Earth's surface, and a secondary field which is due to both induced and remanent magnetization of high susceptibility ferrimagnetic minerals occurring in crustal rocks. The latter field is much weaker than the main field and it is locally variable according to the nature of the underlying crustal rocks. These two components of the emf appear clearly in a power spectrum analysis of the emf intensity along a world encircling profile (Alldredge et al., 1963). The energy of the spectrum is concentrated at wavelengths greater than about 2,000 km and smaller than 400-500 km (Fig.1). In the representation of the emf as an infinite series of spherical harmonic functions, the main field therefore is described by the harmonics up to order and degree about (13, 13) and the field of crustal origin by harmonics higher than (40, 40).

The study of the magnetic anomalies (i.e. the residual field obtained by subtracting the main field from the observed field) is of great interest for understanding the structure of the oceanic and the continental crust. In recent years the longest wavelength components of the anomalous field have received considerable

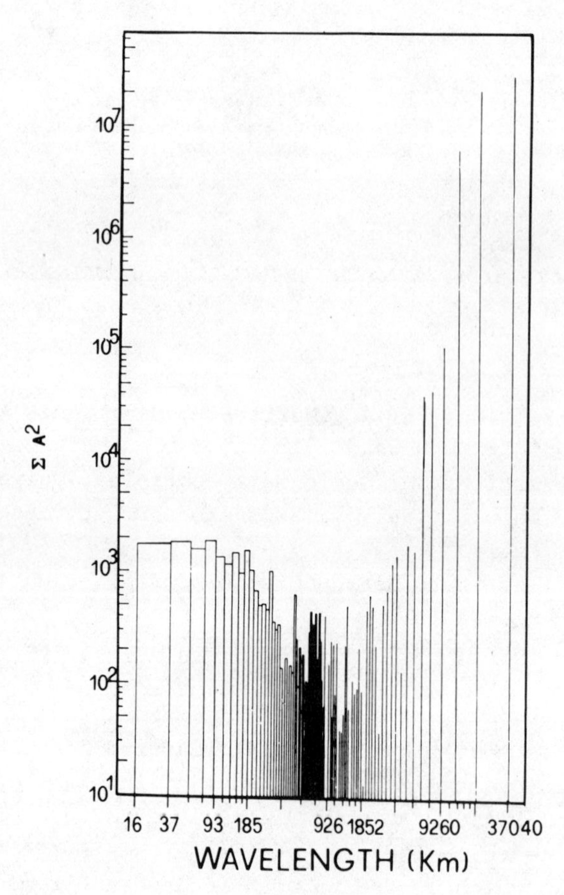


Fig. 1 - Power spectrum of magnetic anomalies along a world-encircling profile (from Alldredge et al., 1963).

attention because:

a) they may be produced by a high susceptibility lower crust or (although less probably) by magnetized horizons within the upper mantle underlying oceanic areas or low heat flow shield areas;

b) The longest wavelength component may represent morphological variations of the bottom of the crustal magnetized layer, i.e. the depth where ferrimagnetic minerals of crustal rocks reach their Curie temperature.

Hence long wavelength magnetic anomalies, when compared with heat flow data, can give informations on the thermal state of deep crust. If the main ferrimagnetic phase occurring in crustal rocks is pure magnetite ($T_C$ = 580 °C), the temperature profiles inferred for different continental provinces indicate that the Curie temperature of magnetite is not reached in a normal continental crust, whereas it is reached at a depth of 25 km in continental areas characterized by a twice as normal heat flow (Fig. 2). Curie temperature of pure magnetite can be exceeded even at shallower depth in active volcanic regions where huge intracrustal magmatic reservoirs exist.

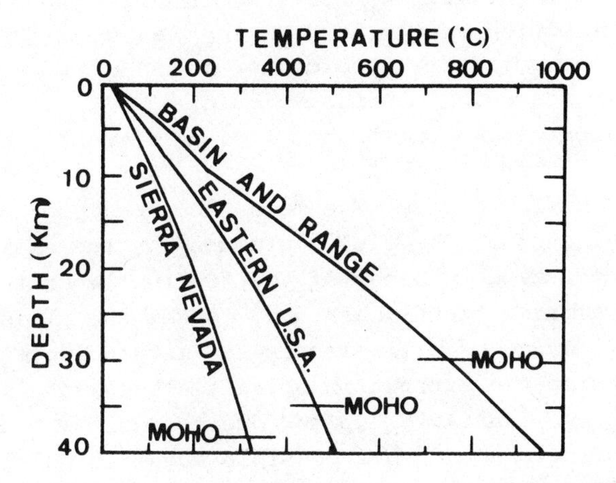


Fig. 2. Temperature variation in the continental crust in a low heat flow (Sierra Nevada), normal heat flow (eastern U.S.A.) and a high heat flow area (Basin and Range)(from Wyllie, 1971).

However, it is to be recalled that Curie temperatures are strongly dependent on the Fe/Ti ratios of "magnetites" (or other ferrimagnetic materials), which in turn depends on the physical-chemical crystallization conditions of the mineral and the composition of the rock. Variations of the thickness of the crustal magnetized layer may therefore reflect, in some cases, lateral compositional variations in the crust (Gasparini et al., 1979).

The identification of the longest wavelength components in the complicated pattern of magnetic anomalies generally observed at or near the Earth surface is not straightforward and cannot be considered a satisfactorily solved problem. The main difficulties to be overcome are:

1) a correct removal of the main field. The subtraction of an erroneous field will generally produce artificial gradients which may alter considerably the longest wavelength components of the field;
2) the removal of the anomalous field components due to magnetized bodies existing in the upper crust. The difficulty of this problem is illustrated in Fig. 3, where a magnetic anomaly profile at the Earth surface in a continental area has been simulated computing the anomalies produced by the magnetized masses distribution represented in the lowest part of Fig. 3. The crust is assumed to be formed by a magnetized layer of varying thickness, whose bottom simulates a "Curie temperature boundary", containing in its interior an irregular and almost continuous high susceptibility layer and a number of discrete magnetized shallow bodies. The magnetic anomaly profile is largely dominated by the fields produced by the shallow masses. The effect of the morphology of the "Curie temperature boundary" appears as a minor trend, masked by larger anomalies.

In the following we will discuss the methodologies which can be used to attempt the extraction of this slowly varying signal from the high level variable "noise" due to shallow sources, an example of application and some possible magnetic models of continental lithosphere.

## EVALUATION OF THE MAIN FIELD

The pattern of the main field can be conveniently represented by determining the coefficients of the spherical harmonic functions representing the emf up to order and degree (13, 13). A first attempt was the elaboration of the International Geomagnetic Reference Field (IGRF), based on all the available survey, satellite and observatory data obtained untill 1968. These data allowed to compute reliable harmonic coefficients only up to

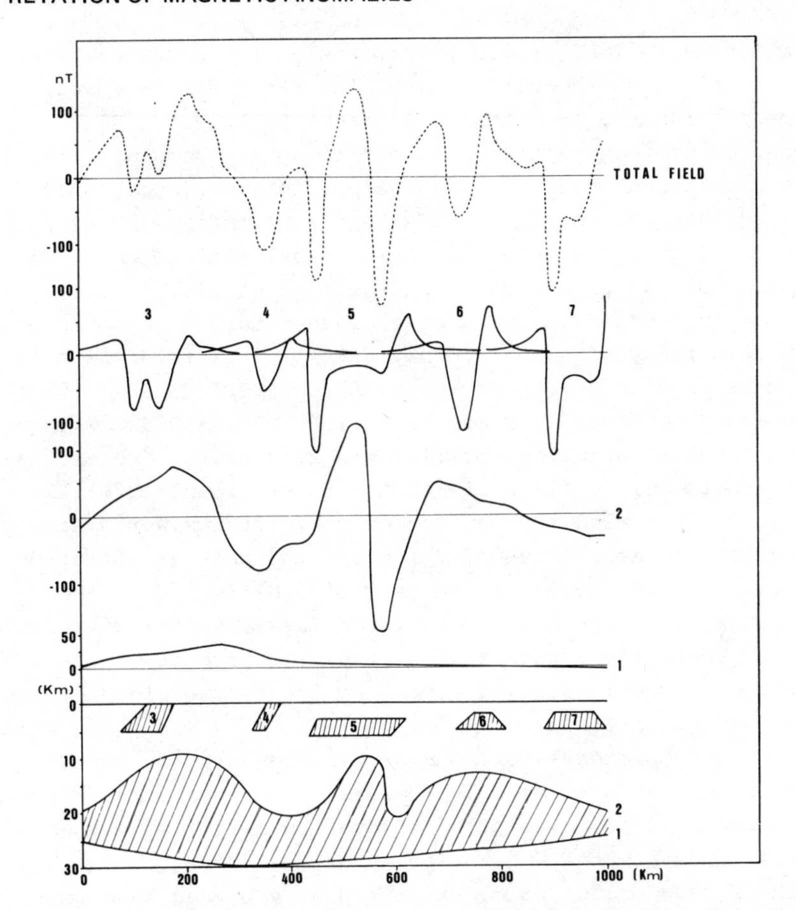


Fig. 3 - Simulated magnetic anomaly profile (total intensity) on a continental crust which has a distribution of magnetized masses as reported in the lower part of the Figure. The dashed area between surfaces 1 and 2 has susceptibility contrast in respect of "normal" crust of $1.2 \times 10^{-2}$, whereas all the shallow masses 3 to 7 have a susceptibility contrast of $8 \times 10^{-3}$ in respect of "normal" crust. The contributions of individual layers to the total field is indicated by continuous curves 1 to 7. The produced field was calculated for the Southern hemisphere, with an inducing field of 24,000 nT and - 19° of inclination.

order and degree (8,8). IGRF contained secular change coefficients in order to permit to predict its variation with time. It was meant to give a representation of the main field for 1965.0 and to predict its variations for a decade (Zmuda, 1971). Despite its uncompleteness, IGRF proved to be reasonably accurate to represent the main field at the date it was elaborated for. However, it became clear quite soon that the weakest point of IGRF was the inaccuracy of the coefficients describing the secular variations, which were based on the few and dispersed observatory data. The bad predictive capability of IGRF was confirmed when the predicted field was compared with a definitive model derived using also all the post-1955 available data (Barraclough et al., 1978). It was observed that the main field at a given date can be be modelled to an accurrancy of about 100 nT, but its variations with time cannot be predicted satisfactorily. A new IGRF (IGRF 1975) ha been recommended to be applied to the period 1975.0-1980.0 with a new set of secular variation coefficients such to fit the secular variation in 1975.0 (IAGA Division 1 Study Group, 1976). It has been shown that also the new set of coefficients were largely inadequate: in many parts of the world the secular variation predicted by IGRF 1975 for the year 1976 differs from the observatory data by 20 to 60 nT . $y^{-1}$ (Dawson and Newitt, 1978). In a few years the errors in the predicted rate of secular variations may completely distort the longest wavelength components of the anomalous field.

The uncompleteness of IGRF as a representation of the main field is demonstrated also by the differences existing between IGRF and regional fields evaluated in some areas from survey data. Typical cases are the Mediterranean area (Corrado et al., 1978) and the Minas Gerais State, Brazil (Mantovani and Shukowsky, in prep.). In both areas regional fields have been evaluated upon the data collected in detailed surveys covering areas of about $10^\circ \times 10^\circ$.. In both cases it has been found that the best fit to the observed anomalous field was given by a second order surface. If local anomalies can be considered as random deviations from a regional field, the best fit second order surface can be assumed to represent this regional field. These best fit surfaces presented systematic deviations from the main field computed by IGRF in both cases. The systematic deviation was greater for Brazil, where the main field pattern is complicated by the existence of a prominent low just off its SE coast. The difference between the two fields in Minas Gerais State range from -40 to +60 nT, producing an apparent anomaly of some thousands km of wavelength corresponding to an harmonic of degree lower than 12 in the planetary representation of the emf. It is therefore likely that these unadequacies in the representation of emf at a given date can be eliminated by the use of the new Magnetic Satellite (MAGSAT) data, which should allow a representation of the main field up to order and degree (14, 14). In the meantime, the possible existence of a spurious gradient due to an improper subtraction of the primary field must be checked and accounted for by utilizing all available data to compute regional fields.

## REMOVAL OF ANOMALIES RELATED TO SHALLOW SOURCES

A classical approach to this problem is through the application of power spectrum analysis and two-dimensional low pass filters. This approach became widely used in gravity and geomagnetism after the work by Spector and Grant (1970). Shuey et al. (1977) and Smith et al. (1977) utilized it specifically for the problem of determination of the depth variations of the bottom of the crustal magnetized layer.

The philosophy of this approach can be illustrated with reference to the simulated magnetic profile of Fig. 3. The problem is to go back to profile 2 (and possibly 1) starting from the observed anomaly field (dashed curve in the upper part of Fig. 3) with the individual contributions by all the magnetized bodies as unknowns. Computions of the magnetic fields associated with bodies of different sizes and shapes, lying at different depths, have shown that the depth to the top of magnetized bodies controls mainly the wavelength and gradient of the anomaly. The amplitude of the anomaly is rather controlled by the magnetization intensity, thickness and inclination of the mass. If it is assumed that the observed anomalies are the resultant of the individual effects of magnetized bodies of comparable size lying at different depths, the spectral characteristics of the anomalies can be used as a discriminating parameter for sources at different depths. Then the first step must be the determination of the frequency intervals where the energy of the spectrum is concentrated. The aim of this step is to identify whether measurable anomalies related to sources at the desired depth exist and to choose the cut off frequency for the low pass filters which are proper to get rid of contributions due to shallow bodies. This is accomplished by carrying out a power spectrum analysis of profiles perpendicular to the main trends of magnetic anomalies. Two widely used methods of power spectrum analysis are based upon the application of the discrete fast Fourier transform method (DFT) and the maximum enthropy method (MESA). They are two radically different approaches to the problem of estimating the power spectrum of a function of time (or space) from a finite set of samples (point values) of the function. The DFT method is a variation of the methods proposed by Blackman and Tukey (1959) for the estimation of power spectra, modified in order to use the advantages of the fast Fourier transform algorithm, as described by Kanasewich (1975). As a very simplified description, the frequency range is divided into a number of intervals and the power spectrum of the time (or space) function is approximated by the average power of the variation of the function in each of these intervals. The average power at the various frequency intervals is estimated by the decomposition of the time (or space) function in a combination of sine and cosine terms by means of the fast Fourier transform algorithm. The frequency range, the number of frequency intervals, the sampling rate of the function and the reliability of the power spectrum estimate are interrelated. A very detailed study of this interdependence is presented in the classical study by Blackman and Tukey (1959).

The MESA method (Burg, 1967) is a data adaptive method. It is based on the predictive aspects of optimum filter theory (see Kanasewich, 1975). The sequence of sampled values of a time (or space) function exhibit an ordering, or pattern, which represents the temporal (or spatial) variation of the sampled function. The MESA is based on the construction of a filter (called the prediction error filter) which destroys all the ordering of the sampled values of the function. Symbolically:

$$\text{(samples of the function)} \xrightarrow{\text{PEF}} \text{(white noise)}$$

The effect of the prediction error filter is to maximize the entropy of the sequence of samples (the concept of entropy is here used as a measure of the disorder of the system). It can be shown that the inverse of the frequency response of the PEF is a good estimate of the power spectrum of the time function. Both DFT and MESA spectral estimations are based on the analysis of data along a finite length profile. The main difference between the two methods is that, while in the DFT method is implicity assumed that the time (or space) function is zero outside the analyzed interval, the MESA method extends the function outside the analyzed interval "in a sensible way". Because of this, MESA operates on greater virtual length of data and gives systematically a higher frequency resolution and best stability than DFT, particularly when it operates on short profiles. As DFT method produce often a distortion at the spectrum extremes, it is less suitable then MESA to be applied to the search of long wavelength anomalies.

Spector and Grant (1970) have shown theoretically that the logarithm of the power of the magnetic intensity spectrum produced by a body of small lateral extension and infinite thickness decreases almost linearly with increasing frequency. The slope of the straight line is proportional to the depth to the top of the body (Fig. 4a). When the thickness of the body is small and the lateral extension is large the logarithm of the power will present a maximum whose position and extension depends on the depth to the top, the thickness and extension of the body (Fig. 4b). The combination of deep seated bodies and shallow bodies of little thickness will produce a spectrum like the one of Fig. 4c.

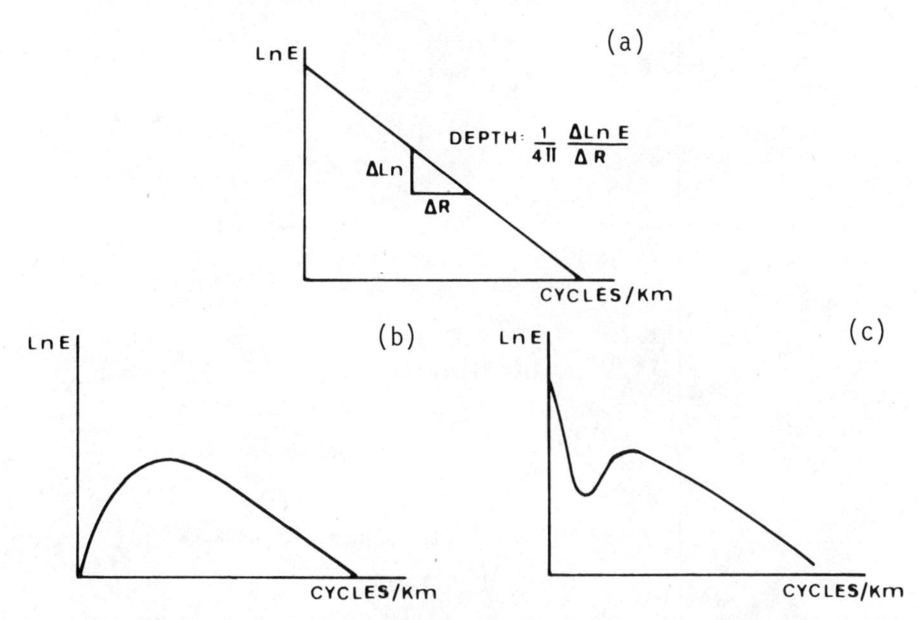


Fig. 4 - Theoretical power spectra of prismatic sources of infinite thickness: (a) monopole source (upper diagram), of small thickness compared to its lateral extension, (b) dipole source (intermediate diagram) and (c) a combination of deep monopole sources and shallow dipole sources (lower diagram). Average depth to the top of the source can be estimated from the slope of the straight line as indicated in the upper diagram (from Spector and Grant, 1970).

DFT and MESA power spectra of the simulated total profile of Fig. 3 are reported in Fig. 5. DFT power spectrum of the field produced by the deep sources (curves 1 and 2 of Fig. 3) is also reported for comparison. At wavelengths longer than 200 km, both power spectra are entirely due to the field related to the deep sources. The contributions by the deep sources is clearly evidenced by the peaks at 280 and 140 km wavelength, particularly resolved in the MESA power spectrum. Cut off to be applied to low frequency pass filters in order to get rid of the undesired contributions by shallow masses are either at 100 or at 200 km.

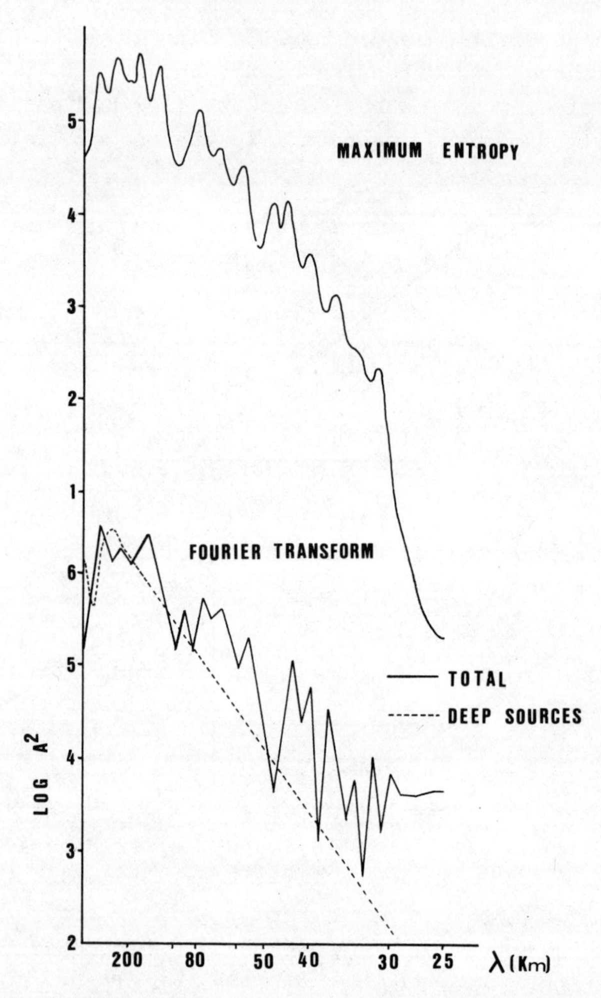


Fig. 5 - MESA and DFT spectra of the simulated anomaly profile of Fig. 3. DFT power spectrum of anomalies related to deep sources is represented by dashed line.

The usual filters applied to geophysical problems are also amenable for use in the present case. All the linear digital filters used in geophysics fall into two categories: non recursive and recursive (also called feedback) filters (see Kulhanek, 1976). The output of a non recursive filter is a weighted average of present inputs, some past inputs and, possibly, some future inputs. The choice of the quantity and values of the weights determines the characteristics of the filter (e.g. high pass, low pass, band pass, rejection, etc.). For a one dimensional linear non recursive filter, the weights are the discrete Fourier tranform of the desired frequency transfer function of the filter. This property allows for the easy calculation of the weights by applying the fast Fourier transform algorithm to the digitized transfer function of the filter. For time sequences, only present and past inputs can meaningfully be used, whereas for space sequences (such as gravity and magnetic anomalies) future inputs are always used in order to ensure zero phase distorsion of the output.

The output of a recursive linear digital filter is a weighted average of the present inputs, some of the past inputs and also some of the past outputs. The weights are also related to the transfer function of the filter, but in a less simple way as compared to non recursive filters.

The non recursive filters have the advantage of zero phase distorsion and unconditional stability. They have the drawback of requiring a great number of terms in order to have a good frequency response when a sharp cut off is required. In this work a non recursive zero phase filter with 33 terms was used.

Recursive filters have a good frequency response with only a few terms, but they invariably introduce phase distorsions in the filtered data. Also the calculation of the weights is not simple as in the former case, and it is further complicated by the conditions imposed to ensure stability of the filter.

Filtered profiles obtained through the applications of low pass filters to total profile of Fig. 3 are reported in Fig. 6. The profile obtained using a cut off at 100 km wavelength contains undistorted informations related to the deep seated sources, but it still contains interferences by shallow sources. This is mostly due to the uncapability of the filter to separate the effects due to shallow bodies of large extension, such as body n. 5 of Fig. 3. This uncapability is inherent to all potential methods and cannot be obviated by any methods. The profile obtained using a cut off at 200 km wavelength is largely unsatisfactory because it includes

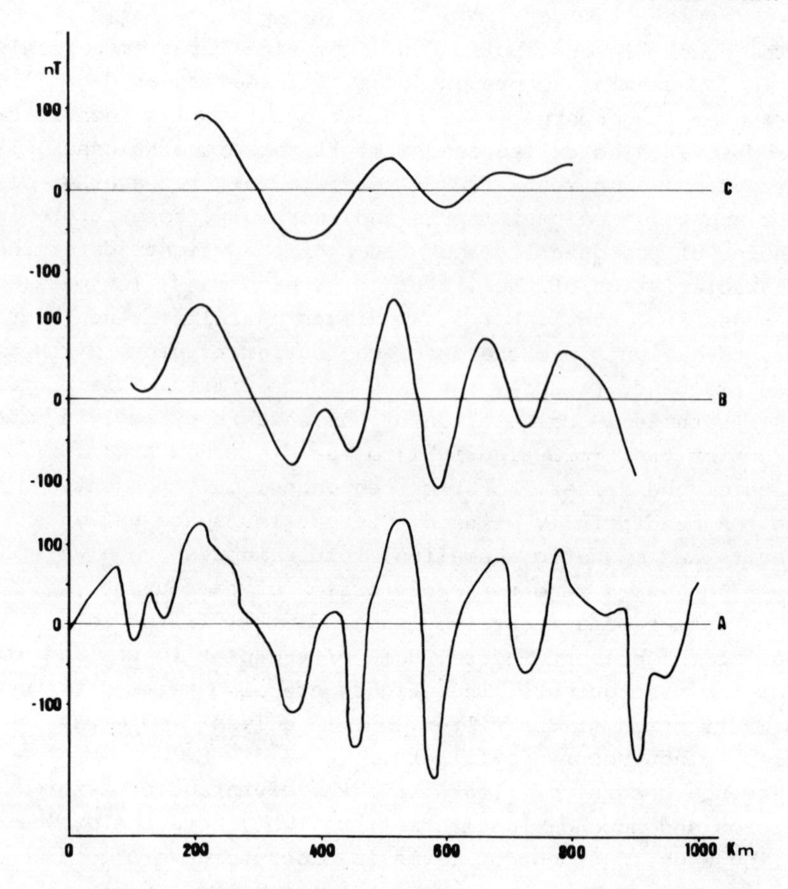


Fig. 6 - Filtered anomaly profiles obtained from the simulated profile of Fig.3: (A) unfiltered profile, (B) filtered profile with low pass filter having a cut off frequency corresponding to a wavelength of 100 km, (C) filtered profile with low pass filter having a cut off frequency corresponding to a wavelength of 200 km.

only a part of the informations due to the deep seated sources, and does not separate curve 1 from curve 2 of Fig. 3. It must be observed that a loss of informations proportional to the cut off wavelength occurs at the edges of the profile. Cut off at wavelength longer than about 250-300 km cannot be used in a 1000 km long profile, because it will reduce the informations to a very narrow band at the center of the profile. Moreover, the profiles obtained in Fig. 6 demonstrate that it is advisable to use cut off frequencies which are quite distant from the frequency characteristic of the desired anomalies in order to prevent any deformation of the desired anomaly. The possible interface by shallow bodies can be removed in the inversion process through the use of inversion models with smoothly varying surfaces.

It must be pointed out that depth variations of the bottom of the magnetized crustal layer cannot be resolved in presence of a very irregular top surface of an overlying lower crustal highly magnetized layer. Therefore a thoughtful choice of the ubication of profiles to be analyzed for the contributions due to deep crustal masses is necessary for improving the reliability of the results.

A method used by Russian geophysicists to separate anomalies of shallow and deep crustal origin is the upward continuation of the field to a given height (Krutikhovskaya and Pashkevich, 1979). The method consists of finding the optimum height where anomalies of little extension are negligibly small in comparison to anomalies of large extension. Aeromagnetic surveys made at different heights and theoretical models indicate, for example, that anomalies of about 5 km of diameter completely disappear at a height of 10-15 km. This method permits a good reconstruction of the regional field when strong local anomalies are present, but it does not allow any improvement in the separation of the contributions due to shallow bodies of large extension from those due to deep sources.

## AN EXAMPLE: LONG WAVELENGTH MAGNETIC ANOMALIES OF MINAS GERAIS AND ESPIRITO SANTO (BRAZIL)

We are presently working on the interpretation of the long wavelength magnetic anomalies obtained by a detailed aeromagnetic survey of the states of Minas Gerais and Espirito Santo (Brazil). The surveyed area is mostly formed by 2,000-2,600 m.y. old high to medium grade metamorphic terrains, which represent an exposed intermediate to lower continental crust sequence. Upper preCambrian to eoCambrian sedimentary formations cover these metamorphic terrains W of 44° W and N of 20° S (see Cordani et al., 1973; Almeida, 1977). Crustal thickness in a part of the surveyed area (between 41°30' - 45° W and 18°-20° S) was estimated in the range 38-45 km on the basis of Bouguer anomalies (Blitzkow et al., 1979). Total intensity magnetic anomalies were obtained by utilizing the IGRF definitive field (Barraclough et al., 1978) as the main field at the year of the survey (1971.5). Total intensity magnetic anomalies are represented in Fig. 7. The magnetic pattern appears

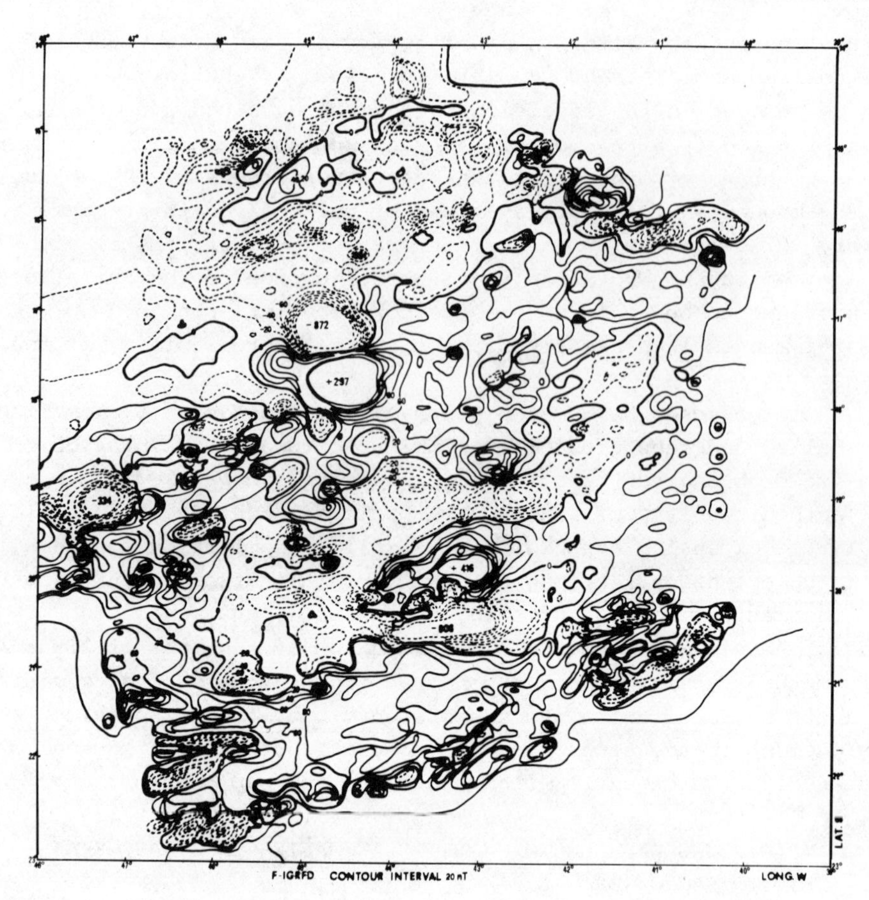


Fig. 7 - Total intensity magnetic anomaly field of Minas Gerais and Espirito Santo States (Brazil) al 1200 m.a.s.l. and reduced at the epoch 1971.5 (from Corrado et al., in prep.).

very complicated, with many anomalies of different wavelength, particularly W of 42° W. The most prominent features are a ENE trending complex anomaly pattern between 19°-21° S and 42°-44° W, and huge almost circular low and high between 17°-18° S and 44°-45° W. The former is related to outcrops of an iron containing metasedimentary formation (itabirite), whereas the latter has no abvious correlations with the geological structures. The identification and interpretation of the longest wavelength components of the anomalous field was attempted by Corrado et al., 1979; Gasparini et al., 1979; Corrado et al., in prep. The interpretation of a profile passing through the less disturbed part of the field is shown as example (Fig. 8). DFT and MESA power spectra are characterized by slow decays and shapes indicating

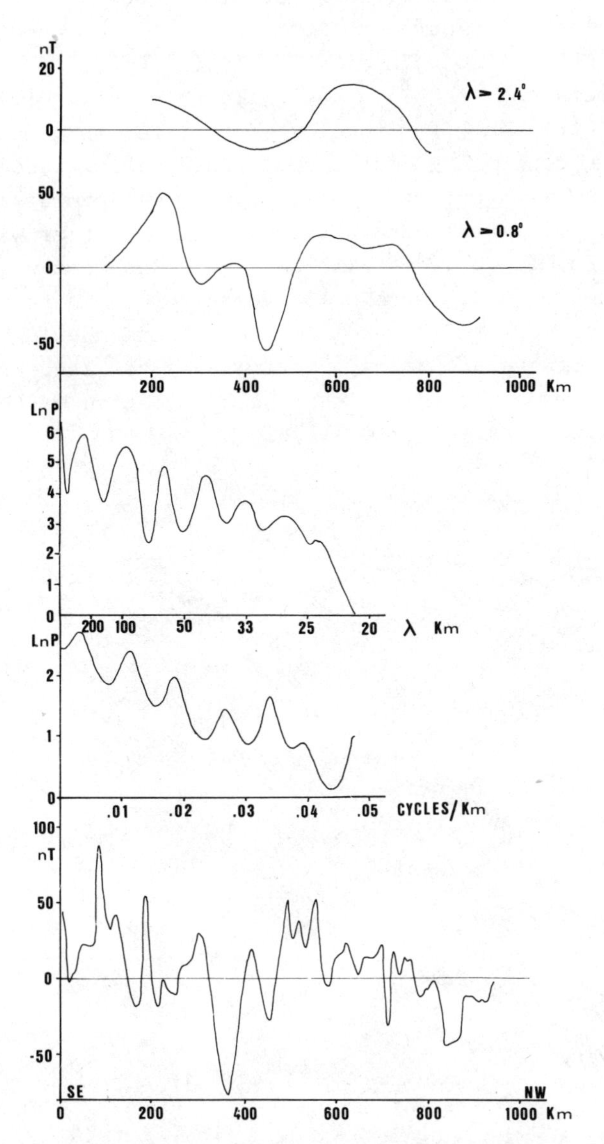


Fig. 8 - Interpretation of a 14° S, 43° W - 23° S, 43°7' W profile. The diagrams from the bottom upward represent: the observed anomalies of total intensity along the profile, the MESA power spectrum, the DFT power spectrum, the filtered profile using a cut off frequency corresponding to $\lambda \geq 88$ km; the filtered profile using a cut off frequency corresponding to $\lambda \geq 266$ km.

that most of the anomalous field is due to masses of large horizontal extension compared to its thickness. Components with longer than 200 km wavelength are clearly indicated by the peaks occurring in both spectra. In a case like this the choice of a proper cut off frequency for the low pass filter is complicated by the practical continuity of the spectrum. Two low pass filters were chosen: one with cut off frequency at 0.011 cycles/km (corresponding to a wavelength of 88 km) and the other with cut off frequency at 0.0038 cycles/km (corresponding to a wavelength of 266 km). The first filter should permit to eliminate the contributions by very shallow bodies, the second should practically eliminate most of the anomalies, leaving only the longest wavelength. The maps obtained after the application of these two dimensional filters are reported in Fig. 9 and 10.

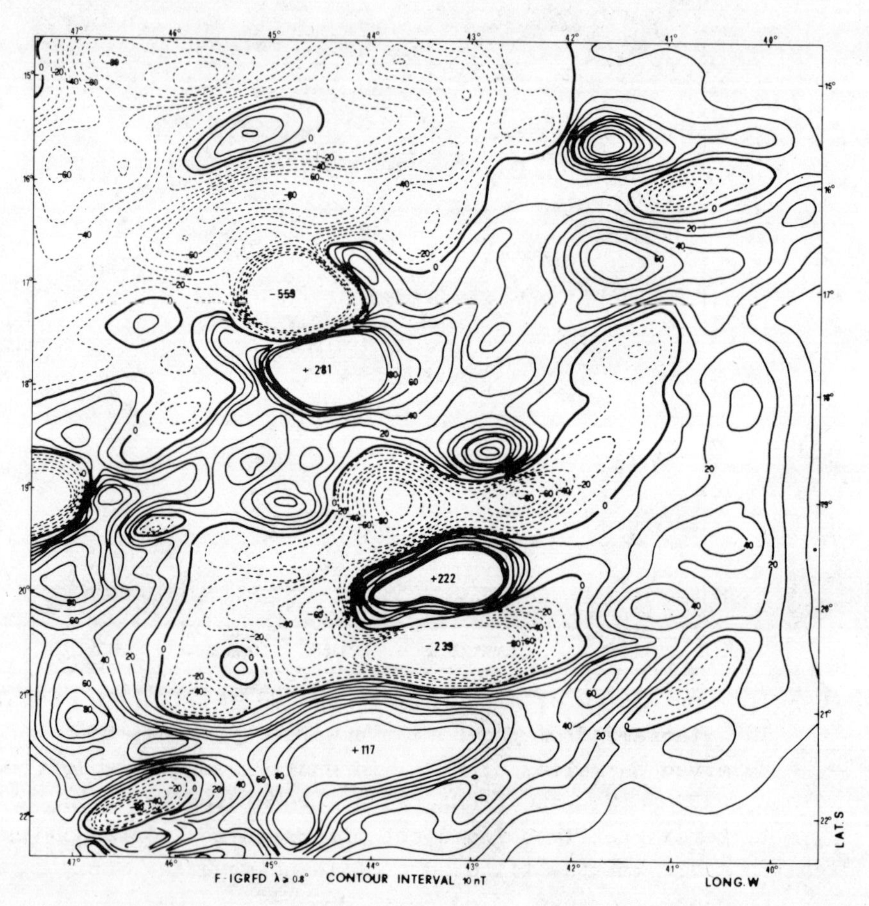


Fig. 9 - Magnetic anomalies at Minas Gerais after having filtered out anomalies with $\lambda < 88$ km (from Corrado et al., in prep.)

Two main features have to be pointed out in the map of Fig. 9:

1) the contributions by outcropping itabirites has not been removed by the filter due to the large extension of the anomaly source;
2) magnetic anomalies have a predominant EWE trend which cuts all the main geological structures, which have a dominant N trend.

Long wavelength Bouguer anomalies also have a trend consistent with that of geological structures, indicating that variations of crustal thickness occur mostly along E-W direction. In the map of Fig. 10 the ENE trend of magnetic anomalies is emphasized.

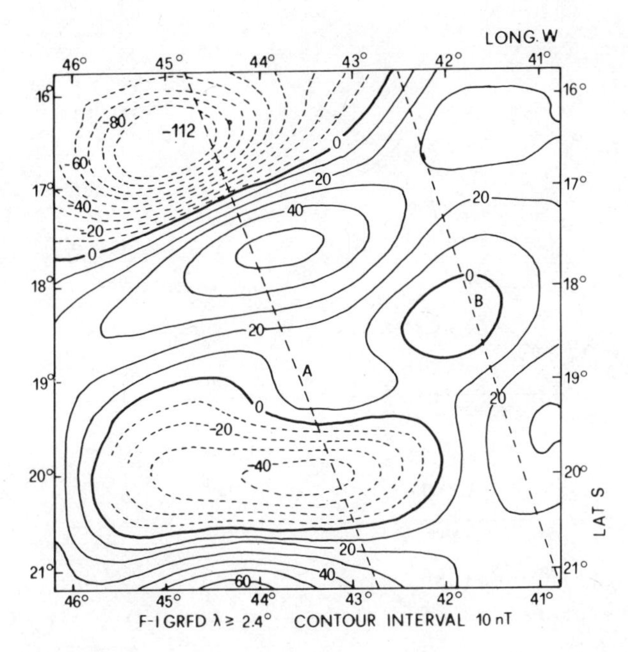


Fig. 10 - Magnetic anomalies at Minas Gerais after having filtered out anomalies with $\lambda < 266$ km (from Corrado et al., in prep.).

Magnetic anomalies appear to have wavelength of the order of 500-600 km, so their characteristics should have not been significantly modified by the used filter. Moreover the oscillating pattern of the field allows to rule out that this anomaly component is due to uncomplete subtraction of the main field. A quantitative interpretation of these anomalies has been attempted by inverting the data of Fig. 10 along two profiles: one passing through the main anomalies (A) and the other through the less disturbed part of the map (B). The used inversion method utilizes two-dimensional models of irregular shape and iterative techniques with least square adjustment of computed to theoretical data (Corrado et al., in prep.).

Profile A was first inverted assuming that the anomalies are due to morphological variations of the top of a lower crustal highly magnetized layer (Fig. 11). The best fit model indicates that the magnetized layer is practically absent in the NW part of the profile, whereas its top varies between 10 and 20 km of depth in most of the profile. This model therefore indicates a strong lateral variation in the magnetic properties of the crust.

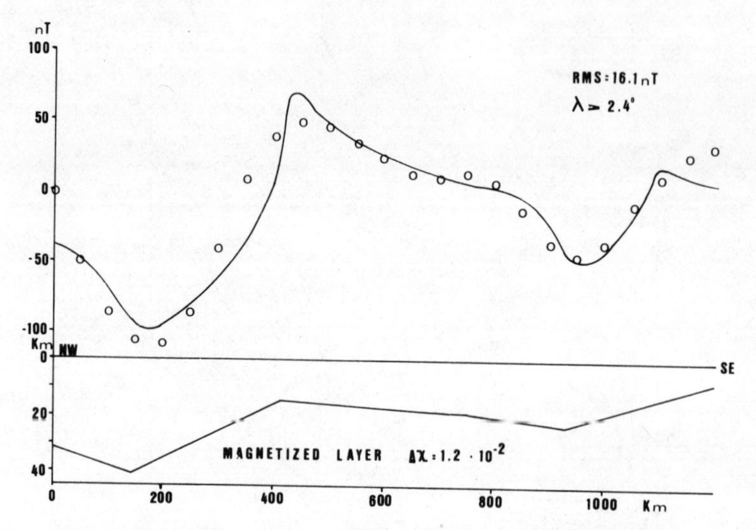


Fig. 11 - Best fit model assuming a magnetized lower crust for the interpretation of profile A of Fig. 10 (Corrado et al., in prep.).

An alternative model has also been used: it assumes a magnetized upper crust layer and considers the morphology of the bottom of this layer as the source of long wavelength anomalies. An interpretation by this model of a profile parallel and slightly displaced eastward in respect of profile A is reported in Fig. 12. The anomalies are fitted by a 15 to 30 km thick upper crustal layer with an average susceptibility of about $4 \times 10^{-3}$. According to this interpretation, the bottom of this layer should correspond to the Curie temperature of ferrimagnetic minerals entering upper crustal rocks. As lateral variations of temperature strong enough to produce the desired effect can be ruled out on the basis of the available heat flow data and the general geological structure of

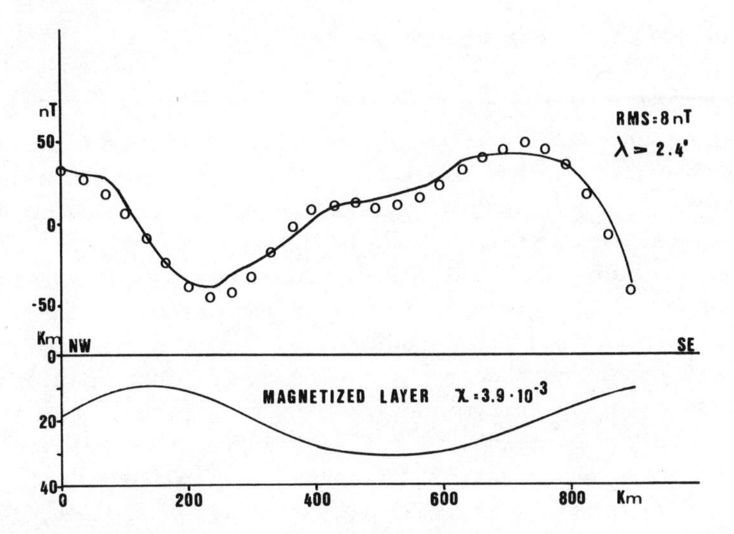


Fig. 12 - Best fit model assuming a magnetized upper crust and non magnetic lower crust (from Corrado et al., in prep.).

the area, this interpretation implies strong lateral variations of Curie temperature as a consequence of compositional variations, as will be discussed in the next paragraph.

Profile B has been interpreted using the magnetized upper crust model (Fig. 13). A smooth bottom surface of the magnetized layer was imposed; the best fit model indicates in this case a depth of Curie temperature slowly varying from 25 to 30 km of depth.

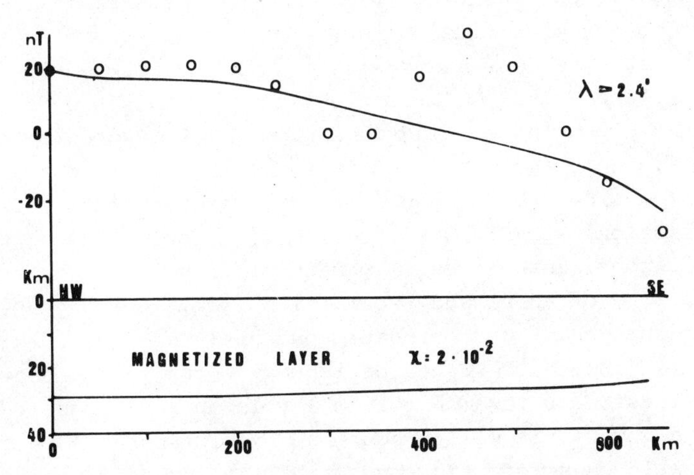


Fig. 13 - Best fit model assuming a non magnetized lower crust and a smooth Curie temperature surface for the interpretation of profile (B) (from Corrado et al., in prep.).

MAGNETIC MODELS OF CONTINENTAL CRUST

The models represented in Figs 11 and 12 represent the two alternative approaches generally used for the interpretation of the long wavelength magnetic anomaly field in continental areas. The former (long wavelength magnetic anomalies are due to a strongly magnetized lower crust and, possibly, upper mantle) has been used by canadian (e.g. Hall, 1974) and russian (e.g. Krutikhovskaya and Pashkevich, 1979) geophysicists in the interpretation of the magnetic anomalies of the Canadian and Ukrainian preCambrian shields. The second model (morphology of the bottom of the upper crustal magnetized layer and non magnetic lower crust) has been often used by geophysicists concerned with the interpretation of magnetic anomalies in young orogenic or recent volcanic areas (e.g. Smith et al., 1977; Shuey et al., 1977; Bhattacharryya and Leu, 1975; Byerly ans Stolt, 1977). The different approaches reflect the implicit assumption that the main ferrimagnetic phase throughout the continental crust is pure magnetite: being its Curie temperature at 580 °C, it saves its ferrimagnetic properties throughout the crust of low heat flow shield areas and looses them at an intracrustal level in high heat flow areas. This assumption however cannot be hold to be valid in general, as already pointed out by Haggerty (1978). Actually the magnetic properties of rocks are mainly controlled by the concentrations of ferrimagnetic minerals of the magnetite ($Fe_3O_4$) - ulvöspinel ($Fe_2TiO_4$) solid solution series. The susceptibility of the members of the series is directly dependent on the magnetite content. Curie temperature of the members of this series change with continuity between 580 °C (magnetite) and - 153 °C (ulvöspinel). The composition of the ferrimagnetic phase occurring in a given rock depends on several parameters, including the composition of the liquid they crystallized from, the oxygen fugacity, the temperature of crystallization, etc. Furthermore, chemical processes occurring within the continental crust may significatly alter the overall susceptibility of an already solidified rock, because they produce important changes of ferrimagnetic phases (Haggerty, 1978). High temperature (about 600 °C) oxydation may occur in the continental crust at least down to amphibolite levels during orogenic or rifting phenomena. It will produce an exsolution of a Fe rich ferrimagnetic phase of high susceptibility and an ilmenite of intermediate composition. In this case an increase of susceptibility will be produced. On the contrary, low temperature oxydation, which may occur in the crust of continental preCambrian shield areas, will produce the formation of maghemite ( $\gamma$ $Fe_2O_3$) from magnetite-ulvöspinel series minerals. Maghemite has a Curie temperature of 545 °C, but it is unstable at temperature between 250 and 550 °C. At this temperature interval it transforms readily into low susceptibility hematite. This effect will therefore produce a noteworthy decrease of susceptibility in the intermediate to lower crust of shield areas. The same effect is produced by metasomatic processes which transform minerals of the magnetite-ulvöspinel series into paramagnetic minerals. On the contrary, very high susceptibilities can be produced by serpentinization of amphibolites or peridotites. A reduction of Fe oxides occurs in fact during this process with the consequent formation of Fe-Ni-Co alloys which have Curie temperatures in the range 600-1100 °C.

Unfortunately, the magnetic characteristics of amphibolite or granulite facies rocks outcropping on the Earth surface cannot be extrapolated with confidence to the physical-chemical conditions existing in the deep crust. Frequently, the ferrimagnetic phases observed in these rocks are the product of exsolutions or even recrystallizations of the phases stable in the deep crust. These transformations mainly occurred during the slow uplift of these formations toward the Earth surface and in most case produced the formation of almost pure magnetite. As a consequence of this, outcropping, or shallow high grade metamorphic rocks have higher susceptibility than equivalent rocks existing at depth within the crust. However one result which can be reasonably extrapolated to lower crust conditions is that high grade metamorphic rocks, regardless of their chemical composition, have a low Königsberger ratio 0.1 to 0.5), (Corrado et al., in prep.) indicating that the contributions by remanent magnetization can be neglected.

Ulvöspinel concentration in magnetites increases with increasing pressure at the crystallization. As a consequence Curie temperature of intrusive rocks at depth will be probably lower than 350 °C.

In conclusion, the following arguments mitigate against the possibility that the lower crust is highly magnetized even in preCambrian shields:

1) low temperature oxidation will produce a strong decrease of susceptibility at temperatures higher than 250 °C which should be reached at a depth of about 30 km (see Fig. 2), because of transformation of maghemite into hematite;
2) metasomatic processes will produce the same effects;

3) also intrusive bodies crystallizing within the intermediate and lower crust will probably have ferrimagnetic minerals with Curie temperature lower than 350 °C.

A linear correlation between intensity of magnetic anomalies and crustal thickness was observed in the Canadian and Ukrainian shields. It was considered as an evidence favouring the hypothesis of a highly magnetized lower crust (Krutikhovskaya and Pashkevich, 1979). However, the correlation may be explained also, if the Moho can be assumed to be an almost isothermal surface in preCambrian areas, as an effect of greater thickness of the upper crustal magnetized layer. Moreover, in the above reported case of the Brazilian shield, long wavelength magnetic anomalies do not appear to be correlated with crustal thickness, as shown also by the almost perpendicular trend that they have in respect of long wavelength gravity anomalies. Therefore, the linear correlation between intensity of long wavelength anomalies and crustal thickness cannot be considered as a convincing evidence that the lower crust is magnetized.

The temperature-depth variation inferred for low heat flow areas would permit to reach Curie temperature of pure magnetite well below the Moho surface (see Fig. 2). Consequently, if the main ferrimagnetic phase is pure magnetite the upper mantle should have a high susceptibility. This consideration has stimulated researches looking for evidences allowing to favour or reject this possibility. Peridotites belonging to ophiolitic formation or oversthrust lower continental crust sequences (Ivrea-Verbano type) have generally very high magnetic susceptibilities (of the order of $10^{-2}$). As for granulite facies rocks, the main ferrimagnetic phase of these peridotites is pure magnetite formed during the slow uplift of these masses. However, a more reliable insight into the real magnetization characteristics of mantle peridotites is furnished by the study of ultramafic xenoliths included in basaltic volcanic rocks. These xenoliths, in many cases, are fragments of mantle rapidly brought to the surface by erupting basaltic magmas. The small residence time within the magma has allowed these xenoliths to save their originary magnetic characteristics. Wasilewski et al. (1979) have studied the magnetization of mantle xenoliths (peridotites, dunite, eclogites) from a number of localities, and Corrado et al. (in prep.) have analyzed peridotitic mantle xenoliths from Afar (Ethiopia). Mantle xenoliths have been found to be only slightly ferrimagnetic (the measured remanent magnetization is equivalent to a concentration of about 0,001% to 0,01% $Fe_3O_4$) and to have a paramagnetic susceptibility of 2 to

20 x $10^{-6}$. The very low magnetization level of these rocks is due to complex Cr, Mg, Al, Fe spinels and magnesian ilmenite. Haggerty (1978) suggested that serpentinization of mantle peridotites may produce exolution of metal Fe and formation of metal alloys with Curie temperature between 600 and 1100 °C. Although this process may theoretically furnish magnetized layers within the mantle, experimental evidence for it has not been yet reported.

## CONCLUSIONS

Power spectrum analyses of total intensity magnetic anomalies on continental areas show the existence of magnetic anomalies with wavelengths of the order of some hundreds km.

These anomalies can be be related to strong lateral variations of magnetization of crustal rocks at a depth of 20-30 km. Two alternative interpretations have been proposed: (1) they can be due to the occurrence of a strongly magnetized lower crust and, possibly, upper mantle. This can hold particularly in low heat flow continental shield areas; (2) they can be due to depth variations of the bottom of a magnetized upper crustal layer. This is very likely to occur in high heat flow areas (recent orogenic and active volcanic areas) where the Curie temperature even of pure magnetic is reached at depths shallower than 25-30 km.

However, our present knowledge on the magnetic characteristics of rocks indicate that it is not safe to assume that the main ferrimagnetic phase through the whole continental crust is pure magnetite. In fact, plutonic rocks which crystallized at lower crustal levels may contain a ulvöspinel-magnetite solid solution as the main ferrimagnetic phase. Metamorphic and igneous rocks having experienced low temperature oxidation or metasomatic processes may contain no more high susceptibility ferrimagnetic phases. Therefore, the bottom of the crustal magnetized layer may represent a compositional boundary between a magnetized upper crust and a non magnetic lower crust even in a low heat flow area (see Gasparini et al., 1979).

The available evidence from magnetic properties of rocks allow to rule out that the lithosperic mantle can produce magnetic anomalies, except the very particular case when serpentinization processes may have produced the exsolution of metal Fe (see Haggerty, 1978). No evidence of magnetic anomalies which can be related to the upper mantle has been so far reported, so that all the informations which can be extracted from long wavelength magnetic anomalies are limited to the crust.

ACKNOWLEDGEMENTS

This work was carried out with the support of International cooperation program between Italian CNR and Brazilian CNPq.

BIBLIOGRAPHY

Alldredge, L.R., Van Vookhis, G.D., and Davis, T.M., 1963, A magnetic profile around the world, J. Geophys. Res., 68:3679.

Almeida, F.F., de, 1977, O craton de Sao Francisco, Rev. Brasil Geoc., 7:349-364.

Barraclough, D.R., Harwood, J.M., Leaton, B.R., and Malin S.R.C., 1978, A definitive model of the geomagnetic field and its secular variation for 1965. I. Derivation of model and comparison with the IGRF, Geophys. J.R. Astr. Soc., 55:111-121.

Bhattacharryya,B.K., Lev,L.K., 1975,Analysis of magnetic anomalies over Yellowstone National Park: mapping of Curie point isothermal surface for geothermal reconnaissance, J. Geophys. Res., 80:4416-4465.

Blackman, R.B., Tuckey, J.W., 1959, The measurement of Power Spectra from the point of view of communication engineering. New York, Dover Publ.

Blitzkow, D., Gasparini,P., Mantovani, M.S.M., Sa, N.C., de, 1979, Crustal structures of Southeastern Minas Gerais, Brazil, deduced by gravity measuments, Rev. Brasil, Geoc., 9:38-47.

Burg, J.P., 1967, Maximum enthropy spectral analysis. Paper presented at 37th Ann. Int. JEG Meeting Oklahoma.

Cordani, U.G., Amaral, G., Kawashita, K., 1973, The Precambrian evolution of South America, Geol. Rund., 62:309:317.

Corrado, G., Gasparini, P., Mantovani, M.S.M., Rapolla, A., 1979, Depth of Curie temperature computed from aeromagnetic anomalies in Southeastern Minas Gerais, Brazil, Rev. Brasileira de Geociencias, 9:33-38.

Corrado, G., Pinna, E., Rapolla, A., 1977, The magnetic field of Italy: description and analysis of the new T, Z and H maps between 40° N and 44° N, Boll. Geof. Teor. e Appl., vol. XX, 75:140-156.

Dawson, E., and Newitt, L.R., 1978, IGRF comparisons, Phys. of the Earth and Plan. Int., 16:P1-P6.

Gasparini, P., Mantovani, M.S.M., Corrado, G., and Rapolla, A., 1979, Depth of Curie temperature in continental shields: a compositional Boundary?, Nature 278 N°5707:845-846.

Haggerty, S.F., 1978, Mineralogical constraints on Curie isotherms in deep crustal magnetic anomalies, Geophys. Res. Letters, 5:105-108.

IAGA Division 1 Study Group, 1976, International geomagnetic reference field 1975, EOS (Trans. Am. Geophys. Union), 57:120-121.

Kanasewich, E.R., 1975, Time sequence analysis in geophysics Univ. of Alberta Press, Edmonton, Canada.

Krutikhovskaya, Z.A., and Pashkevich, I.K., 1979, Long-wavelength magnetic anomalies as a sorce of information about deep crustal structure, J. Geophys., 46:301-317.

Kulhanek, O., 1976, Introduction to digital filtering in geophysics Elsevier Publ. Co. Amsterdam.

Shuey, R.T., Schellinger, D.K., Tripp, A.C., Alley, L.B., 1977, Curie depth determination from aeromagnetic spectra, Geophys. J.R. Astr. Soc., 50:75:101.

Smith, R.B., Shuey, R.T., Pelton, J.R., Bailey, J.P., 1977, Yellostone hot spot: contemporary tectonics and crustal properties from earthquake and aeromagnetic data, J. Geophys. Res., 82:3665-3676.

Spector, A., Grant, F.S., 1970, Statistical models for interpreting aeromagnetic data, Geophysics, 35:293-302.

Wasilewski, P.J., Thomas, H.H., Mayhew, M.A., 1979, The Moho as a magnetic boundary, Geophys. Res. Letters, 6:541-544.

Willye, P., 1971, The dynamic Earth, J. Wiley and Sons, New York.

Zmuda, A.J., 1971, World Magnetic Survey 1957-1969, IAGA Bulletin n. 28. IUGG Publication 206 pp.

# GEOELECTRIC METHODS

# ON THE POSSIBILITY OF A DIRECT AUTOMATIC INVERSION OF D.C. GEOELECTRIC DATA

Luigi Alfano

Istituto di Geofisica
Università di Milano
Italy

ABSTRACT

A method suitable for the direct automatic inversion of D.C. geoelectrical data is proposed. The aim is the interpretation of data pertaining to underground structures of arbitrary shape. Complicated structures present considerable difficulties if they are interpreted by means of indirect procedures, based on the search for a best fit between the field data and a theoretical model. The mathematical problem involved in direct inversion is a nonlinear one; it has been tackled by means of a linearization procedure; based on successive approximations. Some topics of practical interest, concerning the handling of a great number of data, and the influence of too strong geological surface noises are discussed.

## INTRODUCTION

The technique of the "Vertical Electrical Sounding" (V.E.S.) is much used in geophysical prospecting; it is generally carried out by means of a Schlumberger array, or, in the case of deep explorations, by means of dipole-dipole arrays. The interpretation of an electrical sounding alone has the best possibility of success when only horizontal stratifications occur in the subsoil; but when this hypothesis is not applicable, the problem is indeterminate owing possible unknown lateral variations.

Another frequently used technique is horizontal sounding, which is carried out by means of a fixed current and moving potential electrodes. The fixed geoelectrical field is surveyed at the ground surface over a large area, with great variations of the distances between the potential and the current electrodes, in order to make evident in a map the presence of apparent resistivity anomalies (lateral variations). But a possible horizontal stratification may influence the apparent resistivity map, and may simulate nonexistent lateral resistivity changes.

Thus, it appears that both vertical and horizontal soundings, if used alone, give insufficient sets of data, and the interpretation is not a uniquely determinate problem. However, the integration of these two techniques will supply more meaningful data. Another example of a useful procedure is to make a number of vertical soundings distributed over the surveyed area, with different centers and different directions of the spreads. This generates an amount of data which, considered on the whole, has the characteristics of both the vertical and the horizontal exploration. Now, if the lateral variations are not small, it is practically impossible to interpret each vertical sounding separately. But, unfortunately, the simultaneous interpretation of these soundings may be very difficult if only indirect methods are available, i.e., if initial models of the underground resistivity structures must be formulated in advance and then fitted to the field data.

If such complicated structures occur a mathematical direct method of data inversion appears to be desirable, a method which is substantially an operator whose input is given by the measured data and whose output is the interpretation. This is the topic of the present paper.

## THE GEOELECTRICAL FIELD CAUSED BY DISCONTINUITY SURFACES OF ANY SHAPE

### 1. The Forward Problem

Let us assume that the underground half-space is divided into several parts of different resistivity by a number of discontinuity surfaces of arbitrary shape. The problem of computing the values of a geoelectrical field caused by known current sources (we can call this the forward problem) has been carried out (Alfano 1959, 1962). In the first of the two papers, the case of mutually intersecting vertical and horizontal discontinuity planes was considered. In the second, the discontinuity surfaces were assumed to be curved. It was shown that the mathematical formulation of this problem may be reached by means of a distribution of electric charges on the discontinuity surfaces (secondary charge distribution). Thus, if P is a generic point of S, where S is the set of discontinuity surfaces, including the ground surface, let us indicate by $\sigma(P)$ the charge density (electric charge on a unit surface). The values of $\sigma(P)$ are functions of the geometric characteristics of the discon-

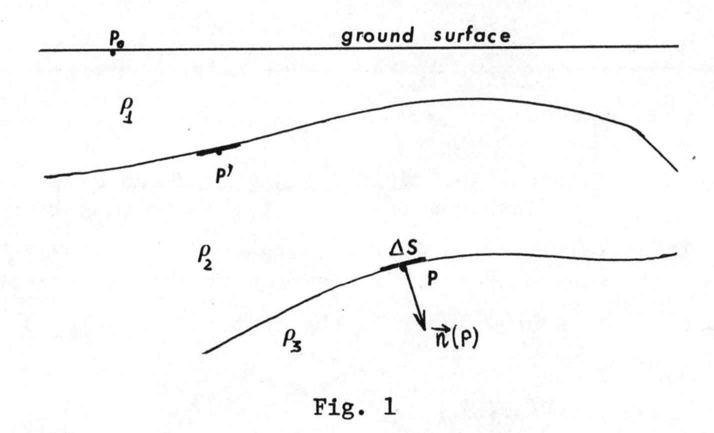


Fig. 1

tinuity surfaces; but they depend also on the positions of the actual current sources and on the resistivity values.

A simple example is shown in Fig. 1, where an actual point source is at the point $P_0$ of the ground surface. Two discontinuity surfaces subdivide the underground half-space into three parts with resistivity values $\rho_1$, $\rho_2$, $\rho_3$, respectively. $S_{ij}$ is the surface separating the regions with the resistivity values $\rho_i$ and $\rho_j$. In the above-mentioned papers, the following formula was shown to be valid:

$$\sigma_W(P) = \frac{K(P)}{2\pi}\,\overrightarrow{E_W(P)}\cdot\overrightarrow{n(P)} = \frac{K(P)}{2\pi}\left[\overrightarrow{E_{pW}(P)} + \overrightarrow{E_{sW}(P)}\right]\overrightarrow{n(P)} \tag{1}$$

$(W = 1, 2, \ldots, W_0)$

This formula satisfies the boundary conditions of a geoelectric D.C. field in a conducting medium. In this formula $E_{pW}(P)$ is the primary field, i.e., the field caused by the true current electrodes. These, in practice, must be at least two (positive and negative) with positions respectively at the points $P_{W+}$ and $P_{W-}$, and surrounded by media with resistivity values $\rho_{W+}$ and $\rho_{W-}$. In this case the expression of the primary field is given by

$$\overrightarrow{E_{pw}(P)} = \frac{\rho_w I \,\overrightarrow{(P - P_{w+})}}{2\pi |P - P_{w+}|^{3/2}} + \frac{\rho_w I \,\overrightarrow{(P - P_{w-})}}{2\pi |P - P_{w-}|} \tag{2}$$

$(w = 1, 2, \ldots, w_o)$

Here I is the current injected into the ground and $w_o$ is the number of the dipoles and consequently of the fields we will consider separately. The term $\overrightarrow{E_{sw}(P)}$ in (1) is the secondary field at P. It has been shown to result from all the secondary charges except those in the element of area of which P is the center. Thus $\overrightarrow{E_{sw}(P)}$ is given by

$$\overrightarrow{E_{sw}(P)} = \int_{(S-dS)} \frac{\sigma_w(P') \cdot \overrightarrow{(P - P')}}{|P - P'|^{3/2}} \, dS' \qquad (w = 1, 2, \ldots, w_o) \tag{3}$$

where S is represented by all the discontinuity surfaces, and dS is the element of area surrounding P.

Now, substituting (2) and (3) in (1) and setting

$$\overrightarrow{E_{pw}(P)} \cdot \overrightarrow{n(P)} = \Gamma_w(P) \cdot \frac{\rho_w I}{2\pi} \tag{4}$$

$$\Gamma(P, P') = \frac{\overrightarrow{(P - P)}}{|P - P|^{3/2}} \cdot \overrightarrow{n(P)} \tag{5}$$

we obtain

$$\sigma_w(P) = \left[K(P)/2\pi\right] \cdot \left[\Gamma_w(P) \frac{\rho_w I}{2\pi} + \int_{(S-dS)} \Gamma(P - P') \cdot \sigma_w(P') \cdot dS'\right] \tag{6}$$

$(w = 1, 2, \ldots, w_o)$

which is an integral equation where $\sigma_w(P)$ is the unknown function.

Now, the value of the geoelectrical potential in a generic point M is the sum of two components: the first, $V_{pw}(M)$, caused by the actual current sources on $P_{w+}$ and $P_{w-}$ (primary potential), and the second, $V_{sw}(M)$, caused by the charge distribution $\sigma_w(P)$ (secondary potential). Thus:

$$V_w(M) = V_{p,w}(M) + V_{s,w}(M)$$

$$V_{p,w}(M) = C_w(M, P_{w+}, P_{w-}, \rho_{w+}, \rho_{w-}) \tag{7}$$

$$V_{s,w}(M) = \int_S \frac{\sigma_w(P)}{|MP|} dS = \int_S C(M,P) \cdot \sigma_w(P) \cdot dS$$

where the meaning of $C_w(M)$ and $C(M,P)$ is obvious. We are interested in the cases where M and eventually other potential electrodes are on the ground surface. The theory can easily be extended to cases with more than two discontinuity surfaces and more than one current source dipole.

Formulas (6) and (7) represent the problem completely, whatever the number, position, and shape of the discontinuity surfaces may be.

## 2. Practical Formulas

Formulas (6) and (7) must be modified in order to carry out the numerical calculations. The discontinuity surfaces must be subdivided into a finite number of surface elements $\Delta S_i$ (where i = 1, 2, 3, ..., N); on each $\Delta S_i$ the charge density is assumed to have a constant value. It follows that the integrals turn into sums, and that the integral equation (6) turns into the following system of linear equations:

$$\sigma_{i,w} = \frac{K_i}{2\pi} \cdot \left[ \Gamma_{i,w} \frac{\rho_w I}{2\pi} + \sum_{j=1}^{N} \Gamma_{i,j} \cdot \sigma_{j,w} \cdot \Delta S_j \right] \tag{8}$$

$$\Gamma_{ii} = 0 \; ; \qquad (w = 1, 2, \ldots, w_0) \; ; \; (i = 1, 2, \ldots, N)$$

In (8) for each surface element $\Delta S_i$ one equation expresses the value of the charge density $\sigma_i$ as a linear function of the primary field and the other $N-1$ values $\sigma_j$. $K_i$ corresponds to the K(P) of (6), and has the same value, constant on the same discontinuity surface. Moreover, since $P_i$ and $P_j$ are the centers of $\Delta S_i$ and $\Delta S_j$, respectively, $\Gamma_w(P)$ and $\Gamma(P,P')$ turn into $\Gamma_{io}$ and $\Gamma_{ij}$, so that

$$\Gamma_{iw} = \frac{1}{|P_i - P_w|^2} \cdot \frac{\overrightarrow{(P_i - P_w)}}{|P_i - P_w|} \tag{9}$$

$$\Gamma_{ij} = \frac{1}{|P_i - P_j|^2} \cdot \frac{\overrightarrow{(P_i - P_j)}}{|P_i - P_j|} \tag{10}$$

$$(i = 1, 2, \ldots, i_o) \qquad (w = 1, 2, \ldots, w_o)$$

where $\overrightarrow{n_i}$ is the unit vector perpendicular to $\Delta S_i$ on $P_i$ and

$$\frac{\overrightarrow{(P_i - P_w)}}{|P_i - P_w|} \cdot \overrightarrow{n_i} = \cos \alpha_{iw} ; \qquad \frac{\overrightarrow{(P_i - P_j)}}{|P_i - P_j|} \cdot \overrightarrow{n_i} = \cos \alpha_{ij} ; \tag{11}$$

where $\alpha_{ij}$ and $\alpha_{iw}$ are the angles defined by the vector $\overrightarrow{n_i}$ with the vectors $\overrightarrow{(P_i - P_j)}$ and $\overrightarrow{(P_i - P_w)}$ respectively.

The ground surface $S_o$, which may or may not be assumed to be a plane, must also be considered a discontinuity surface of the underground half-space, having an infinite value of the resistivity. Consequently on $S_o$ too we find a secondary charge distribution.

Generally the points M for which we compute the values of the potential are on this ground surface. Let us consider one of these points, $M_h$, the center of an elementary area $\Delta S_h$, and its potential value $V_h$, which is expressed, according to (7), by

$$V_{h,w} = C_{h,w} \frac{\rho_w I}{2\pi} + \sum_{i=1}^{N} C_{hi} \cdot \sigma_i \cdot \Delta S_i \tag{12}$$

$$C_{h,w} = \frac{1}{|M_h - P_w|} ; \qquad C_{hj} = \frac{1}{|M_h - P_j|} \tag{13}$$

$$(h = 1, 2, \ldots, \mu) \qquad (w = 1, 2, \ldots, W_o)$$

and $C_{hh} \cdot \sigma_h$ is the contribution of the elementary area $\Delta S_h$, of which $M_h$ is the center. $V_h$ depends on the value of $\sigma_h$ in $\Delta S_h$ and on the shape of this area.

The solution of the system (8) in terms of the unknowns $\sigma_i$ will be introduced into expression (12), and the potential $V_h$ on $M_h$ may be computed.

Formulas (8)-(12) are basic for the computation of the potential values $V_h$ at each point of the ground surface if the positions and the shapes of the discontinuity surfaces and the resistivity values in the underground half-space are known. In particular, formula (8) may be written in matrix form:

$$\begin{Vmatrix} (2\pi/K_1) & \Gamma_{12}\Delta S_2 & \Gamma_{13}\cdot\Delta S_3 & \ldots & \Gamma_{1N}\cdot\Delta S_N \\ \Gamma_{21}\cdot\Delta S_1 & (2\pi/K_2) & \Gamma_{23}\cdot\Delta S_3 & \ldots & \Gamma_{2N}\cdot\Delta S_N \\ \Gamma_{31}\cdot\Delta S_1 & \Gamma_{32}\cdot\Delta S_2 & (2\pi/K_3) & \ldots & \Gamma_{3N}\cdot\Delta S_N \\ \cdot & \cdot & \cdot & \cdot & \cdot \\ \Gamma_{N1}\cdot\Delta S_1 & \Gamma_{N2}\cdot\Delta S_2 & \Gamma_{N3}\cdot\Delta S_3 & \ldots & (2\pi/K_N) \end{Vmatrix} \cdot \begin{Vmatrix} \sigma_1 \\ \sigma_2 \\ \sigma_3 \\ \cdot\cdot \\ \sigma_N \end{Vmatrix} = \begin{Vmatrix} \Gamma_{1w} \\ \Gamma_{2w} \\ \Gamma_{3w} \\ \cdot\cdot \\ \Gamma_{Nw} \end{Vmatrix} \frac{\rho_w I}{2\pi} \tag{14}$$

and in a similar way we can write (12) as

$$Q_w = \frac{\rho_w I}{2\pi}$$

$$\begin{vmatrix} C_{12} & C_{12} & C_{13} \dots & C_{1N} \\ C_{21} & C_{22} & C_{23} \dots & C_{2N} \\ C_{31} & C_{32} & C_{33} \dots & C_{34} \\ \dots & \dots & \dots & \dots \\ C_{n1} & C_{n2} & C_{n3} & C_{nN} \end{vmatrix} \cdot \begin{vmatrix} \sigma_1 \cdot \Delta S_1 \\ \sigma_2 \cdot \Delta S_2 \\ \sigma_3 \cdot \Delta S_3 \\ \dots \\ \sigma_N \cdot \Delta S_N \end{vmatrix} = \begin{vmatrix} V_1 - C_{1w} \cdot Q_w \\ V_2 - C_{2w} \cdot Q_w \\ V_3 - C_{3w} \cdot Q_w \\ \dots \\ V_n - C_{nw} \cdot Q_w \end{vmatrix} \tag{15}$$

This form may be convenient for the discussion of some aspect of the inversion procedure. But first, some comments on (14) and (15) are appropriate. In particular, let us point out that the values $K_i$ in (14), in general, are not all different, since many elements $\Delta S_i$ pertaining to the same discontinuity surface must have the same value of $K_i$; moreover, the number of $K_i$ is $\mu$, where $\mu$ is the number of discontinuity surfaces. Consequently, $N \geqslant \mu$, where N is the number of $\sigma_i$ and also of $\sigma_i$ and also of $\Delta S_i$. Further, $\mu \geqslant m$, where m is the number of subdivisions of the underground half-space defined by the discontinuity surfaces, and consequently is also the number of resistivity values $\rho_i$. In (12) and (15), as in (8) and (13), the number n of the points $M_h$ has been chosen so that $n \cdot w = m$. There is no particular reason for this in the forward problem, but there is in the inversion procedure, for which the preceding formula will be utilized later.

Finally, let us point out that the true resistivity ratios, $\rho_2/\rho_1$, $\rho_3/\rho_1$, ..., $\rho_m/\rho_1$, do not appear explicitly in (14) and (15), but in their place the coefficients $K_1$, $K_2$, ..., $K_\mu$ are present; these latter parameters are not independent, so that a set of $\mu - m$ bounding equations of the following type must be written:

$$f(K_1, K_2, \dots, K_\mu) = 0 \tag{16}$$

## 3. Successive Approximations

The computing procedure is relatively simple and consists of the following steps:

(a) Some initial approximate values $\sigma_{io}$ are introduced in the second members of the system (8), in the place of the unknowns $\sigma_i$; then (8) is solved with respect to the unknowns $\sigma_i$ contained in the first member. A first approximation $\sigma_{i1}$ is so obtained.

(b) The values $\sigma_{i1}$ are substituted in the second members of (8) and the previous operation is repeated, so that a second approximations $\sigma_{i2}$ is obtained.

(c) We continue in this manner. We substitute in the second members of the system (8) the values obtained by means of the (p - 1)th approximation in order to compute the values of the pth approximation.

It has been shown in the previously cited papers that this procedure is convergent even if $\sigma_{io}$ is zero. But clearly a better start may be obtained by the use of closer initial values. This forward procedure, which concerns the calculation of the geoelectric potentials on the basis of the shapes and positions of underground resistivity values, may be carried out without too much difficulty, but sometimes the great number of data to be handled may cause some trouble. We have reviewed it here only with the aim of introducing the main topic of this paper, the direct inversion of the D.C. geoelectric data, which is more difficult, because of the nonlinearity of the equations and the instability of the mathematical problem when randomly distributed geological irregularities influence the field data.

## THE DIRECT INVERSION

### 4. General Considerations

The methods for the interpretation of geoelectrical data may be divided into two classes: indirect methods and direct methods. The indirect ones which are based on the best fit between field data and a previously assumed theoretical model present an important advantage: if the field data are disturbed by geological (especially superficial) irregularities, we do not meet with serious difficulties. In effect, if the field data do not agree with an assumed, perhaps too simple model, we can simply ascertain this fact, without any trouble in the execution of the automatic program, which concerns only calculations of the above-mentioned forward type.

In contrast, the direct methods involve mathematical operators for which the field data are the input and the interpreation is the output. They are hard to handle.

Other advantages of the indirect methods are the possibility of trying again with other, more suitable models, and the linearity of the mathematical problem resulting from (14) and (15), which are linear with respect to $\sigma_i$ and $V_h$. On the other hand, a severe drawback is presented by the necessity of developing beforehand suitable models for cases of complicated structures.

As far as the direct methods are concerned, the principal advantage lies in the possibility of handling simultaneously and according to some fixed rules great numbers of data when complicated underground structures occur. The most important drawback is the nonlinearity of the problem. This fact does not introduce many difficulties when synthetic data, deriving from theoretically prepared models, are to be handled. But when actual problems with data measured in the field are tackled, serious troubles can be expected. The output of the automatic programs may present strange solutions, for instance, with negative resistivity values. Results of this type, devoid of meaning, may occur in cases of wrong input data, or when structural irregularities not included in the interpretive scheme are present in the surveyed area. Notwithstanding these difficulties, we think it is worthwhile to make this direct inversion procedure practically available. Obviously, the present paper cannot exhaust the problem, since a considerable amount of practical experience is needed.

## 5. Procedure for Direct Inversion

Let us suppose the subsoil is divided into m homogeneous parts by $\mu$ discontinuity surfaces ($\mu \geqslant m$). As already pointed out, the values $K_i$ in equations (14), (15), and (16) are not all different since the elements $\Delta S_i$ relating to the same discontinuity surface must have the same value of the resistivity ratio. It follows that there are only $\mu$ different values of $K_i$, i.e., a number equal to the number of discontinuity surfaces. Consequently, as will be shown later in more detail, the m resistivity values $\rho_j$ and the N charge values $\sigma_i$ are the unknowns of the problem represented by the N + $\mu$ equations (14), (15), and (16).

The parameters $\Gamma_{ij}$ and $C_{hi}$, which represent the shape and the position of the discontinuity surfaces (and also their positions with respect to the current and potential electrodes) in these equations, should be unknowns in the inversion problem, and in fact, the geometric characteristics of the discontinuity surfaces are unknown when we attempt an interpretation of the data. But the procedure presented here is planned in such a way as to handle them as known parameters. This is accomplished by subdividing the underground volume covered by the data into elementary little volumes $H_i$ of known position and shape, with generally different and unknown resistivity values. Now, if these elementary volumes are sufficiently small, any type of underground structure may be represented. Consequently, the resistivity values relating to all the $H_i$ remain the only unknowns of the problem. The fact that the positions of the elementary volumes are predetermined implies that $\Gamma_{ij}$ and $C_{hi}$ are known parameters of the inversion problem. Let us now examine the procedure in more detail by means of the following steps.

(a) The part of the underground half-space which we consider to be connected with the surveyed area on the ground surface is assumed to be a parallelepiped $\Omega$ whose top side is coincident with the ground surface. It is subdivided by three sets of planes perpendicular respectively to the x, y, and z axes of a rectangular system (the first two axes are horizontal and the third is vertical). Consequently, the planes with equations $x = x_i$ and $y = y_j$ are vertical, and the planes with equation $z = z_k$ are horizontal. It follows that the underground half-space is divided into a number of partial parallelepipeds $H_1, H_2, \ldots, H_m$, having unknown resistivity values $\rho_1, \rho_2, \ldots, \rho_m$. $H_0$ is the part of the underground half-space which is outside the parallelepiped $\Omega$. If the partial volumes $H_i$ are sufficiently small any possible structure may be approximated by means of proper resistivity values.

(b) Each side of a given $H_i$ is a possible discontinuity surface; it may be denoted by $S_{ab}$ if it divides $H_a$ from $H_b$, with resistivity values $\rho_a$ and $\rho_b$. Moreover, each side, characterized by a value $K_{ab} = (\rho_b - \rho a)/(\rho_b + \rho_a)$, may be subdivided into some elementary areas $\Delta S_{ij}$, which obviously all have the same value of $K_{ab}$.

(c) The number of parallelepipeds, equal to the number of true unknowns $\rho_i$, is m; the number of $K_{ab}$ is equal to $\mu > m$; the number of elementary areas $\Delta S_i$ is $N > \mu$. In practice it is expedient to assume the parallelepipeds different in size, those nearer the surface being smaller and the deeper-lying ones larger. In fact, because of the low resolution of the geoelectric methods, it is useless to assume parallelepipeds which are too small with respect to their depth.

(d) As far as the balance between equations and unknowns is concerned, let us point out that in (14) we have N equations and N unknowns $\sigma_i$. In (15) the number of equations is n, which is the number of the measured potential data for each of $w_0$ positions of the current electrodes (and consequently of the different geoelectrical fields taken into consideration), so that the total number of the measured data is equal to $n \cdot w_0 = m$, equal to the number of the unknowns $\rho_h$. But it must be pointed out that the unknowns actually appearing in (15) are the $K_{ij}$, whose number is $\mu > m$; the difference between the $(N + \mu)$ unknowns $K_{ij}$ and $\sigma_i$ and the $(N + m)$ equations (14) and (15) is balanced by the $(\mu - m)$ equations (16), which bound the values $K_{ij}$. The use of these bounding equations agrees with the above-noted fact that the $K_{ij}$ values cannot be independent.

(e) The resolution of the system (14) may be expressed by

$$\sigma_{i,w} = \psi_i\left(\rho_w, K_{01}, K_{02}, \ldots, K_{ab}, \ldots, K_{m-1,m}\right) \tag{17}$$
$$(w = 1, 2, \ldots, w_0)$$

which unfortunately is nonlinear with respect to the $K_{ij}$. The substitution of (17) into (15) or (12) yields the system

$$V_{h,w} = \varphi_{h,w}\left(K_{01}, K_{02}, \ldots, K_{ab}, \ldots, K_{m-1,m}\right) \quad (h = 1, 2, \ldots, h_0) \tag{18}$$

which contains $\mu$ unknowns $K_{ij}$ and at least m equations, where $m < w_0 \cdot h_0$. As has already been pointed out, the bounding equations between the values $K_{ij}$ are $\mu - m$ in number. Consequently, equations (18), together with the $\mu - m$ bounding equations (16), form a system of at least $\mu$ nonlinear equations in the unknowns $K_{ij}$. It must be remarked that knowledge of the $K_{ij}$ values is not sufficient, since at least one resistivity value is needed if we are to know all the other resistivity values. However, all the surface values $\rho_w$ relating to the surface parallelepipeds surrounding the current electrodes may be considered as known by means of direct measurements.

(f) In the scheme just described the bounding equations (16) may assume a particular form which may be deduced by means of the following considerations. The edges of the main parallelepiped $\Omega$, which are parallel to the x, y, and z axes, have been divided into p, q, and r parts respectively. Consequently the number of the partial parallelepipeds, and also of $\rho_i$, is given by the product $m = pqr$. The number of sides, and consequently of $K_{ab}$, is given by

$$\begin{aligned} \mu &= pq(r+1) - pr(q+1) - qr(p+1) \\ &= 3pqr + pq + pr + qr \end{aligned} \tag{19}$$

so that the number $\mu - m$ of the bounding equations is

$$\mu - m = 2pqr + pq + pr + qr =$$

$$= pq(r+1) + pr(q+1) + qr \quad (20)$$

As far as the form of the bounding equations is concerned, let us consider four contiguous elementary parallelepipeds $H_a$, $H_b$, $H_c$, $H_d$, with resistivity values $\rho_a$, $\rho_b$, $\rho_c$, $\rho_d$, and having a corner in common. Now, the sides $S_{ab}$, $S_{bc}$, $S_{cd}$, $S_{da}$ having this corner in common separate respectively $H_a$ from $H_b$, $H_b$ from $H_c$, $H_c$ from $H_d$, and $H_d$ from $H_a$. If we cross the four sides successively, we can write

$$\rho_b/\rho_a = (1 + K_{ab})(1 - K_{ab})$$

$$\rho_c/\rho_b = (1 + K_{bc})(1 - K_{bc})$$

$$\rho_c/\rho_d = (1 + K_{cd})(1 - K_{cd})$$

$$\rho_d/\rho_a = (1 + K_{da})(1 - K_{da})$$

so that, multiplying these four equations, the following relation may be obtained:

$$\frac{1+K_{ab}}{1+K_{ab}} \; \frac{1+K_{bc}}{1+K_{bc}} \; \frac{1+K_{cd}}{1+K_{cd}} \; \frac{1+K_{da}}{1+K_{da}} = 1 \quad (21)$$

which shows clearly that the $K_{ab}$ values are not independent. Equations (21) must be written for $\mu - m$ corners, divided into three sets parallel, respectively, to the x, y, and z axes. Obviously all the equations must be independent of each other.

## 6. The Linearization of the Problem

First, a formal modification must be applied in equations (8) or (14), and in (12) or (15), by means of the following substitutions:

$$\sigma_i \cdot \Delta S_i \quad \text{is replaced by} \quad \sigma_{\varepsilon i} \cdot \Delta S_{\varepsilon i}$$
$$\sigma_j \cdot \Delta S_j \quad \text{is replaced by} \quad \sigma_{\eta j} \cdot \Delta S_{\eta j} \tag{22}$$

where the indexes $\varepsilon$ and $\eta$, which may assume values from 1 to $\mu$, refer to one of the $\mu$ discontinuity surfaces $\Delta S_\varepsilon$ or $\Delta S_\eta$, i.e., the sides $S_{ab}$ introduced in Section 5. On the other hand, the indexes i and j refer to the position of each charge element $\sigma_i \cdot \Delta S_{\varepsilon i}$ or $\sigma_j \cdot \Delta S_{\eta j}$ in its own discontinuity surface, so that, for instance, in a side characterized by a particular value of $\varepsilon$ (or $\eta$), the elementary charges are indicated by means of the index i (or j), where $i = 1, 2, \ldots, i_o$ and $j = 1, 2, \ldots, j_o$. Consequently (8) turns into

$$\sigma_{\varepsilon i,w} = (K_\varepsilon/2\pi) \cdot \left[ \Gamma_{i,w} \frac{\rho_w I}{2\pi} + \sum_{\eta=1}^{\mu} \sum_{j=1}^{j_o} \Gamma_{\varepsilon i,\eta j} \cdot \sigma_{\eta j} \cdot \Delta S_{\eta j} \right] \tag{23}$$

$$\left[ \begin{array}{l} i = 1, 2, \ldots, i_o \\ \varepsilon = 1, 2, \ldots, \mu \\ w = 1, 2, \ldots, w_o \end{array} \right.$$

and (13) becomes

$$V_{h,w} = C_{h,w} \frac{\rho_w I}{2\pi} + \sum_{\varepsilon=1}^{\mu} \sum_{j=1}^{i_o} C_{h,\varepsilon i} \cdot \sigma_{\varepsilon i,w} \cdot \Delta S_{iw} \tag{24}$$

$$(h = 1, 2, \ldots, m)$$

Now, a first attempt at the resolution of the problem represented by equations (23) and (24) and the bounding equations may be carried out by resolving first the linear system (23) with respect to the N unknowns $\sigma_{\varepsilon iw}$ and substituting in (24) the nonlinear expressions of $\sigma_{\varepsilon iw}$ as functions of the $K_\varepsilon$ so obtained. Thus,

$$\sigma_{\varepsilon i,w} = f_{\varepsilon i} (\rho_w, K_1, K_2, \ldots, K_\mu) \tag{24a}$$

or

$$V_{h,w} = \left[ C_{h,w} \frac{\rho_w I}{2\pi} + \sum_{\varepsilon=1}^{\mu} \sum_{j=1}^{j_o} C_{h,\varepsilon i} \cdot f_{\varepsilon i,w}(K_1, \ldots, K_\mu) \right] \tag{25}$$

Expression (25), together with (21), forms a system of $\mu$ equations in the $\mu$ unknowns $K_i$; this is a nonlinear system.

For linearization, as is customary, it is sufficient to consider small variations of the unknowns and of the parameters with respect to a known solution. For instance, let us consider in (23), (24), and (21) the variations $\Delta\sigma_{\varepsilon iw}$, $\Delta K_{\varepsilon w}$, $\Delta V_{hw}$, and let us write

$$\left(\sigma_{\varepsilon i,w} + \Delta\sigma_{\varepsilon i,w}\right) = \left(K_\varepsilon + \Delta K_\varepsilon\right)\left[\Gamma_{\varepsilon i,w}\frac{\rho_w I}{2\pi} + \right.$$

$$\left. + \sum_{\varepsilon=1}^{\mu}\sum_{i=1}^{i_o}\Gamma_{\varepsilon i,\eta j}\cdot\left(\sigma_{\eta j,w} + \Delta\sigma_{\eta j,w}\right)\cdot\Delta S_{\eta j}\right]; \tag{26}$$

$$\left(V_{h,w} + \Delta V_{h,w}\right) = \left[C_{hw}\frac{\rho_w I}{2\pi} + \sum_{\varepsilon=1}^{\mu}\sum_{i=1}^{i_o}C_{h\varepsilon,i}\cdot\left(\sigma_{\varepsilon i,w} + \Delta\sigma_{\varepsilon i,w}\right)\cdot\Delta S_{\varepsilon i}\right] \tag{27}$$

$$f_\lambda\left(K_1, K_2, \ldots K_\mu\right) = 0 \; ; \quad \begin{matrix}(\lambda = 1,2, \ldots, \mu - m)\\ (w = 1,2, \ldots, w_o)\\ (h = 1,2, \ldots, n)\end{matrix} \tag{28}$$

These equations may be transformed into the following:

$$\Delta\sigma_{\varepsilon i,w} = \frac{\Delta K_\varepsilon}{2\pi}\left[\Gamma_{\varepsilon iw}\frac{\rho_w}{2\pi} + \sum_{\eta=1}^{\mu}\sum_{j=1}^{j_o}\Gamma_{\varepsilon i,\eta j}\cdot\sigma_{\eta jw}\Delta S_{\eta j}\right] +$$

$$+ \frac{K_\varepsilon}{2\pi}\left[\sum_{\eta=1}^{\mu}\sum_{j=1}^{j_o}\Gamma_{\varepsilon i,\eta j}\Delta\sigma_{\eta jw}\Delta S_{\eta j}\right] \tag{29}$$

or more synthetically,

$$\Delta\sigma_{\varepsilon i, w} = A_{\varepsilon i, w}\,\Delta K_\varepsilon + \Delta B_{\varepsilon i, w}\,K_\varepsilon \tag{29a}$$

Moreover,

$$\Delta V_{hw} = \sum_{\varepsilon=1}^{\mu}\sum_{i=1}^{i_0} C_{h\varepsilon i}\cdot\Delta\sigma_{\varepsilon i, w}\cdot\Delta S_{\varepsilon i} \tag{30}$$

$$(\partial f_\lambda/\partial K_1)\Delta K_1 + (\partial f_\lambda/\partial K_2)\Delta K_2 + \ldots + (\partial f_\lambda/\partial K_\mu)\Delta K_\mu = 0 \tag{31}$$

The following procedure may be suggested. In a first approximation instead of solving the system with respect to $\Delta\sigma_{\varepsilon i}$, let us set $\Delta\sigma_{\eta j} = 0$ in the second term of (29) or (29a); then let us substitute (29a) into (30) with the aim of solving the linear system formed by (30) and (31) with respect to the unknowns $\Delta K_\varepsilon$. We now have the first approximation values of $\Delta\sigma_{\varepsilon j}$ and $(\Delta K_\varepsilon)_1$. These may be substituted in the second term of (29), and the operation may be repeated, reaching the second approximation value $(\Delta K_\varepsilon)_2$. By iteration, it is possible to reach the approximation $(\Delta K_\varepsilon)_i$.

A demonstration of the convergence of this procedure is not difficult. As has been pointed out in the previous sections, the resistivity values can be obtained since we know not only the values $\Delta K$, but also the $\rho_w$ relating to the surface parallelepipeds surrounding the current poles, which may be measured directly.

As far as the balance between unknowns and data is concerned, let us remember that the volume was divided into $m = pqr$ parallelepipeds with unknown but distinct resistivity values. On the other hand the number of areas in each horizontal plane, and also on the ground surface, is $n = pq$, so that for each couple of current electrodes, and consequently for each distinct geoelectric field, we can obtain only n useful distinct potential values. Now, since $m > n$, it follows that the problem is not defined with the use of only one geoelectric field, and that at least $w_0 = r$ different fields (current electrode positions) are needed.

## 7. The Procedure When No Initial Values are Available

The previous considerations are based on the knowledge of some initial, sufficiently close, approximate values. Frequently, however, these are not available. But the described procedure may also be utilized in such cases. When no initial values of K are available, they may be assumed to have the value zero, i.e., the underground half-space is assumed to be homogeneous at the start of the successive approximation procedure. The initial secondary component of $V_{hw}$ in (27) and (30) is also assumed to be zero. However, the measured potentials, which are the known terms of the equations, contain a secondary component different from zero, which may be very large, in complete disagreement with the assumption of a homogeneous space. In this case convergence is not certain.

A solution may be suggested based on the following steps which involve a procedure of approximations to the true solution:

(a) The measured data, i.e., the potential values $V_h$, may be changed in order to obtain good convergence; later the original values are restored. In particular, the mean value

$$\overline{V} = \left( \sum_{h=1}^{m} V_h \right) \Big/ m \qquad (32)$$

and the difference

$$(DV_h) = (V_h - \overline{V})$$

may be considered.

(b) A positive number $\beta < 1$ may be chosen (for instance, $\beta = 0.9$) to set up the series

$$\beta^t (\Delta V_h),\ \beta^{t-1}(\Delta V_h),\ \ldots,\ \beta(\Delta V_h),\ (\Delta V_h)$$

in which the last terms are close to the actual value $\Delta V_h$, and the first few terms very small, depending on the used value of t. Based on this series, a number of potential values may be written as follows:

$$(\Delta V_h)_t = \overline{V} + (V_h - \overline{V})\beta^t \qquad (33)$$

$$(t = 1, 2, \ldots, t_0)$$

For high values of t, these values refer to quasi-homogeneous media, correlated to the mean value of formula (36a), and characterized by very slight inhomogeneities. Low values of t refer to electrical media closer in character to the actual one.

(c) Assuming $t_o$ to be the maximum value of t, we represent by means of (33) potential values having very little variation with respect to the main value and referable to a quasi-homogeneous medium. Consequently, if we start with a known (zero approximation) value of $\sigma_{\varepsilon iw} = 0$ and $K_\varepsilon = 0$, the procedure described in Section 6 may be applied, with a good convergence rate. We can then repeat the calculations, assuming $(\Delta K_\varepsilon)_t$ and $(\sigma_{\varepsilon iw})_t$ obtained by means of $t = t_o$ as initial conditions, and the potentials $(\Delta V_h)_{(t_o-1)}$ may be computed for $t = t_o - 1$ as index of the known terms. The calculations are repeated, assuming in the ith approximation as initial conditions

$$(\Delta K_\varepsilon)_{t_o}, (\Delta K_\varepsilon)_{t_o-1}, \ldots, (\Delta K_\varepsilon)_{t_o-i}$$

and as known terms

$$(\Delta V_h)_{t_o}, (\Delta V_h)_{t_o-1}, \ldots, (\Delta V_h)_{t_o-i-1}$$

until the solution is reached.

## 8. Considerations of Practical Interest

In the assumed scheme the part of the underground half-space involved in the inversion procedure is, as we know, a parallelepiped $\Omega$, and the resistivity values inside it are the unknowns. But the resistivity values of the space outside $\Omega$ are also unknown, without any possibility of knowing them. Thus, ignorance must necessarily influence our knowledge of the inside of $\Omega$, and in particular, of those of its parts which are near to the bounding surfaces. An intuitive consequence of this fact is that the area effectively explored must extend well beyond a central zone for which the best results are required. Here the interpretation is more reliable, while for the bordering area it is at best doubtful.

Another difficulty to be overcome lies in the great number of data and parameters to be handled. It may be helpful to carry out a first approximation procedure, subdividing $\Omega$ into very large parts, so that a relatively small number of unknowns and parameters is present in this first attempt. In a second step, each of these large volumes in turn is subdivided into the smaller parallelepipeds discussed in this paper and interpreted separately. During this last operation only one part is considered in its fine structure, while the others remain undivided. This can be done since the fine structure of each major subdivision of $\Omega$ is not influenced very much by the fine structure of the others. Some successive approximations of this type may bring us to the solution.

Geoelectrical inhomogeneities very near the ground surface may cause some difficulties in the application of the direct inversion procedure. Measured potential values, influenced by these geological features, may be incompatible with the assumed, often simplified scheme of the underground space. As already noted, such cases may give rise to results without physical meaning, such as, for instance, negative resistivities. The goal of validating the output of the inversion may be reached by including in the scheme close to the surface some very small volumes in positions near the current and particularly the potential electrodes. The presence of these volumes may justify possible irregular characteristics of the measured data, since they simulate the behavior of the true inhomogeneities. Thus, good or at least acceptable interpretive results may be obtained.

These and other problems may be discussed from a theoretical point of view; but only by means of a sufficient amount of experimentation on synthetic and on true field cases will the method of direct inversion become effective in the practice of the interpretation of D.C. geoelectric data.

REFERENCES

Alfano, L., 1959, Introduction to the Interpretation of Resistivity Measurements for Complicated Structural Conditions. Geophysical Prospecting, 311, 366.

Alfano, L., 1962, Sulla Interpretazione dei Dati Geoelettrici in Presenza di Strutture di Firma Qualunque. Quaderni di Geofisica Applicata, XXIII, 3, 25.

# METHODS FOR GEOELECTRICAL DATA INVERSION

Marcello Bernabini

Istituto di Geofisica Mineraria

Università di Roma

## INTRODUCTION

Professor Alfano has faced the wide problem of interpretation of resistivity measurements in the presence of tridimentional distribution. This is probably the line of research that will produce remarkable progress in the application of resistivity measurements. However, we will only carry out a schematic examination of the methods which are now employed in the automatic interpretation of vertical electrical soundings. A description, even approximate, of the various methods adopted by different authors would have required even more space than that at our disposal.

Fortunately at the end of 1979 Professor Koefoed published his last book, "resistivity sounding measurements", essentially dedicated to the interpretation of vertical electrical soundings. In this book the methods proposed by the various authors are described and many programs for various calculators are reported. Our task has therefore been much simplified.

We refer to this book for details on the various methods that we will point out and for the notations used.

In order to simplify the treatment, we will always consider the Schlumberger arrangement unless a different arrangement is cleary specified.

The potential on the surface of a layered horizontal soil is defined by

$$V = \frac{\rho_1 I}{2\pi} \int_0^\infty K(\lambda)\, J_0(\lambda r)\, d\lambda$$

and the apparent resistivity for the Schlumberger arrangement is given by

$$\rho_a = \rho_1 S^2 \int_0^\infty K(\lambda)\, J_1(\lambda S) \lambda\, d\lambda$$

with S equal to half the distance between the current electrodes AB.

Note that the function K ($\lambda$) called Kernel function introduced by Slichter (1933) differs from the Kernel function $\theta(\lambda)$ introduced by Stefanescu (1930). The relationship between the two functions is given by

$$K(\lambda) = 1 + 2\,\theta(\lambda)$$

It is common knowledge that the Kernel function is determined by imposing boundary conditions once the parameters, thickness and resistivity, of each layer have been chosen.

Two main methods have been proposed to obtain the expressions of the Kernel function: the Flathe and the Pekeris methods. The Kernel function can be obtained, according to Flathe (1955) with the recurrence relations.

$$K_{i+1} = \frac{N_{i+1}}{D_{i+1}} = \frac{N_i(u) + k_i u_i^2 N_i(1/u)}{D_i(u) - k_i u_i^2 D_i(1/u)}$$

where

$$u_i = e^{-\lambda h_i} \qquad k_i = \frac{\rho_{i+1} - \rho_i}{\rho_{i+1} + \rho_i} \qquad h_i = \text{depth of the boundary.}$$

From these relations we can start from the first layer and sucesively add the lower layers. The Pekeris (1940) recurrence relation is given by

$$K_i = \frac{K_{i+1} + P_i \tanh(\lambda t_i)}{P_i + K_{i+1} \tanh(\lambda t_i)}$$

where

$t_i = h_i - h_{i-1}$ = thickness of the layer

$P_i = \rho_i/\rho_{i+1}$ = ratio between the resistivities

To calculate the Kernel function by such a method, we start from the last layer n considered isolated and we add the higher layers (n - 1) up to the first layer.

We are now going to consider the sequence

$$(h, \rho) \rightarrow K(\lambda) \rightarrow \rho_a(S)$$

This is the process to obtain the apparent resistivity curve when the layers parameters are known.

The inversion of the electrical vertical soundings follows the inverse proceeding.

$$\rho_a(S) \rightarrow K(\lambda) \rightarrow (\rho, h)$$

The first step is possible by applying the Hankel theorem

$$\int_0^\infty \int_0^\infty f(x)\, J_\nu(yx)\, J_\nu(yu)\, x\, y\, dx\, dy = f(u)$$

By integrating the Stefanescu integral and applying such a theorem we obtain

$$K(\lambda) = \frac{1}{\rho_1} \int_0^\infty \rho_a(S)\, J_1(\lambda S)\, \frac{1}{S}\, dS$$

From this $K(\lambda)$ function, resistivity and thickness of the various layers can be obtained by applying, for instance, the Pekeris relation which can be inverted to give

$$K_{i+1} = P_i \frac{K_i - \tanh(\lambda t_i)}{1 - K_i \tanh(\lambda t_i)}$$

Such a relation means that the effects of the shallowest layer can be removed from the Kernel function. This can be repeated for each successive layer until we reach the substratum.

## LINEAR FILTERING

Up until 1970 the calculation of the two integrals to obtain the apparent resistivity from the Kernel function and viceversa, was long and complex.

Let us consider the exponential forms

$$S = e^{x} \quad \text{and} \quad \lambda = e^{-y}$$

The Stefanescu integral becomes

$$\rho_a(x) = \int_{-\infty}^{+\infty} K(y)\, J_1(e^{-(x-y)})\, e^{2(x-y)}\, dy$$

which can be considered as convolution integrals between the two function K(y) and $J_1(y)\, e^{2y}$.

Also, the inverse integral

$$K(\lambda) = \frac{1}{\rho_1} \int_{o}^{\infty} \rho_a(S)\, J_1(\lambda S)\, \frac{1}{S}\, dS$$

can be transformed, by putting

$$S = e^{x} \quad \text{and} \quad \lambda = e^{-y}$$

we then have

$$K(y) = \frac{1}{\rho_1} \int_{-\infty}^{+\infty} \rho_a(x)\, J_1(e^{-(y-x)})\, dx$$

which is the convolution integral between the two functions $\rho_a(x)$ and $J_1(x)$.

This form due to Kunetz (1966) and Ghosh (1971), has made the calculations of the two integrals much easier, because the linear filters theory could be applied to the interpretation of the electrical soundings.

The first consideration due to Ghosh is that:if we apply the Fourier transform to Kernel or apparent resistivity curves expressed in logarithmic scale, the amplitude spectrum approaches zero asymptotically as the frequency increases. We can say that the harmonic components with wave length smaller than half decade (if we use the decimal logarithmic scale) are practically negligible.

If we go into the frequency domain through the Fourier transform, we obtain the spectra of the filtered function (output) by the product of the amplitude spectra of the input function and of the filter, and by the addition of the phase spectra.

Let us now consider the Bessel function of the first order $J_1$. It can be written in asymptotic form

$$J_1(\lambda) = \sqrt{\frac{2}{\pi\lambda}} \cos\left(\lambda - \frac{3\pi}{4}\right)$$

that in logarithmic argument becomes:

$$J_1(-y) = \sqrt{\frac{2}{\pi e^{-y}}} \cos\left(e^{-y} - \frac{3\pi}{4}\right)$$

We can also see from figure 1 the frequencies approach infinity as x increases.

Since we have to multiply two amplitude spectra of which the first is limited, all the high frequencies of the Bessel function are truncated. Therefore, we can eliminate all frequencies higher than a given frequency from the Bessel function $J_1$ in the second integral and from the function $(J_1 . e^{2x})$ in the first integral, before making the convolution. Obviously the cut-off frequency must be higher than the highest frequencies contained in the apparent resistivity curve, otherwise the filtered curve obtained would be distorted.

The cut-off operation has been done by Ghosh in the frequency domain for both functions $J_1$ and $J_1 e^{2x}$; for the first function it can also be made in the distance domain by the convolution of the

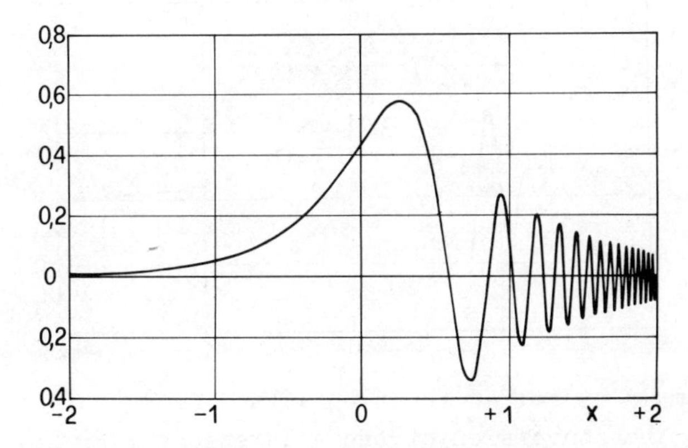


Fig. 1. Bessel function of first order of logarithmic argument

$J_1$ with the sinc function of proper period. It must be pointed out that the cut-off frequency must be within the highest frequency of the apparent resistivity curve and a frequency equal to half the sampling frequency.

The filtered function, thus obtained, approaches zero on both sides with an oscillation of frequancy equal to the cut-off frequency (fig. 2). Therefore such a function constitutes an operator of finite length.

From the given filtered functions it is possible to obtain digital operators with a small number of elements.

The smallest number of elements, that can be considered, is evidently that which is obtained by a sampling frequency equal to twice the cut-off frequency: moreover we should choose the position of the samples in order that they coincide with the zero points of the oscillations on the right side of the filtered function.

In this way, Ghosh obtained 9 elements filters with a sampling frequency of three per decade. Other filters, with a higher number of samples, have been obtained by various authors either for the conversion of the Kernel function in the apparent resistivity or for the inverse conversion. The coefficients of such filters can be found in the bibliography.

To end this very short reminder as to the possibility of application of linear filters, it must be noted that such a possibility allowed other methods used in the deconvolution process, for example, in reflection seismic to be applied in filter calculations.

Koefoed and Dirks (1979) have, for example, consider two methods. In the first one we suppose the input function and the corresponding "output" function already known and we wish to determine

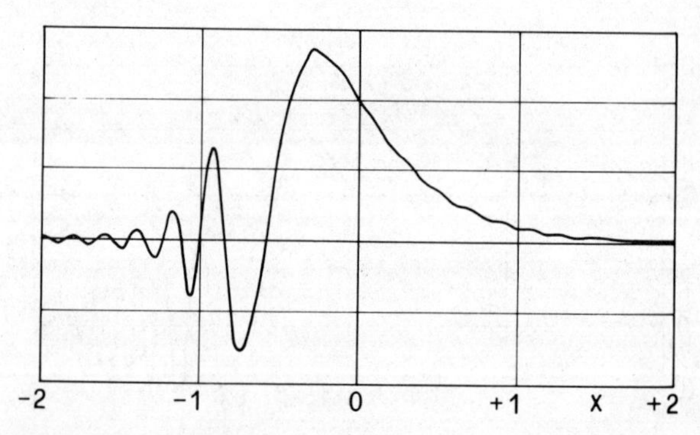


Fig. 2. Example of filter function with 5 Hz cut-off frequency for the conversion of the apparent resistivity function in the Kernel function.

the filter operator of a given length. The operator must be such that, when the input function is introduced into the filter, we obtain an out-put function which is the most approximate to the desired output function. The criterion adopted is that of least squares.

In the other method we consider the Z transform of the output and input functions and we obtain a ratio between the two polynomials. The quotient so obtained is the Z transform of the filter. The problem, in this case, is in the convergence to zero of the remainder.

Coming back to the utilization of the linear filters, whichever method is used to obtain the filter coefficient, the conversion of the Kernel function in the apparent resistivity and the inverse conversion can be done in digital form by the equations.

$$\rho_{a(J)} = \rho_1 \sum_{i=-\infty}^{+\infty} K_i \, f_{J-i} \quad \text{and} \quad K_J = \frac{1}{\rho_1} \sum_{i=-\infty}^{+\infty} \rho_{a(i)} \, f^{-1}_{J-i}$$

It is also well known that if the length of the filter is q and if we want an out-put of m values, the length of the input function will have to be m+q-1 and therefore for the function that has to be transformed m+q-1 values will have to be calculated, measured or extrapolated.

Therefore to obtain the above mentioned conversions, the addition of q products will have to be carried out, for each of the m values , an operation that even a pocket calculator can make in a very short time.

## INTERPRETATION METHODS

Coming back now to the methods of interpretation, we can point out that in the sequence

$$(h,\rho\ ) \rightarrow K\ (\lambda) \rightarrow \rho_a$$

the second step that 10 years ago needed large computers and long computing time, has now, by applying the linear filters theory, become simpler and shorter than the first one.

Similarly in the inverse process

$$\rho_a \rightarrow K\ (\lambda) \rightarrow (h,\rho\ )$$

the first step is also very easy.

Let us now consider the methods utilized for the inversion of the resistivity data. The methods, theoretically more rigorous and elegant, are those that follow the last indicated process which allows the direct conversion of the apparent resistivity to the thicknesses and resistivities of the layers.

We are not considering here the Kunetz-Rocroi method, which was published in 1970 when the filters theory had not yet been applied, this method is much more complicated and leads to results not easily interpretable from the geological point of view.

As we have already said before, the Koefoed's method, theoretically easier, applies, on the first step, the linear filters hence leading to the Kernel function or the socalled resistivity transform of the first layer:

$$T_1 = K_1 \rho_1$$

Note that the method of the digital linear filters can be applied when the sampling interval is constant. In the more general case the method proposed by Patella (1975) can be followed; with such a method we have:

$$K(\lambda) = 1 + \frac{1}{\rho_1} \sum_{J=1}^{K} \bar{\rho}_a (r_J, r_{J+1})\, J^{+}(\lambda r_J)$$

in which

$$\bar{\rho}_a (r, r_{J+1}) = \frac{1}{2} \left[ \rho_a (r_{J+1}) + \rho_a (r_J) \right]$$

is the arithmetic mean of two consecutive values of apparent resistivity and $J^{+}(\lambda r_J)$ is given by

$$J^{+}(\lambda r_J) = \int_{r_J}^{\infty} J_1 (\lambda S) \frac{1}{S} dS$$

a function which is known.

With such a method we have different coefficients for each value of $\lambda$.

With the above mentioned methods it is therefore possible to obtain either the Kernel function or the resistivity transform.

Koefoed (1979) proposed to obtain from the latter, the parameters ρ and h of the layers through the Pekeris formula which takes the form

$$T_{i+1} = \frac{T_i - \rho_i \tanh(\lambda t_i)}{1 - T_i \tanh(\lambda t_i)/\rho_i}$$

The process is the following:

1) we determine the values of $\rho_1$ and $h_1$ of the first layer;
2) we introduce those values in the previous equation and we obtain the function $T_2(\lambda)$ concerning the second layer;
3) from function $T_2$ we obtain resistivity and thickness of the second layer and so on.

In the first expression the method was not automatic; the values of resistivity and the thicknesses were obtained with two layers standard curves and the application of the equation (25) was done with graphic methods.

Later Koefoed (1976-1979) proposed an automatic method which also considers the confidence limits of the observed values and the magnification of the limits during the process. The method is as follows

1) The resistivity value $\rho_1$ is obtained from the first value of the apparent resistivity introduced. We must therefore extrapolate the curve towards the left hand side until we obtain the asymptotic value within the confidence limits.
2) The program controls, proceeding from the right hand towards the left, which values are equals to $\rho_1$ within the confidence limits.
3) On the following points we calculate the modified Kernel function given by

$$G_i(\lambda) = \frac{T_i(\lambda) - \rho_i}{T_i(\lambda) + \rho_i}$$

In the case of two layers such a function is expressed by

$$G_i = k_i\, e^{-2\lambda t_i} \qquad k_i = \frac{\rho_{i+1} - \rho_i}{\rho_{i+1} + \rho_i}$$

which in logaritmic form becomes

$$\ln G_i = \ln k_i - 2 \lambda t_i$$

Putting in diagram the logaritm of G in function of $\lambda$, we obtain a straight line from which we get $k_i$, and then $\rho_2$, and $t_i$. These values are determined by interpolation with the least squares method, taking into consideration the points following the first one until no interpolation within the confidence limits is possible.

4) We calculate the new resistivity transform $T_2$ with the Pekeris formula and we apply the same sequence assuming the resistivity value $\rho_2$ previously determined. It is possible to introduce in the program the values of thickness or resistivity obtained from geological information.

A different method has been proposed by Szaraniec ( 1979). We consider the Kernel function obtained from the measured values of apparent resistivity and we try to follow such a function with a theoretical Kernel calculated by determining, step by step, the thickness and resistivity of the layers beneath.

We suppose the Kernel experimental function and the theoretical Kernel $K_i$, calculated on a model of i layers, coinciding until the point $\lambda_j$. After the point $\lambda_j$ the two Kernels diverge.

We now consider the Flathe recurrence relations

$$K_{i+1} = \frac{N_{i+1}}{D_{i+1}} = \frac{N_i(u) + k\, u_i^2\, N_i\,(1/u)}{D_i(u) - k\, u_i^2\, N_i\,(1/u)}$$

Solving such an equation for $k_i\, u_i^2$ we obtain

$$k_i\, u_i^2 = \frac{K_{i+1}\, D_i(u) - N_i(u)}{K_{i+1}\, D_i\,(1/u) + N_i(1/u)}$$

The values of $D_i$ and $N_i$ are known as the thicknesses and resistivities of the first i layers have been determined. The values of $K_{i+1}$ are assumed from the experimental Kernel.

The unknown quantity $k_i$ and $u_i$ can be determined by considering a pair of subsequent points $\lambda_i$ and $\lambda_{i+1}$ and applying the previous relation expressed in logaritmic form. We again apply the previous relation to the pair $\lambda_{j+1}$ and $\lambda_{j+2}$ and so on until the values of $h_i$ and $k_i$ are obtained which deviate from the previous ones. The values of $h_i$ and $k_i$ are assumed to be the arithmetic mean of the previous values. We then consider the layer (i+1) in the same way and so on.

Let us now consider another approach to the inversion problem: the indirect automatic methods.

As in the old graphic methods, we consider a model, we calculate the theoretical curve, we compare such a curve with the experimental one; if it fits well, the model is assumed; otherwise we change the model and so on. Various possibilities have been considered in the automatic methods:

1) the comparison can be done either on the apparent resistivity or on the Kernel function.

   In the first case the scheme is the following:

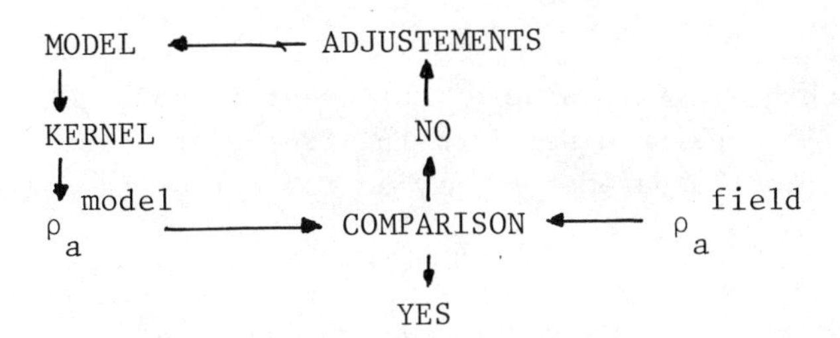


We may see that for each step we must calculate the theorical apparent resistivity.

In the second case we have

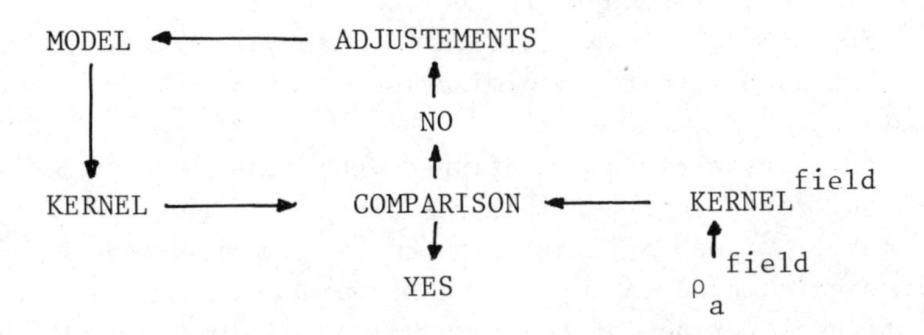


We must obtain the experimental Kernel from the measured $\rho_a$ values, and for each step only the theorical Kernel must be calculated.

2) the kind of comparison, or the error criterion, may be based either on the mean square deviation between the theoretical and

experimental values of $\rho_a$ or of K or on the mean square deviation between the logarithms of the same values (relative deviations).
3) the system utilized to vary the model parameters.

Among the various systems proposed, the most rigorous are the steepest descent method applied by Vozoff (1958), Bichara and Lackshamanan (1976) and Koefoed (1979) and the matrix equations method applied by Meinhardus (1970), Inman (1975), Johansen (1977) and Cecchi, Marchisio and Montana (1979).

Let us now consider the steepest descent method.
Firstly , values of the layers parameters, h and $\rho$, are chosen on the basis of a preliminary interpretation and we calculate a first value of the error and of the derivative of the error with respect to the parameters.

We consider

$$E = \Sigma \left[ Q^m - Q^f \right]^2$$

in which $Q^m$ is the value of the quantity obtained from the model and $Q^f$ the one obtained from the field data; the variations of the error with the parameters may be taken from the previous one and we have

$$\frac{\partial E}{\partial P_K} = 2 \Sigma \left[ Q^m - Q^f \right] \frac{\partial Q^m}{\partial P_K}$$

On the basis of such derivatives it is possible to obtain parameter variations such as to reduce the error.

The subsequent application of such variations, conveniently modified, leads to the minimization of error and therefore to the solution.

As we have already mentioned, the computing criteria of the error differs. Vozoff (1958) has considered the differences between the Kernel values arising from the field measurements and the Kernel values calculated on the model; Bichara and Lakshmanan (1976) the relative difference of the same values; Koefoed (1979) the relative difference between the apparent resistivity values obtained from the field data and those obtained from the model.

Similar differences are present also in the computing of the parameter changes; Koefoed for instance considers the following equation

$$\Delta P_K = - \frac{E - \varepsilon}{|\text{grad } E|^2} \frac{\partial E}{\partial P_K} P_K$$

in which $P_K$ is the considered parameter (resistivity or thickness) and $\varepsilon$ is a constant.

The derivatives from the Kernel function with respect to the parameters may be obtained from the Pekeris recurrence relations. For the Koefoed's E function in which the apparent resistivities appear, the partial derivative of the apparent resistivities is obtained by applying to the K derivatives the same linear filter used for computing the apparent resistivities.

The process is applied until the error is reduced; when the error increases, the variation of the parameters is reduced by a preassigned factor; the process stops when the error has fallen below a preassigned limiting value, or the variation in the parameters is smaller than a preassigned value, or the number of executed steps is too high.

In the program it is possible to limit the interval within which the parameters may vary.

The second method, more rigorous, is based on the solution of matrix equations.

Also, in this case, we minimize the squares of differences between the values obtained from field and values obtained from a model.

A first model $M_1$ is obtained in first approximation for instance with graphic methods.

Let the values obtained from field data be $y_i$ and those obtained from the model be $x_i$.

We have

$$E = \sum_{i=1}^{n} ( y_i - x_i)^2 = \min$$

Expanding $x_i$ by the Taylor Series around the $M_1$ model values, we obtain

$$E = \sum_{i=1}^{n} \left[ y_i - x_i(M_1) - \sum_{k=1}^{m} \left( \frac{\partial x_i}{\partial P_k} \right)_{M_1} \Delta P_k \right]^2 = \min$$

in which $x_i$ are the values of the quantity considered computed on the basis of the model $M_1$. $P_k$ are the model parameters (resistivity and thickness), and the derivatives of the quantity, considered with respect to the parameter, are computed in point $M_1$.

Minimizing the previous equation, we obtain a set of correction of $P_k$.

We correct the model and therefore obtain a new value of E. The process is repeated until there is no further decrease in E.

By the differentiation of the equation with respect to $\Delta P_J$ and equalizing to zero, we have

$$\sum_K \left[ \Delta P_k \sum_J (D_{J,K} \; D_{J,n}) \right] = \sum_J (e_J D_{J,n}) \qquad (n = 1....m)$$

in which

$$e_J = y_J - x_J \qquad \text{and} \qquad D_{iJ} = \left(\frac{\partial x}{\partial P_J}\right)_i$$

The equations system may be put in the form of matrix equations

$$D^T . D \Delta P = D^T e$$

where D is the matrix of the derivatives $D_{iJ}$ and $D^T$ is the transpose of this matrix.

The solution of the equations system can lead to too high values of $\Delta P$ for which the system does not converge, particularly in the presence of some noise.

To avoid this, Marquardt (1963) has modified the previous equation by putting

$$(D^T . D + \alpha^2 I) \Delta P = D^T e$$

in which I is the unit matrix and $\alpha$ a positive constant. Johansen (1977) considers the eigenvalues of the $D^T D$ matrix and, at first, assumes for $\alpha$ the smallest of such values. When the system does not converge anymore, it is substituted with the next largest eigenvalue until the solution is obtained.

On the contrary, Cecchi, Marchisio and Montana (1979) multiply the value of $\alpha$ by a factor lower than 1 after each step. If the system does not converge, $\alpha$ is increased again.

There are other differences among the processes of the various authors. Meinardus (1970) considers the Kernel function as a quantity to compare , Inman (1975) the apparent resistivity, Johansen (1976) the logarithms of the apparent resistivity values.

As a general criterion of the comparison between the steepest descent method and the matrix equation, we think it is useful to repeat what Inman (1975), Cecchi, Marchisio and Montana (1979) have said. They affirm that generally the steepest descent method converges in shorter time than the second when the starting model is far from the solution, while the second one would be preferable when the model is nearer to the true layering.

The matrix equations method also permits the carrying out of the confidence limits of the obtained parameters.

Inman (1975) and Johansen (1977) showed, for instance, that small eigenvalues indicate equivalence condition between the parameters in the direction of the corresponding eigenvector.

We are not considering here the Marsden (1973) and the Zohdy (1974) methods as they are less rigorous than the previous ones and are not included in the scheme indicated at the beginning of our lecture.

## CONSIDERATIONS ON THE INTERPRETATION METHODS

We recalled that in the spectrum of the theoretical apparent resistivity curves and of the Kernel function frequencies higher than 1.5-2 are not present in a significant way, the abscissa is considered expressed in decimal logarithms.

On the contrary, in the curves obtained in the field higher frequencies are very often present. They represent noises.

Such noises can be of three kinds
1) errors of measurement 2) inhomogeneities in the various layers and particularly in the first layer 3) boundaries not flat and not horizontal.

The noise due to measurements errors can be considered as a stationary random process and therefore as a white noise. Such a signal has a spectrum in which all frequencies are present.

Some unhomogeneities can also be considered as a white noise, some produce distortion in the apparent resistivity curves with a band of limited frequency, either towards the high frequencies or towards the low frequencies. To take an example, we can refer to the italian geological formation indicated with the name of "flysch". In such a formation either small random unhomogeneities or layers and piles of layers more or less conductive than the surrounding medium are present; the apparent resistivity curves determined on such a formation are often so irregular that the interpretation in the case of two layers is also often very difficult and uncertain (see fig. 3).

Inclined or curved boundaries produce, in the curves, distortions that often have frequencies included in the same band of the curves for a stratified horizontal soil.

We measure, in the field, the apparent resistivity at preassigned electrodic distances, that is to say, we sample the apparent resistivity curve with a preassigned sampling frequency.

From the sampling theory, we know that if $f_{max}$ is the highest frequency present in the signal spectrum, to represent the signal completely, we must choose a sampling frequency $f_s$ equal or higher

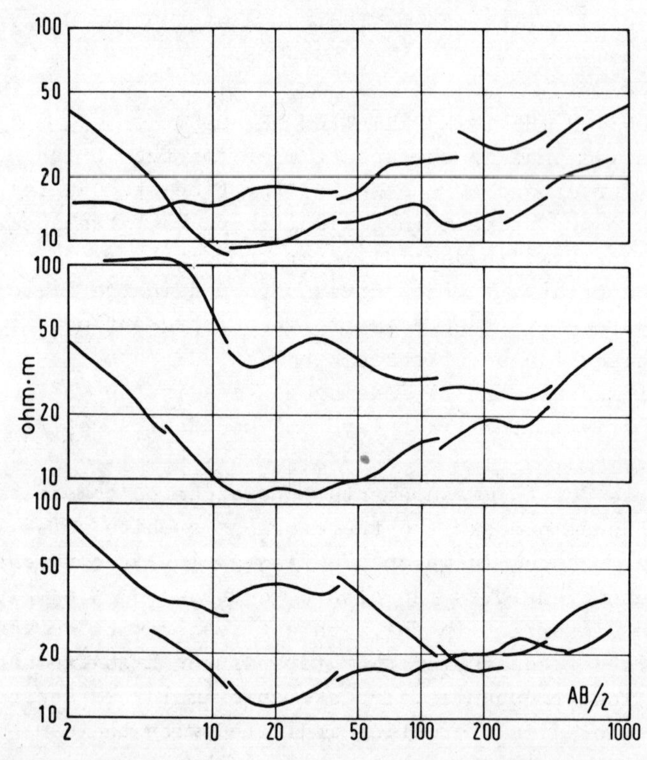


Fig. 3. Some apparent resistivity curves obtained on a flysch area in Latium (Italy).Noises due to inhomogeneities are evident

than $2f_{max}$. If we choose a sampling frequency $f_s$ lower than $2f_{max}$, the frequencies higher than $f_s/2$ will be converted into frequencies lower than $f_s/2$. This is the well known aliasing effect.

Since the spectrum of field curves of apparent resistivity also contain very high frequencies, we should choose an infinite sampling frequency: that is to say make infinite measurements.

For practical reasons we adopt an interval of measurements (and therefore a sampling frequency) compatible from an economical point of view.

But in such a way we do not represent the field curve completely.

Let $f_{max}$ be the maximum frequency of the theorical curve calculated for an horizontal stratified soil that approaches the real soil and $f_s$ be the sampling frequency. The noise frequencies lower or equal to $f_{max}$ cannot be separated from those due to the layers; the frequencies included between $f_{max}$ and $f_s/2$ can still be separated; on the contrary frequencies higher than $f_s/2$, that have been converted by aliasing effect into frequencies lower than $f_s/2$, are partly transformed into frequencies lower than $f_{max}$ and cannot be separated anymore from the frequency due to the layers.

A second consideration is over the equivalence principle. The equivalence among curves with an equal number of layers is well known and the studies of Maillet (1947), Rocroi (1975), Koefoed (1979), Johansen (1977) have indicated the validity areas of the equivalence according to the measurements accuracy and to the noise.

In our opinion, the equivalence among curves obtained with models having a different number of layers is much more important. It is well known that it is possible to fit a two layers curve with a five or ten layers curve.

While in the first case it may not be essential to determine the resistivity and therefore the thickness of a layer, the change in the kind of model (higher or lower number of the layers) can lead to solutions which are not acceptable from the geological point of view.

Let us now analyze the two groups of methods that have been recalled before. With the direct method, first the Kernel is obtained and then from the Kernel the parameters of the layers.

In the first step, because of the aliasing effect, the use of the filter produces some noises that are difficult to be distinguished from the signal. This occurs particularly when using filters sampled with a frequency close to the highest frequency of the signal.

As an example, in figure 4 A we see a three layers theoretical curve in which three different noises have been added. If we utilize the Ghosh filter with nine coefficients with a sampling of 3 points for decade, we obtain Kernel curves (fig. 4 B) which are very distorted. Utilizing a filter with a higher number of values (fig. 4 C), the aliasing effect is lower.

Another element to be considered is the extrapolation of the field curve on both sides, necessary to apply the filter, as we have already said. The error influence in the extrapolation is not easy to individualize in the obtained Kernel function.

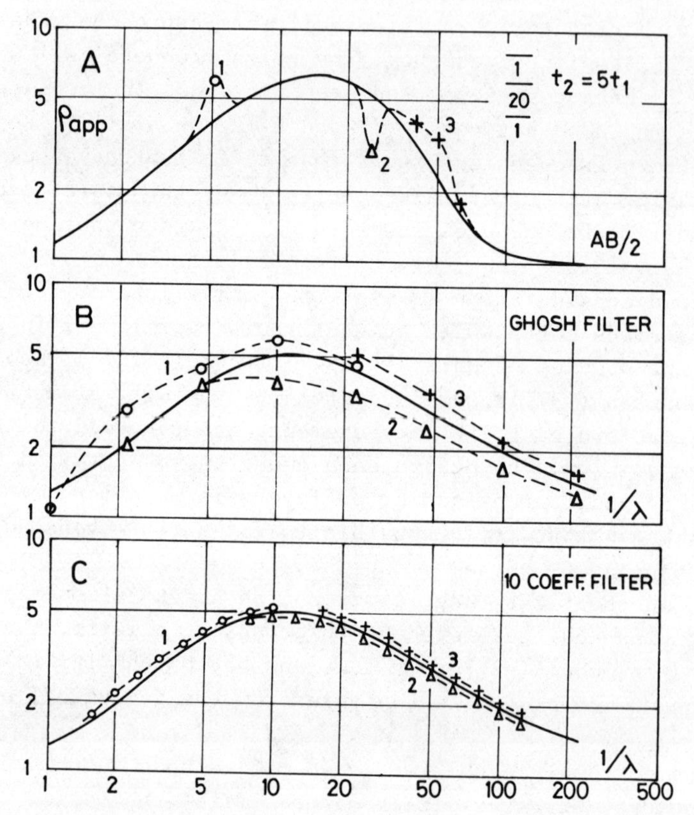


Fig. 4. Effect of two different filters in the conversion of a apparent resistivity curve in Kernel curves.

A) A three layers theoretical curves with three different noises

B) Kernel curves obtained with the 9 element Ghosh filter

C) Kernel curves obtained with a 41 element filter with the same cut-off frequencies.

In the next step, from the Kernel to the parameters, the noises introduced lead to incorrect solutions and in a much wider way as number of the layers is increased. In fact in the elaboration of the Kernel function the errors increase as long as the elaboration carries on. Moreover, it is difficult to impose a model and it is therefore likely to obtain solutions unacceptable from the geological poin of view.

The indirect method is undoubtly heavier as it needs a series of very long computing cycles. On the other hand it has the great advantage of allowing an easy introduction of the model type and of the values of some parameters which may be known.

We have seen that the convergence check, that is to say the error criterion, can be carried out either on the values of the Kernel function or on the values of the apparent resistivity.

In the first instance each cycle is shorter because the step from the Kernel to the apparent resistivity is missing. Then we make the comparison on the transformed values: the considerations previously made on such a transform are valid in this case too.

If we make a comparison on the values of resistivity, we consider the values measured without filtering or extrapolation.

The computing time are in this case longer but not so much, thanks to the applications of linear filters.

Moreover we believe that it would always be better to consider the relative values either of the apparent resistivities or of the parameters, that is to say to work always on a logarithmic scale.

Finally, we would like to add that, generally, any method works well on theoretical curves or on curves obtained on ideal areas such as those that can be found in Northern Europe.

In areas where the noise level is very high, the continuous intervention of man is necessary and therefore the methods that allow such an intervention are to be preferable.

This intervention will be based on the geological information that will show which model is the best one, on the evaluation and identification of the noises which will allow some anomalous values to vary and finally on the comparison among near electric soundings.

In fact the analysis of all data obtained in an area allows us to eliminate some indeterminations and will show the most probable solution.

We believe that the interpretation of single electrical sounding could in future be refined, shorter and simpler, but it has now reached such limits that do not allow a decisive improvement.

Such an improvement will be possible when we will be able to determine which models and which values of the parameters of some layers are compatible, within certain confidence limits, for all a series of electrical soundings.

This will be perhaps one of the main lines of research on the resistivity method.

## REFERENCES

Bernabini, M.and Cardarelli, E., 1978. The use of filtered Bessel functions in direct interpretation of geoelectrical soundings, Geophys. Prospect., 26: 841-852.

Bichara, M. and Lakshmanan, G., 1976. Fast automatic processing of resistivity soundings, Geophys. Prospect., 24: 354-370.

Flathe, H., 1955. A practical method of calculating geoelectrical model graphs for horizontally stratified media, Geophys. Prospect., 3: 268-294.

Ghosh, D.P., 1971. The application of linear filter theory to the direct interpretation of geoelectrical resistivity sounding measurements, Geophys. Prospect., 19: 192-217.

Inman, J.R., 1975. Resistivity inversion with ridge regression, Geophysics, 40: 798-817.

Johansen, H.K., 1977. A man/computer interpretation system for resistivity soundings over a horizontally stratified earth. Geophys. Prospect., 25: 667-691.

Koefoed, O., 1976. Progress in the direct interpretation of resistivity soundings: an algorithm. Geophys. Prospect., 24: 233-240.

Koefoed, O., 1979. Resistivity sounding measurements, Elsevier, Amsterdam.

Koefoed, O. and Dirks, F.J.H., 1979. Determination of resistivity sounding filters by the Wiener Hopf least square method. Geophys. Prospect., 27: 245-250.

Kunetz, G., 1966. Principles of Direct Current Resistivity Prospecting . Borntraeger, Berlin.

Kunetz, G. and Rocroi, J.P., 1970. Traitement automatique des sondages électriques. Geophys. Prospect., 18: 157-198

Marquardt, D.W., 1963. An algorithm for least-squares estimation of nonlinear parameters. J.Soc.Ind.Appl.Math., 11: 431-441.

Marsden, D., 1973. The automatic fitting of resistivity sounding by a geometrical progression of depths. Geophys. Prospect., 21: 266-280.

Meinardus, H.A., 1970. Numerical interpretation of resistivity soundings over horizontal beds. Geophys. Prospect., 18: 415-433.

Patella, D., 1975. A numerical computation procedure for the direct interpretation of geoelectrical soundings. Geophys. Prospect., 23: 335-362.

Pekeris, C.L., 1940. Direct method of interpretation in resistivity prospecting. Geophysics, 5: 31-46.

Slichter, L.B., 1933. The interpretation of resistivity prospecting method for horizontal structures. Physics, 4:307-322.

Stefanescu, S.S. and Schlumberger, C. and M., 1930. Sur la distribution électrique potentielle autour d'une prise de terre ponctuelle dans un terrain à couches horizontales, homogènes et isotropes. J. Phys.Radium, 7:132-141.

Szaraniec, E., 1980. Direct resistivity interpretation by accumulation of layers. Geophys. Prospect., 28:257-268.

Vozoff, K., 1958. Numerical resistivity analysis: horizontal layers. Geophysics, 23:536-566.

# A GENERAL TRANSFORMATION SYSTEM OF DIPOLE GEOELECTRICAL SOUNDING INTO SCHLUMBERGER'S AS AN APPROACH TO THE INVERSION

Domenico Patella

Istituto di Geodesia e Geofisica
Università degli Studi
70100 Bari, Italy

## ABSTRACT

A synthesis of recent studies on the transformation of electrical dipole sounding curves into equivalent Schlumberger curves is given. The transformation is here considered as a powerful mean for attempting quantitative interpretation of generally noise degraded field dipole curves, obtained over arbitrary underground structures.

## INTRODUCTION

The ever increasing demand for geophysical exploration of the earth's crustal structures for the search of exploitable energy and mineral resources has involved also the dc geoelectrical probing methods. In fact, as is known, the resistivity physical parameter may undergo significant deviations and produce visible anomalies of the surface measures in correspondence with buried mineral targets and fluid filled permeable reservoirs.

Among the various geoelectrical techniques, vertical sounding is the most largely used, since a proper choice of the number and position of the sounding centers can provide very useful informations about both vertical and lateral resistivity variations. As for the field sounding technique, the classical well studied Schlumberger array still remains the preferred one for easy and quick use in shallow and moderately deep investigations. Moreover, in recent years, conspicuous developments have been reached in the domain of quantitative interpretation of Schlumberger apparent resistivity diagrams, and very accurate models of the investigated underground volumes can be drawn from generally poorly noise degraded field sounding curves.

However, in very deep exploration the Schlumberger array suffers heavily from the following two disadvantages: (i) large sounding spreads with consequently large cable length, and (ii) problematic inductive coupling between the cables of the current and measuring circuits.

In order to overcome these strong disadvantages, Alpin (1950) introduced the dipole arrays, the generic one being represented in Fig. 1. It soon appeared a good practical solution, and extensive theoretical studies were performed first on the behaviour of the electric field generated by a dipolar source, and then on the construction of theoretical dipole sounding curves for some simple one-dimensional layered structures and two-dimensional models with plane horizontal and vertical contacts.

In the last decade also in Italy there has been a strong increase of interest in the use of the dipole techniques, with the aim of giving contributions to the solution of geodynamic and geothermal problems. But since the beginning, in the real geological context of the survey areas, the morphology of the sounding curves appeared very often as irregular to be in practice hardly interpretable.

Detailed analyses revealed that, as a rule, the significant information content of highly noise degraded dipole field curves is critically masked by disturbing effects very likely provoked by recurrent situations of uneven topography and mainly by shallow and narrow geological disomogeneities. This was to be expected, since, as from theory, the dipole arrays, owing to their intrinsic "differential" character, are strongly sensitive to any sharp physical discontinuity close to the sounding line. On account of

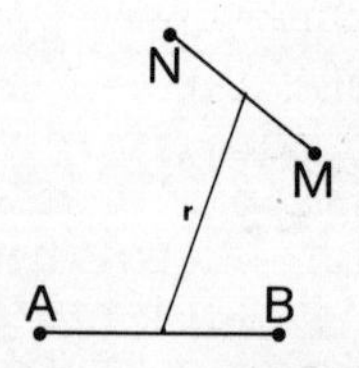


Fig. 1. The generic dipole electrode array.

the above experimental evidence, the following conclusions have been drawn about dipole soundings: (i) they are easy to be put in practice during field surveys, being possible to bypass any sort of obstacles with relatively short dipoles, and (ii) they produce, very frequently, highly noise perturbed apparent resistivity curves, any attempt at quantitative interpretation of which may be exposed to high uncertainty.

A close examination of the above conclusions reveals that these are exactly opposite to those previously drawn about the Schlumberger array. Therefore it appears that a possible way for gaining resolution in deep geoelectrics may be characterized by the following two steps: (i) execution of dipole soundings in the field, and (ii) quantitative interpretation in the Schlumberger domain. This sounding scheme, that joins the main advantage of each array, implies the intermediate passage of the "transformation" of the original field dipole sounding curve into an equivalent Schlumberger curve.

Alfano (1974) and Patella (1974) were the first to introduce the concept of the transformation. Starting both from the above basic ideas, but following different mathematical schemes, they were able to suggest also particular techniques for carrying out the transformation. Moreover, very recently, Alfano (1980) has studied in detail the problem of the transformation in the general case of arbitrary underground structures.

In the following sections we shall give a synthesis of these studies, which may be considered also as a preliminary approach to the problem of the "inversion" of geoelectrical dipole data.

## THE RELATIONSHIP BETWEEN DIPOLE AND SCHLUMBERGER APPARENT RESISTIVITY IN THE CASE OF LAYERED STRUCTURES

Consider at first the half Schlumberger array MNP of Fig. 2.

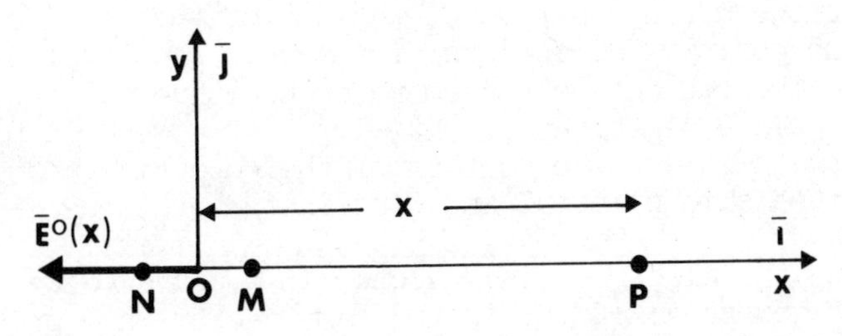


Fig. 2. Schematic representation of the half Schlumberger array.

P is a positive current pole with strength +I; $\overline{MN}$ is the potential dipole, directed along the x-axis of a plane rectangular coordinate system (x,y) placed on the ground surface, with the origin at the center of the potential dipole. It follows:

$$\overline{MN}/|\overline{MN}| = \overline{i},$$

where $\overline{i}$ is the unit vector of the x-axis. OP=x is the variable distance of the current pole from the origin of the system.

The electric field $\overline{E}^o$ in O, provoked by the positive source in P, has always the direction of the x-axis, whatever be the position of P along the same axis, which is assumed to correspond with the sounding expansion line. This property is due to the radial symmetry of layered structures. The modulus of the electric field depends however on the distance x.

Thus, if $\Delta V_{MN}^P$ is the potential difference measured between M and N and provoked by the current pole P, and $\overline{E}^o(x)$ the corresponding electric field evaluated at O, then, under the hypothesis $|\overline{MN}| \ll OP$, we have:

$$\Delta V_{MN}^P/|\overline{MN}| \simeq \overline{E}^o(x) \cdot \overline{i} = E^o(x). \quad (1)$$

By definition, the half Schlumberger apparent resistivity $\varrho_{a,S}(x)$ is given by:

$$\varrho_{a,S}(x) = (2\pi x^2)\, E^o(x)/I. \quad (2)$$

For simplicity and without loss of generality, we assume I constant whatever the position of the current pole P along the x-axis.

Consider now the generic dipole array of Fig. 3. $\overline{MN}$ is the potential dipole the direction of which is defined by the unit vector $\overline{u}$; $\overline{AB}$ is the current dipole the direction of which is defined by the unit vector $\overline{n}$. $\varphi$ and $\vartheta$ are the angles, positive counterclockwise, between the positive directions of the unit vectors $\overline{i}$ and $\overline{u}$ and the positive directions of the unit vectors $\overline{i}$ and $\overline{n}$, respectively. The current pole A emanates positive current +I, while the current pole B negative current -I. C is a generic point running along the n-axis. The centers of the potential and current dipoles are the same points O and P as in the previous Fig. 2.

$\overline{E}^o(A)$ is the electric field vector in O provoked by the positive current source A; $-\overline{E}^o(B)$ is the same vector in O provoked by the negative current source B (the minus sign specifies this).

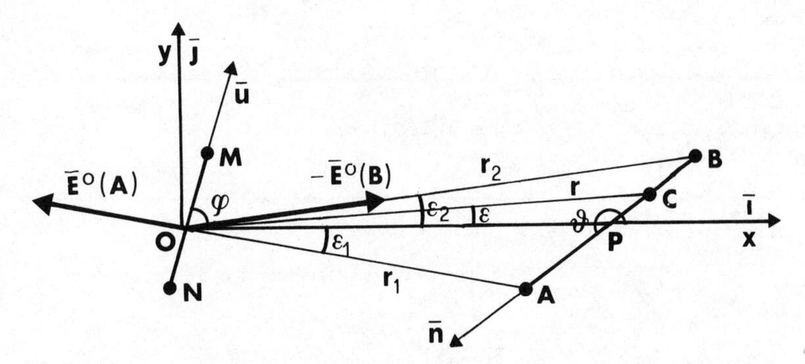


Fig. 3. Schematic representation of a generic dipole array on the surface of a layered structure.

Again, owing to the radial symmetry of the assumed underground structure, the electric fields $\overline{E}^\circ(A)$ and $\overline{E}^\circ(B)$ have always the directions $\overline{OA}$ and $\overline{OB}$, respectively, whatever the position of the current dipole $\overline{AB}$. Their moduli depend on the distances $r_1$ and $r_2$, respectively. In general, for a generic current pole C, distant r from O and running along the n-axis, it is $r=r(n)$ and, consequently, $\overline{E}^\circ(C)=\overline{E}^\circ[r(n)]$. In fact, for layered structures, the modulus of the electric field depends only on the distance r, but never on the position of the source in the (x,y) plane.

From Fig. 3 we have:

$$E_x^\circ(A) = |\overline{E}^\circ(A)| \cos \varepsilon_1 = E^\circ(A) \cos \varepsilon_1 \tag{3a}$$

$$E_y^\circ(A) = |\overline{E}^\circ(A)| \sin \varepsilon_1 = E^\circ(A) \sin \varepsilon_1 \tag{3b}$$

$$E_x^\circ(B) = |\overline{E}^\circ(B)| \cos \varepsilon_2 = E^\circ(B) \cos \varepsilon_2 \tag{3c}$$

$$E_y^\circ(B) = |\overline{E}^\circ(B)| \sin \varepsilon_2 = E^\circ(B) \sin \varepsilon_2 . \tag{3d}$$

Thus, if $\Delta V_{MN}^{AB}$ is the potential difference between M and N, provoked by the current dipole $\overline{AB}$ and it is $MN \ll OP$ and $AB \ll OP$, then with (3a-d) we have:

$$\Delta V_{MN}^{AB}/|\overline{MN}| \simeq [\overline{E}^\circ(A) - \overline{E}^\circ(B)] \cdot \overline{u} =$$

$$\left([E_x^\circ(A)-E_x^\circ(B)]\,\overline{i} + [E_y^\circ(A)-E_y^\circ(B)]\,\overline{j}\right) \cdot \overline{u} =$$

$$\left([E^\circ(A)\cos\varepsilon_1 - E^\circ(B)\cos\varepsilon_2]\cos\varphi + [E^\circ(A)\sin\varepsilon_1 - E^\circ(B)\sin\varepsilon_2]\sin\varphi\right) =$$

$$E^\circ(A)\cos(\varphi - \varepsilon_1) - E^\circ(B)\cos(\varphi - \varepsilon_2) \simeq$$

$$\left.\frac{\delta E^\circ(C)\cos(\varphi - \varepsilon)}{\delta n}\right|_{\substack{C\equiv P\\ \varepsilon=0}} \cdot \left|\overline{AB}\right| =$$

$$\left.\frac{\delta\left(E^\circ\left[r(n)\right]\cos(\varphi - \varepsilon)\right)}{\delta n}\right|_{\substack{r=x\\ \varepsilon=0}} \cdot \left|\overline{AB}\right| =$$

$$\left(\frac{\delta E^\circ(x)}{\delta n}\cos\varphi + E^\circ(x)\left.\frac{\delta\cos(\varphi - \varepsilon)}{\delta n}\right|_{\varepsilon=0}\right)\cdot\left|\overline{AB}\right| =$$

$$\left(\frac{\delta E^\circ(x)}{\delta x}\frac{dx}{dn}\cos\varphi + E^\circ(x)\sin\varphi\left.\frac{d\varepsilon}{dn}\right|_{\varepsilon=0}\right)\cdot\left|\overline{AB}\right|. \qquad (4)$$

Now from Fig. 3 it is derived:

$$\frac{dx}{dn} = \cos\vartheta \quad \text{and} \quad \left.\frac{d\varepsilon}{dn}\right|_{\varepsilon=0} = \frac{\sin\vartheta}{x}. \qquad (5)$$

Thus, with (5), from (4) we obtain:

$$\frac{\Delta V_{MN}^{AB}}{\left|\overline{MN}\right|\cdot\left|\overline{AB}\right|} \simeq \frac{\delta E^\circ(x)}{\delta x}\cos\varphi\cos\vartheta + \frac{E^\circ(x)}{x}\sin\varphi\sin\vartheta \ . \qquad (6)$$

As is known, for a generic dipole array the apparent resistivity $\varrho_{a,D}(x)$ is given by:

$$\varrho_{a,D}(x) = \frac{2\pi x^3}{\left|\overline{AB}\right|\cdot\left|\overline{MN}\right|}(\sin\varphi\sin\vartheta - 2\cos\varphi\cos\vartheta)^{-1}\cdot\frac{\Delta V_{MN}^{AB}}{I}. \qquad (7)$$

With (6), (7) becomes:

$$\varrho_{a,D}(x) = \frac{2\pi x^3}{I(\sin\varphi\sin\vartheta - 2\cos\varphi\cos\vartheta)}\cdot$$

$$\cdot\left(\left[\delta E^\circ(x)/\delta x\right]\cos\varphi\cos\vartheta + \left[E^\circ(x)/x\right]\sin\varphi\sin\vartheta\right). \qquad (8)$$

Eq.(8) explicitely contains, at the right-hand side, the electric field $E^{\circ}(x)$ and its derivative with respect to x, i.e. two quantities which can be measured and deduced, respectively, by the previously described half Schlumberger sounding. In fact, from (2) we obtain:

$$E^{\circ}(x) = I\,\varrho_{a,S}(x)/2\pi x^2 \tag{9}$$

$$\frac{\delta E^{\circ}(x)}{\delta x} = \frac{I}{2\pi x^2}\frac{\delta \varrho_{a,S}(x)}{\delta x} - \frac{I}{\pi x^3}\varrho_{a,S}(x). \tag{10}$$

Substituting (9) and (10) into (8) and putting:

$$\alpha = \frac{2\cos\varphi\cos\vartheta - \sin\varphi\sin\vartheta}{\cos\varphi\cos\vartheta}, \tag{11}$$

we finally get:

$$\varrho_{a,D}(x) = \varrho_{a,S}(x) - \frac{x}{\alpha}\frac{\delta \varrho_{a,S}(x)}{\delta x} \tag{12}$$

which is the required relationship between dipole and half Schlumberger apparent resistivities, when considering layered structures. It is possible to deduce some particular cases as follows:

| | | | | |
|---|---|---|---|---|
| azimuthal | array | $\varphi=270°$ | | $\varrho_{a,az}(x)=\varrho_{a,S}(x)$ |
| equatorial | " | $\varphi=270°$ | $\vartheta=270°$ | $\varrho_{a,eq}(x)=\varrho_{a,S}(x)$ |
| radial | " | $\varphi=0°$ | | $\varrho_{a,r}(x)=\varrho_{a,S}(x)-\frac{x}{2}\frac{\delta\varrho_{a,S}(x)}{\delta x}$ |
| axial | " | $\varphi=0°$ | $\vartheta=180°$ | $\varrho_{a,ax}(x)=\varrho_{a,S}(x)-\frac{x}{2}\frac{\delta\varrho_{a,S}(x)}{\delta x}$ |
| perpendicular | " | $\varphi=\vartheta+90°$ | | $\varrho_{a,per}(x)=\varrho_{a,S}(x)-\frac{x}{3}\frac{\delta\varrho_{a,S}(x)}{\delta x}$ |
| parallel | " | $\varphi=\vartheta$ | | $\varrho_{a,par}(x)=\varrho_{a,S}(x)-\frac{x\cos^2\vartheta}{3\cos^2\vartheta-1}\frac{\delta\varrho_{a,S}(x)}{\delta x}$ |

Relation (12) can be treated as a linear differential equation of the first order; the general solution is given by:

$$\varrho_{a,S} = -e^{\int(\alpha/x)dx}\left(\int(\alpha/x)\varrho_{a,D}\, e^{-\int(\alpha/x)dx} dx + c\right), \qquad (13)$$

which reduces easily to:

$$\varrho_{a,S} = -\alpha x^{\alpha}\int(\varrho_{a,D}/x^{\alpha+1})dx \ , \qquad (14)$$

since the constant c must be zero to take account that for x going to zero or to infinity (according to the sign of $\alpha$) $\varrho_{a,S}(x)$ does not reach always an infinite value.

For practical purposes, in order to carry out easily the integration at the right-hand side of (14), we must introduce, instead of the complicated methematical expression of $\varrho_{a,D}(x)$, a rather simple algebraic expression which interpolates with good approximation at least successive portions of the $\varrho_{a,D}(x)$ curve. Let us go on as follows:

(i) we split up the whole interval $0 \leqslant x \leqslant +\infty$ into a set of contiguous partial intervals with spreads even different from one another, of the type:

$$s_j \leqslant x \leqslant s_{j+1} \qquad (j=1,2,3,\ldots\ldots,n-2,n-1); \qquad (15)$$

(ii) we interpolate in each of the above partial intervals the function $\varrho_{a,D}(x)$ with the following algebraic expression:

$$\varrho_{a,D}(x) \simeq a_j x^{b_j}. \qquad (16)$$

Eq.(16), if placed on a log-log scale, is represented in each partial interval by a straightline segment with a slope $b_j$, passing through the points with coordinates $(\log s_j, \log\varrho_{a,D}(s_j))$ and $(\log s_{j+1}, \log\varrho_{a,D}(s_{j+1}))$.

Let us impose now the following condition: in each point $s_j$ on the positive axis x the value $\varrho_{a,D}(s_j)$ is known. Such condition is in the field, of course, accounted for whatever be the number of discrete points $s_j$, since $\varrho_{a,D}(s_j)$ is the dipole apparent resistivity measured at every predetermined spacing between the dipole centres. On the other hand no restrictive condition is imposed on the choice of the partial intervals (15); it is advisable, however, to select them as small possible, compatibly with field difficulties, in order to improve the approximation given by (16).

After substitution of (16) into (14) and integration, one obtains in each partial interval:

$$\varrho_{a,S} = \frac{\alpha}{\alpha - b_j} a_j x^{b_j} - \alpha a_j k x^{\alpha},$$

where k is a new constant deriving from the integration and required to satisfy the condition that in each common point of two contiguous partial intervals $\varrho_{a,S}$ must be continuous.

If we put:

$$c_j = \alpha a_j k,$$

it is finally obtained:

$$\varrho_{a,S} = \frac{\alpha}{\alpha - b_j} a_j x^{b_j} - c_j x^{\alpha}, \tag{17}$$

which can be rewritten more concisely:

$$\varrho_{a,S} = \frac{\alpha}{\alpha - b_j} \varrho_{a,D} - c_j x^{\alpha} \tag{18}$$

valid for $s_j \leqslant x \leqslant s_{j+1}$ and with $j=1,2,\ldots,n-2,n-1$.

In $s_j$ we have:

$$\varrho_{a,S}(s_j) = \frac{\alpha}{\alpha - b_j} \varrho_{a,D}(s_j) - c_j s_j^{\alpha}. \tag{19}$$

In (18) $\alpha$, $s_j$ and $\varrho_{a,D}(s_j)$ are all known quantities; to obtain the value of $\varrho_{a,S}(s_j)$, $b_j$ and $c_j$ must be determined.

For the determination of $b_j$ it is sufficient to consider that, from the starting hypotheses, we have:

$$\begin{cases} \varrho_{a,S}(s_j) = a_j s_j^{b_j} \\ \varrho_{a,S}(s_{j+1}) = a_j s_{j+1}^{b_j} \end{cases} \tag{20}$$

which in logarithmic form becomes:

$$\begin{cases} \log\varrho_{a,D}(s_j) = \log a_j + b_j \log s_j \\ \log\varrho_{a,D}(s_{j+1}) = \log a_j + b_j \log s_{j+1} \end{cases} \tag{21}$$

Solving the system (21) with $\log a_j$ and $b_j$ unknowns, we obtain at once:

$$b_j = \log\left(\varrho_{a,D}(s_{j+1})/\varrho_{a,D}(s_j)\right) \Big/ \log(s_{j+1}/s_j) \tag{22}$$

To determine $c_j$, proceed as follows: consider the interval $s_{j-1} \leqslant x \leqslant s_j$, contiguous and preceding $s_j \leqslant x \leqslant s_{j+1}$, where:

$$\varrho_{a,S} = \frac{\alpha}{\alpha - b_{j-1}} a_{j-1} x^{b_{j-1}} - c_{j-1} x^{\alpha}. \tag{23}$$

Since $\varrho_{a,S}$ must be continuous in $s_j$, $\varrho_{a,S}(s_j)$ from (17) must be equal to $\varrho_{a,S}(s_j)$ from (23). It follows that:

$$c_j = c_{j-1} + \gamma_j - \beta_{j-1}. \tag{24}$$

where:

$$\gamma_j = \frac{\alpha}{\alpha - b_j} a_j s_j^{(b_j - \alpha)} = \frac{\alpha}{\alpha - b_j} \varrho_{a,D}(s_j) s_j^{-\alpha} \tag{25}$$

and

$$\beta_{j-1} = \frac{\alpha}{\alpha - b_{j-1}} a_{j-1} s_j^{(b_{j-1} - \alpha)} = \frac{\alpha}{\alpha - b_{j-1}} \varrho_{a,D}(s_j) s_j^{-\alpha} \tag{26}$$

Through the recurrence formula (24) it is possible to obtain $c_j$ from $c_{j-1}$ and so on backward since $\beta_{j-1}$ and $\gamma_j$ are known quantities for every j. In practice, it is sufficient to find out the first value $c_o$ to obtain all following $c_j$.

Accordingly, if the interval $s_{j+1} \leqslant x \leqslant s_{j+2}$, contiguous and following directly $s_j \leqslant x \leqslant s_{j+1}$, is considered and the above reasoning is repeated, one obtains another recurrence formula:

$$c_j = c_{j+1} + \beta_j - \gamma_{j+1}, \tag{27}$$

which allows to calculate $c_j$ from $c_{j+1}$ and so on forward. In practice, it is now sufficient to find out the last value $c_n$ to obtain all preceding $c_j$.

Practical determination of the constants $c_j$

Let us see when the recurrence formula (24) must be applied and when the recurrence formula (27).

It is necessary at this moment to study the behaviour of the function $\varrho_{a,S}(x)$, given by (17), for shortest spacings ($x \to 0$) and for largest spacings ($x \to +\infty$) at varying values of the array parameter $\alpha$.

The parameter $\alpha$ is surely positive for radial (axial) and perpendicular arrays. In the case of parallel array $\alpha$ may be positive, zero or negative according to the value of the azimuth $\vartheta$ (Alpin, 1950).

A) First case: $\alpha < 0$

Since:

$$\lim_{x \to +\infty} \varrho_{a,D}(x) = \varrho_m \tag{28}$$

where $\varrho_m$ is the true resistivity of the last layer (substratum), which we suppose finite and different from zero, it is always possible to find a spacing $s_n$, sufficiently large, so that we may retain:

$$\varrho_{a,D}(s_n) \simeq \varrho_m \tag{29}$$

and, consequently:

$$\varrho_{a,D}(x) \simeq \varrho_m \tag{30}$$

in the whole interval $s_n \leqslant x < +\infty$.

Putting (30) into (14) and integrating, we have:

$$\varrho_{a,S}(x) = \varrho_m - c_n x^{\alpha} \tag{31}$$

for $s_n \leqslant x < +\infty$. Accordingly:

$$\lim_{x \to +\infty} \varrho_{a,S}(x) = \varrho_m \tag{32}$$

whatever be the finite value of $c_n$. Therefore, with such boundary condition it is not possible to obtain the value of $c_n$ and, consequently, one cannot use recurrence formula ((27) when $\alpha$ is negative.

On the other hand, since:

$$\lim_{x \to 0} \varrho_{a,D}(x) = \varrho_1 \tag{33}$$

where $\varrho_1$ is the true resistivity of the first layer (overburden), finite and different from zero, it is always possible to find a spacing $s_1$, sufficiently small, so that we may retain:

$$\varrho_{a,D}(s_1) \simeq \varrho_1 \tag{34}$$

and, consequently:

$$\varrho_{a,D}(x) \simeq \varrho_1 \tag{35}$$

in the whole interval $0 \leqslant x \leqslant s_1$.

Putting (35) into (14) and integrating, we have:

$$\varrho_{a,S}(x) = \varrho_1 - c_o x^{\alpha} \tag{36}$$

in $0 \leqslant x \leqslant s_1$. Accordingly:

$$\lim_{x \to 0} \varrho_{a,S}(x) = \varrho_1 \tag{37}$$

if and only if:

$$c_o = 0. \tag{38}$$

Therefore, starting from (38) it is possible to obtain all the following $c_j$ by using recurrence formula (24).

In practice, when $\alpha < 0$, it is necessary, in order to have a $\varrho_{a,D}$ field graph transformable into the equivalent in $\varrho_{a,S}$, to determine directly on the field the left asymptotic branch of the curve, or to carry out extrapolations to the left of the field curve.

B) <u>Second case</u>: $\alpha > 0$

By the same reasoning as for the first case, we may easily verify that in the interval $0 \leqslant x \leqslant s_1$ it is again:

$$\varrho_{a,S}(x) = \varrho_1 - c_o x^{\alpha} \tag{39}$$

and therefore, when $\alpha > 0$, we have:

$$\lim_{x \to 0} \varrho_{a,S}(x) = \varrho_1, \tag{40}$$

whatever be the finite value of $c_o$. From such boundary condition it is not possible to obtain the value of $c_o$ and, as a consequence, we cannot use recurrence formula (24).

On the other hand, in the interval $s_n \leqslant x < +\infty$, where we have again:

$$\varrho_{a,S}(x) = \varrho_m - c_n x^{\alpha}, \tag{41}$$

with $\varrho_m$ finite and different from zero, it may be easily verified that:

$$\lim_{x \to +\infty} \varrho_{a,S}(x) = \varrho_m \tag{42}$$

if and only if:

$$c_n = 0. \tag{43}$$

The same result is obtained when $\varrho_m$ is zero. In fact, it is always possible to find a spacing $s_n$, sufficiently large, so that in the interval $s_n \leqslant x < +\infty$ we have:

$$\varrho_{a,D}(x) \simeq a_n x^{b_n} \tag{44}$$

with

$$b_n < 0. \tag{45}$$

By the condition (45) we easily verify that:

$$\lim_{x \to +\infty} \varrho_{a,D}(x) = 0. \tag{46}$$

From (17) one obtains:

$$\varrho_{a,S}(x) = \frac{\alpha}{\alpha - b_n} a_n x^{b_n} - c_n x^{\alpha}, \tag{47}$$

and:

$$\lim_{x \to +\infty} \varrho_{a,S}(x) = 0 \tag{48}$$

if:

$$c_n = 0. \tag{49}$$

Finally, the result remains unchanged even when $\varrho_m$ is infinity. In such event, it must be remembered that as for $\varrho_{a,S}$ and whatever be the dipole array, $\varrho_{a,D}$ has an asymptotic branch for the largest spacings rising with a slope +1 on log-log scales (Alpin, 1950). Therefore it is always possible to find a point $s_n$ so that in the interval $s_n \leqslant x < +\infty$, $\varrho_{a,D}$ and $\varrho_{a,S}$ are represented by parallel straight lines. Since now:

$$\varrho_{a,D}(x) = a_n x \tag{50}$$

($b_n$ is obviously +1) in $s_n \leqslant x < +\infty$, and from (17):

$$\varrho_{a,S}(x) = \frac{\alpha}{\alpha-1} a_n x - c_n x^{\alpha}, \tag{51}$$

the condition of parallellism is satisfied only if:

$$c_n = 0. \tag{52}$$

Again also when $\alpha > 0$, it is necessary, in order to have a dipole sounding graph transformable into the equivalent Schlumberger sounding graph, to determine directly on the field the right asymptotic branch of the curve (i.e. for large spacings), or to carry out extrapolations to the right of the field curve.

For all the cases here discussed, some examples of transformation of theoretical and field sounding curves can be found in Patella (1974), Nardi et al. (1979) and Patella et al. (1979).

## DIPOLE AND SCHLUMBERGER SOUNDINGS OVER ARBITRARY UNDERGROUND STRUCTURES

Let us consider again at first the half Schlumberger scheme of Fig. 4. In general, for arbitrary underground structures, the direction of the electric field $\overline{E}^{o}(x)$, generated by the positive current pole P and evaluated at O, is unknown whatever be the position of point P along the x-axis, which is assumed again to correspond with the sounding expansion line.

If, as usual, $\Delta V_{MN}^{P}$ is the potential difference measured between M and N, and MN $\ll$ OP, then:

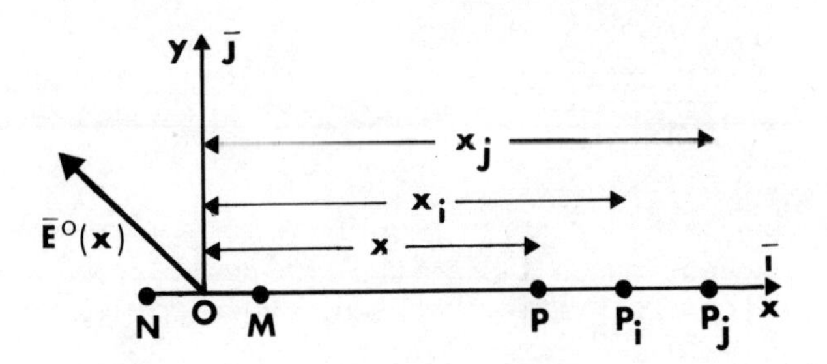


Fig. 4. The half Schlumberger array over an arbitrary underground structure.

$$\frac{\Delta V^P_{MN}}{|\overline{MN}|} \simeq \bar{E}^o(x) \cdot \frac{\overline{MN}}{|\overline{MN}|} = \bar{E}^o(x) \cdot \bar{i} = E^o_x(x), \tag{53}$$

where $E^o_x(x)$ is the x-component of the electric field. Relation (53) is valid provided that no outcropping discontinuity in the resistivity crosses the potential dipole.

The half Schlumberger apparent resistivity is of course given by:

$$\varrho_{a,S}(x) = (2\pi x^2)\, E^o_x(x)/I. \tag{54}$$

Again, for simplicity and without loss of generality, we maintain constant the intensity of current I whatever the position of the current pole along the x-axis.

Consider now the two current pole positions $P_i$ and $P_j$ with $OP_j > OP_i$, i.e. $x_j > x_i$. We have:

$$\varrho_{a,S}(x_j) = (2\pi x_j^2)\, E^o_x(x_j)/I \tag{55}$$

$$\varrho_{a,S}(x_i) = (2\pi x_i^2)\, E^o_x(x_i)/I \tag{56}$$

from which:

$$\frac{\varrho_{a,S}(x_j)}{x_j^2} - \frac{\varrho_{a,S}(x_i)}{x_i^2} = (2\pi/I)\,[\,E^o_x(x_j) - E^o_x(x_i)] =$$

$$= (2\pi/I)\int_{x_i}^{x_j} \frac{\delta E_x^o(x)}{\delta x}\,dx. \qquad (57)$$

The last equality of (57) is valid provided that the electric field $E_x^o(x)$ is continuous inside the interval $(x_i, x_j)$, i.e. no discontinuity in the resistivity crosses, outcropping, the same interval as in Fig. 5.

Taking the limit for $x_j$ going to infinity, we obtain from (57):

$$\varrho_{a,S}(x_i) = -(2\pi x_i^2/I)\int_{x_i}^{+\infty} \frac{\delta E_x^o(x)}{\delta x}\,dx, \qquad (58)$$

since:

$$\lim_{x_j \to +\infty} \varrho_{a,S}(x_j)/x_j^2 = 0 \qquad (59)$$

in all realistic cases (Alfano, 1980).

Now we shall study if it is possible to obtain $\delta E_x^o(x)/\delta x$, needed for the calculation of the half Schlumberger apparent resistivity in $x_i$, by means of a generic dipole sounding, expanded along the same x-direction as in Fig. 6.

In general, for arbitrary structures, the directions of the electric fields $\overline{E}^o(A)$ and $\overline{E}^o(B)$, generated by the dipole source $\overline{AB}$ and evaluated in O (see Fig. 6), are unknown.

If $MN \ll OP$ and $AB \ll OP$, and no outcropping discontinuity in the resistivity crosses the dipoles $\overline{MN}$ and $\overline{AB}$, then the following mathematical passages are possible for the evaluation of the potential difference between M and N, provoked by the current dipole $\overline{AB}$:

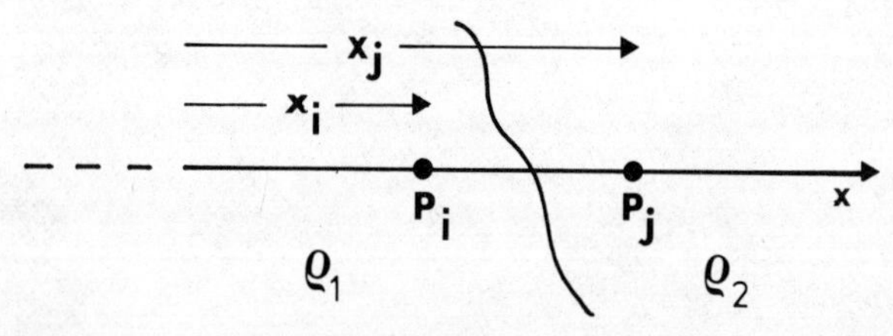


Fig. 5. Schematic depiction of an outcropping discontinuity in the resistivity, crossing the sounding expansion line.

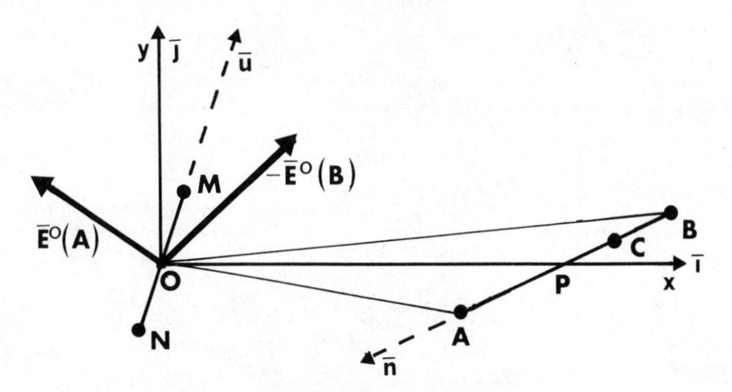


Fig. 6. The generic dipole array over an arbitrary underground structure.

$$\frac{\Delta V_{MN}^{AB}}{|\overline{MN}|} \simeq [\overline{E}^{\circ}(A) - \overline{E}^{\circ}(B)] \cdot \overline{u} =$$

$$\{[E_x^{\circ}(A)-E_x^{\circ}(B)]\overline{i} + [E_y^{\circ}(A)-E_y^{\circ}(B)]\overline{j}\} \cdot \overline{u} =$$

$$\left( \frac{\delta E_x^{\circ}(C)}{\delta n}\bigg|_{C\equiv P} |\overline{AB}|\overline{i} + \frac{\delta E_y^{\circ}(C)}{\delta n}\bigg|_{C\equiv P} |\overline{AB}|\overline{j} \right) \cdot \overline{u} =$$

$$\left( \frac{\delta E_x^{\circ}[x(n),y(n)]}{\delta n}\bigg|_{C\equiv P} \cos\varphi + \frac{\delta E_y^{\circ}[x(n),y(n)]}{\delta n}\bigg|_{C\equiv P} \sin\varphi \right) |\overline{AB}| =$$

$$\left( \frac{\delta E_x^{\circ}(x)}{\delta x}\frac{dx}{dn}\cos\varphi + \frac{\delta E_x^{\circ}(x)}{\delta y}\frac{dy}{dn}\cos\varphi + \right.$$

$$\left. \frac{\delta E_y^{\circ}(x)}{\delta x}\frac{dx}{dn}\sin\varphi + \frac{\delta E_y^{\circ}(x)}{\delta y}\frac{dy}{dn}\sin\varphi \right) |\overline{AB}|. \qquad (60)$$

But:

$dx/dn = \cos\vartheta$ and $dy/dn = \sin\vartheta$,

so from (60) we obtain:

$$\frac{\Delta V_{MN}^{AB}}{|\overline{MN}|\,|\overline{AB}|} = \frac{\delta E_x^o(x)}{\delta x}\cos\varphi\cos\vartheta + \frac{\delta E_x^o(x)}{\delta y}\cos\varphi\sin\vartheta + \frac{\delta E_y^o(x)}{\delta x}\sin\varphi\cos\vartheta + \frac{\delta E_y^o(x)}{\delta y}\sin\varphi\sin\vartheta \,. \tag{61}$$

From (61) we see that the required quantity $\delta E_x^o(x)/\delta x$ cannot be obtained by measurements made with the generic dipole array of Fig. 6. Each measurement of this type contains the four unknowns:

$$\delta E_x^o(x)/\delta x,\ \delta E_x^o(x)/\delta y,\ \delta E_y^o(x)/\delta x,\ \delta E_y^o(x)/\delta y.$$

Thus, in conclusion, for arbitrary underground structures, we cannot obtain half Schlumberger apparent resistivity data from generic dipole measurements, or, in other words, we cannot transform dipole sounding data into equivalent half Schlumberger data, unless we make some changes of the sounding procedure.

One possible practical solution consists in using the double dipole array of Fig. 7. By making all possible combinations between the dipoles, from (61) we obtain:

$$\begin{aligned}
P_1^1 &= A\cos\varphi_1\cos\vartheta_1 + B\cos\varphi_1\sin\vartheta_1 + C\sin\varphi_1\cos\vartheta_1 + D\sin\varphi_1\sin\vartheta_1\\
P_2^1 &= A\cos\varphi_2\cos\vartheta_1 + B\cos\varphi_2\sin\vartheta_1 + C\sin\varphi_2\cos\vartheta_1 + D\sin\varphi_2\sin\vartheta_1\\
P_1^2 &= A\cos\varphi_1\cos\vartheta_2 + B\cos\varphi_1\sin\vartheta_2 + C\sin\varphi_1\cos\vartheta_2 + D\sin\varphi_1\sin\vartheta_2\\
P_2^2 &= A\cos\varphi_2\cos\vartheta_2 + B\cos\varphi_2\sin\vartheta_2 + C\sin\varphi_2\cos\vartheta_2 + D\sin\varphi_2\sin\vartheta_2
\end{aligned} \tag{62}$$

where:

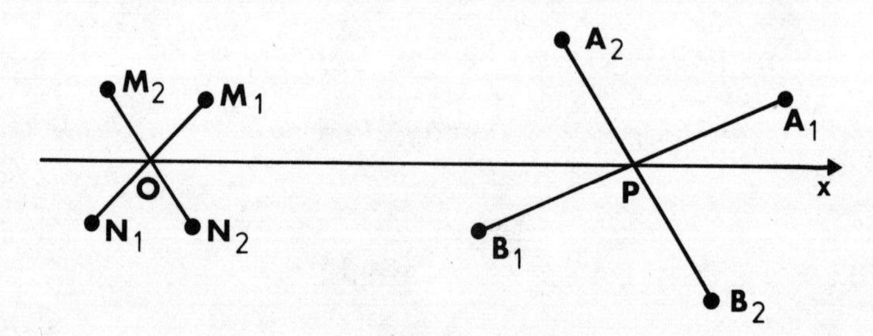


Fig. 7. The double dipole array.

$$P_n^m = \frac{\Delta V_{M_nN_n}^{A_mB_m}}{|\overline{M_nN_n}|\ |\overline{A_mB_m}|}, \quad \text{with} \quad m=1,2 \ \text{and} \ n=1,2 \tag{63}$$

and

$$A=\delta E_x^o(x)/\delta x,\ B=\delta E_x^o(x)/\delta y,\ C=\delta E_y^o(x)/\delta x,\ D=\delta E_y^o(x)/\delta y. \tag{64}$$

(62) is a system of four equations in four unknowns, which allows to calculate the required quantity $A=\delta E_x^o(x)/\delta x$. The use of the double dipole array of Fig. 7 is however too complicated in the field.

A second practical solution consists in using the reduced double dipole array of Fig. 8. In this case we have from (61):

$$\begin{cases} P_1 = -A\cos\varphi_1 - C\sin\varphi_1 \\ P_2 = -A\cos\varphi_2 - C\sin\varphi_2 \end{cases} \tag{65}$$

where:

$$P_n = \frac{\Delta V_{M_nN_n}^{AB}}{|\overline{M_nN_n}|\ |\overline{AB}|}, \quad \text{with} \quad n=1,2 \tag{66}$$

and A and C are given by (64).

(65) is a system of two equations in two unknowns, which allows again to calculate the required quantity A.

A third practical solution consists in using the known axial dipole array, obtained from the generic dipole array by setting $\varphi=0°$ and $\vartheta=180°$ (see Fig. 9).

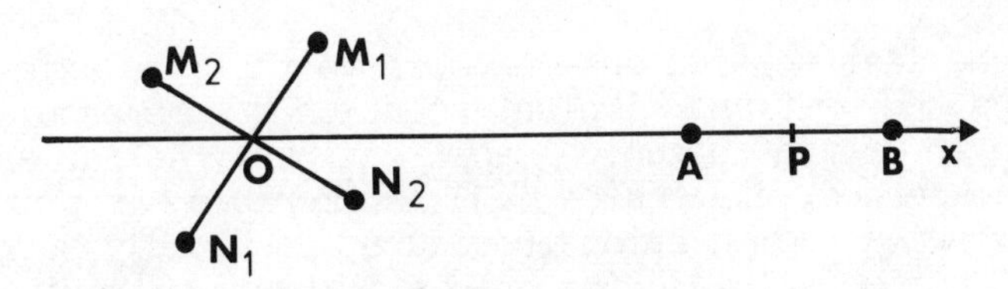


Fig. 8. The reduced double dipole array.

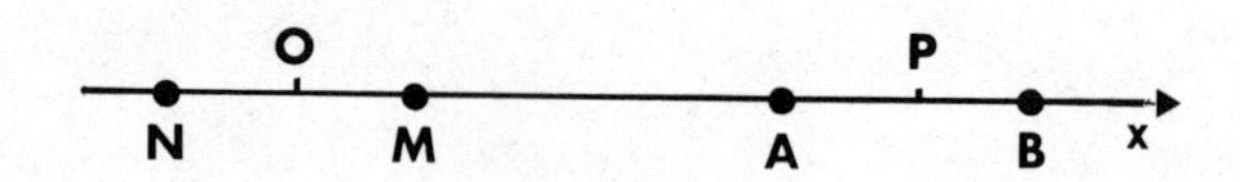


Fig. 9. The axial dipole array.

In this case we have:

$$\frac{\Delta V_{MN}^{AB}}{|\overline{MN}|\,|\overline{AB}|} = -\frac{\delta E_x^{\circ}(x)}{\delta x} \tag{67}$$

which solves in the simplest way our problem.

The axial dipole apparent resistivity is given from (7) by setting $\varphi=0^{\circ}$ and $\vartheta=180^{\circ}$:

$$\varrho_{a,ax}(x) = \frac{\pi x^3}{|\overline{MN}|\,|\overline{AB}|}\frac{\Delta V_{MN}^{AB}}{I} = -\frac{\pi x^3}{I}\frac{\delta E_x^{\circ}(x)}{\delta x}\,. \tag{68}$$

Obtaining from (54) $E_x^{\circ}(x)$ and evaluating its derivative with respect to x, after substitution into (68), we get:

$$\varrho_{a,ax}(x) = \varrho_{a,S}(x) - \frac{x}{2}\frac{\delta \varrho_{a,S}(x)}{\delta x} \tag{69}$$

which is exactly the known equation relating the two apparent resistivities, already obtained during the study about layered structures.

## THE TRANSFORMATION OF DIPOLE TO SCHLUMBERGER SOUNDING CURVES IN THE CASE OF ARBITRARY UNDERGROUND STRUCTURES

From the last section one can observe that the axial dipole array offers the best and simplest solution of the problem under study, also in view of practical applications. As far as transformation is concerned, expression (69) is amenable, in principle, to the same numerical computation procedure, previously suggested in the case of horizontal layering. This is true, without any limitation, when no outcropping discontinuity in the resistivity crosses the whole sounding expansion line. However, in this case, the required asymptotical extrapolation to the right of the field curve must be not necessarily connected with the presence of an horizontal substratum.

The simplest model which can be easily dealt with to check the validity of what above expressed, is characterized by a half-space divided into two regions with different resistivity by a vertical discontinuity plane. When the sounding expansion line is taken parallel to the outcropping vertical plane, or perpendicular to it but with both the fixed and mobile dipoles always placed above the same region of the half-space, then the numerical computation procedure for the transformation, previously described, can be used without any restriction.

If we consider now the presence of some discontinuities in the resistivity, which outcrop and cross the sounding expansion line, from the above theory we can derive that the proposed method of transformation is still applicable, provided that the following condition be satisfied. It is essential that, during the execution of the sounding, no current dipole $\overline{AB}$, supposed for instance mobile, overlap the strike line of the outcropping discontinuities, as in Fig.10. This is, in fact, on the basis of the validity of formula (69).

In field practice, this condition may be satisfied when the current dipole positions are previously programmed on the geological topographic maps of the survey area, so that overlapping with geological contacts, with presumable strong resistivity contrasts, can be avoided.

Let us finally consider the case in which overlapping of dipoles with surface outcrops of resistivity contrasts cannot be in no way avoided. In this situation the previously described theory, and, in particular the dual expressions (69) and (58) are no longer applicable.

To solve this problem Alfano (1974, 1980) introduced the concept of "the continuous axial dipole sounding", during which the current dipole, supposed again mobile, is progressively displaced

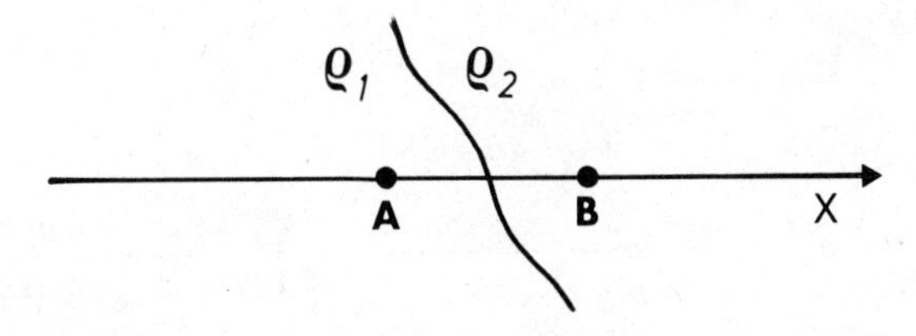


Fig.10. A current dipole overlapping the surface outcrop of a discontinuity in the resistivity.

as in Fig.11. One can easily observe that, after each measurement, the farthest current electrode from the receiving dipole becomes the nearest in the subsequent measurement, i.e. $OB_i = OA_{i+1}$ (i=1,2, 3,....).

With this particular field operative condition in mind, let us take into consideration the first equality of formula (57), which was derived regardless of the real conditions of the ground surface with respect to the mobile current dipole position, i.e.:

$$\frac{\varrho_{a,S}(x_j)}{x_j^2} - \frac{\varrho_{a,S}(x_i)}{x_i^2} = (2\pi/I)\left[E_x^o(x_j) - E_x^o(x_i)\right]. \qquad (70)$$

By adding and subtracting equal quantities, (70) can be rewritten as:

$$\frac{\varrho_{a,S}(x_j)}{x_j^2} - \frac{\varrho_{a,S}(x_i)}{x_i^2} = (2\pi/I)\sum_{k=i}^{j-1}\left[E_x^o(x_{k+1}) - E_x^o(x_k)\right]. \qquad (71)$$

Each difference

$$\Delta E_x^o(x_k, x_{k+1}) = E_x^o(x_k) - E_x^o(x_{k+1}) \qquad (72)$$

is just the quantity measured in the field by the receiving dipole $\overline{MN}$, i.e. the electric field at the midpoint of $\overline{MN}$ provoked by the current dipole $\overline{A_k B_k}$. The sum in (71) is extended over analogous terms relative to continuous displacements of the current dipole.

Taking the limit for $x_j$ going to infinity, (71) becomes:

$$\varrho_{a,S}(x_i) = (2\pi x_i^2/I)\sum_{k=1}^{\infty}\Delta E_x^o(x_k, x_{k+1}), \qquad (73)$$

which is the required solution giving the half Schlumberger apparent resistivity at a generic spacing from continuous axial dipole measurements, in the most general situation of arbitrary underground structures.

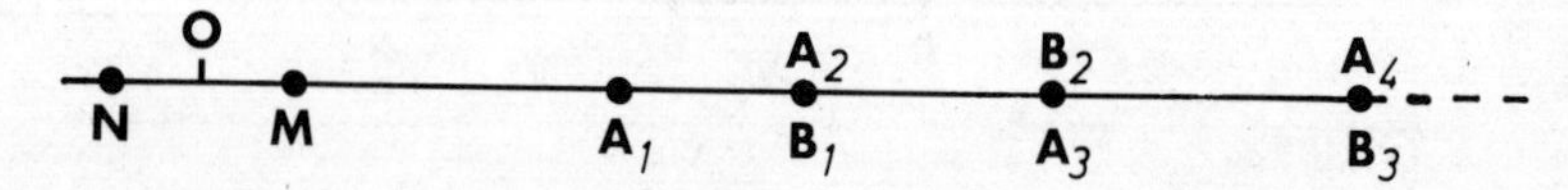


Fig.11. The continuous axial dipole sounding.

The remaining problem concerns the presence in (73) of a sum of infinite terms, which would imply the execution of an infinitely long axial dipole sounding. In practice, however, a finite length is sufficient; the residual sum in (73) may be evaluated by means of an extrapolation procedure similar to that previously discussed (see, also, Alfano, 1974).

Thus, we may consider Alfano's continuous array as the one which gives, at least from a theoretical point of view, the best and most correct solution to the problem of carrying out deep dipole soundings in real geological situations, with the aim of obtaining field diagrams easily transformable into equivalent Schlumberger curves.

## REFERENCES

Alfano, L., 1974, A modified geoelectrical procedure using polar-dipole arrays, Geophys. Prosp., 22: 510.
Alfano, L., 1980, Dipole-dipole deep geoelectric soundings over geological structures, Geophys. Prosp., 28: 283.
Alpin, L. M., 1950, "The Theory of Dipole Soundings", Gostoptekhizdat, Moscow.
Nardi, R., Puccinelli, A., and Patella, D., 1979, Applicazione del metodo del sondaggio dipolare profondo lungo una sezione dalle Alpi Apuane all'Appennino Pistoiese, Atti Soc. Tosc. Sci. Nat., Mem., Serie A, 86:1.
Patella, D., 1974, On the transformation of dipole to Schlumberger sounding curves, Geophys. Prosp, 22: 315.
Patella, D., Rossi, A., and Tramacere, A., 1979, First results of the application of the dipole electrical sounding method in the geothermal area of Travale-Radicondoli (Tuscany), Geothermics, 8: 111.

SHORT NOTES

# CONTINENTAL CHARACTER OF THE LITHOSPHERE BENEATH THE IONIAN SEA

P. Farrugia and G.F. Panza

Istituto di Geodesia e Geofisica
Università di Bari
70100 Bari, ITALY

SUMMARY

Group velocity data in the region of the Ionian Sea have been inverted using the Hedgehog procedure. The range of acceptable solutions shows that only a crustal thickness exceeding 30 km is consistent with the data. Furthermore, the obtained models indicate a clearly defined velocity layering, including velocity inversions, within the crust, as well as a crust-mantle transition zone in place of a sharp Moho discontinuity.

## INTRODUCTION

The region of the Ionian Sea in the eastern Mediterranean has long been an intriguing subject for investigation mainly because of the complexity of the tectonic patterns in the surrounding areas and also because of the several, often conflicting, interpretations that have been given as regards the character of the underlying crust and upper mantle on the basis of various geophysical measurements. In this paper, results are presented of the inversion of Rayleigh wave data along profiles which sample the Ionian bathyal plain for a large part of their length.

## DATA

Three events in Salonica, Crete and the Southern coast of Turkey (Table 1) were recorded on a long-period vertical-component seismograph at Malta. The profiles are shown in Fig.1. A frequency filtering technique (Levshin et al.,1972) was used to obtain the group velocity dispersion data from the three seismograms.

Table 1. Events used

| Event number | Location | Date | Origin Time | Depth (km) | Recording Station |
|---|---|---|---|---|---|
| 1^ | 40.7N 23.2E | 23 05 78 | 23: 34: 11.4 | 10 | Malta |
| 2^ | 35.7N 26.4E | 23 07 79 | 11: 42: 2.8 | 89 | Malta |
| 3^ | 36.4N 31.9E | 28 05 79 | 09: 27: 33.2 | 99 | Malta |
| 4^^ | 34.7N 24.7E | 07 04 74 | 14: 22: 47.1 | 29 | Bari |
| 5^^ | 33.6N 22.9E | 13 11 75 | 03: 07: 26.6 | 33 | Bari |

^This study
^D'Ingeo(1978)

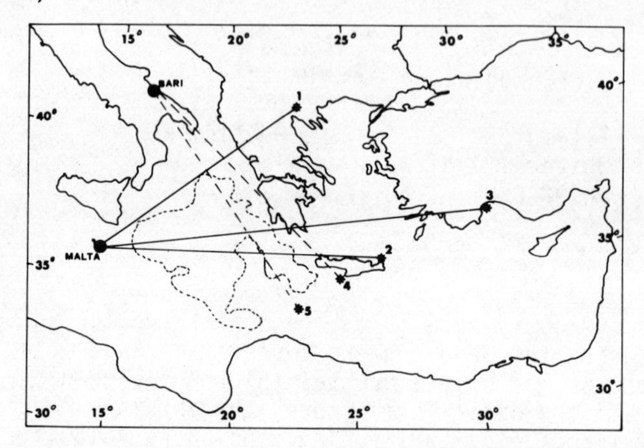


Fig.1. Profiles investigated by Rayleigh wave group velocity dispersion. Numbers refer to Table 1. The dashed contour represents the boundary of the Ionian bathyal plain at 3000 m.

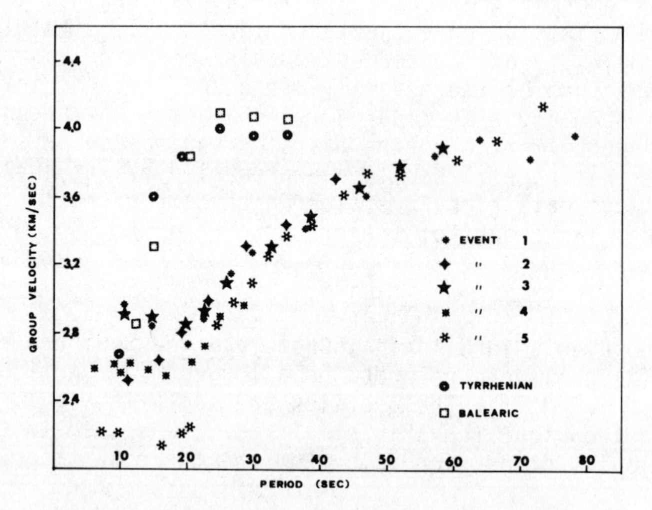


Fig.2. Rayleigh Wave Group Velocities.

The quality of the amplitude matrix allowed the reliable extraction of group velocities over the period range 10s to 80s for event no. 1, 10s to 45s for event no.2 and 10s to 58s for event no.3 (Fig.2). The three curves are very similar at periods longer than 25s.

## INVERSION

A Hedgehog inversion of the data sets was carried out. The parametrization used in the inversion of the three dispersion curves is listed in Table 2. A 5-layer crust was considered, with the 2 upper crustal layers fixed in each inversion. The differences in these two layers were introduced to comply with the trends of the dispersion curves at short periods. The upper mantle structure beneath the lid was taken equal in all 3 cases, with the lithospheric thickness fixed at 90 km. The variable parameters are then the thicknesses and S-wave velocities of the 3 lower crustal layers. However, by assuming always a 1:1 ratio of the thicknesses of the first two layers, the number of variable parameters was further constrained to five. In the inversion, a model was rejected if the difference between computed and observed group velocities exceeded 0.12 $km.s^{-1}$ at any period, or if the r.m.s. difference $\sigma$ exceeded 0.08 $km.s^{-1}$

Table 2. Hedgehog model parametrization

| Depth (km) | Thickness (km) | Shear Velocity ($km.s^{-1}$) |
|---|---|---|
| 0 | 2 | 0.0 |
| 2 | 2 | $3.0^{\circ}$ $3.0^{\wedge}$ $3.1^{+}$ |
| 5 | 3 | $3.6^{\circ}$ $3.1^{\wedge}$ $3.5^{+}$ |
| 8 | P1 | P3 |
| 8+P1 | P1 | P4 |
| 8+P1 | P2 | P5 |
| 8+P1+P2 | $h_{LID}(P_1,P_2)$ | 4.5 |
| 90 | 160 | 4.4 |
| 260 | 120 | 4.8 |
| 370 | 100 | 5.0 |
| 470 | | 5.4 |

° Event no.1
^ Event no.2
+ Event no.3

Parameter Ranges (Values in parentheses indicate the step used)

P1 = 3.0 (3.0) 15.0
P2 = 5.0 (3.0) 20.0
P3 = 2.7 (0.2) 4.1
P4 = 2.7 (0.2) 4.1
P5 = 3.3 (0.2) 4.1

RESULTS

The inversion of data from event no.1 yielded 9 acceptable solutions (Table 3). The solutions may be classified into two categories: a) solutions 1 to 6 which exhibit a clear S-velocity discontinuity between crust and upper mantle in a depth range of 37 km to 43 km; b) solutions 7,8,9 which are characterized by a rather high lowermost crustal velocity of 4.1 km $s^{-1}$. Such velocities may be interpreted in terms of a transition region between crust and mantle material (Prodehl, 1977). It is interesting to note that even if we assign the 4.1 km $s^{-1}$ layer to the upper mantle, the crustal thickness is not less than 32 km. We are therefore in the presence of a continental-like crust.

Another interesting result of this inversion is the presence, in all the acceptable models but one (no.4) of a distinctly stratified crust, containing velocity inversions. Such a structure may be considered indicative of an old stable crust (Mueller, 1977).

Since for this profile, we have been able to determine dispersion values at periods as long as 80 s, a separate inversion was carried out involving the lid velocity as the 5th parameter and fixing the thickness of the third and fourth crustal layers at the rather popular value of 9 km (see Table 3). In such a way, we have been able to confirm that the only value of the lid S-wave velocity consistent with the data is 4.5 km.$s^{-1}$ . Recalling that $\Delta$ P5 = 0.15 km.$s^{-1}$, this means of course, that the lid velocity differs from 4.5 km.$s^{-1}$ by less than + 0.15 km.$s^{-1}$.

The inversion of the dispersion data from event no.2 yielded five possible solutions (Table 4). The different combinations of

Table 3. Acceptable Solutions from the data inversion of event no.1

| Model Number | P1 (km) | P2 (km) | P3 (km.$s^{-1}$) | P4 (km.$s^{-1}$) | P5 (km.$s^{-1}$) |
|---|---|---|---|---|---|
| 1 | 9 | 14 | 3.7 | 3.4 | 3.7 |
| 2 | 9 | 17 | 3.7 | 3.4 | 3.7 |
| 3 | 9 | 17 | 3.9 | 3.2 | 3.9 |
| 4 | 9 | 14 | 3.5 | 3.6 | 3.5 |
| 5 | 9 | 11 | 3.5 | 3.8 | 3.3 |
| 6 | 9 | 11 | 3.7 | 3.6 | 3.3 |
| 7 | 12 | 17 | 3.7 | 3.4 | 4.1 |
| 8 | 12 | 14 | 3.7 | 3.4 | 4.1 |
| 9 | 12 | 11 | 3.7 | 3.4 | 4.1 |

Table 4. Acceptable solutions from data inversion of event no.2

| Model Number | P1 (km) | P2 (km) | P3 (km.$s^{-1}$) | P4 (km.$s^{-1}$) | P5 (km.$s^{-1}$) |
|---|---|---|---|---|---|
| 1 | 7 | 17 | 3.2 | 3.7 | 3.7 |
| 2 | 10 | 14 | 3.2 | 3.9 | 3.7 |
| 3 | 7 | 14 | 3.0 | 4.1 | 3.5 |
| 4 | 10 | 8 | 3.2 | 4.1 | 3.3 |
| 5 | 10 | 11 | 3.2 | 4.1 | 3.5 |

Table 5. Acceptable solutions from data inversion of event no.3

| Model Number | P1 (km) | P2 (km) | P3 ($km.s^{-1}$) | P4 ($km.s^{-1}$) | P5 ($km.s^{-1}$) |
|---|---|---|---|---|---|
| 1 | 9 | 14 | 3.7 | 3.4 | 3.7 |
| 2 | 9 | 17 | 3.7 | 3.4 | 3.9 |
| 3 | 12 | 8 | 3.7 | 3.4 | 3.9 |
| 4 | 6 | 17 | 3.9 | 3.4 | 3.7 |
| 5 | 9 | 14 | 3.7 | 3.6 | 3.5 |
| 6 | 9 | 11 | 3.7 | 3.6 | 3.5 |
| 7 | 6 | 17 | 3.7 | 3.6 | 3.5 |
| 8 | 12 | 8 | 3.7 | 3.6 | 3.5 |
| 9 | 12 | 14 | 3.7 | 3.6 | 3.9 |
| 10 | 12 | 11 | 3.7 | 3.6 | 3.9 |
| 11 | 9 | 14 | 3.7 | 3.6 | 3.7 |
| 12 | 12 | 11 | 3.7 | 3.6 | 3.7 |
| 13 | 9 | 17 | 3.7 | 3.6 | 3.7 |
| 14 | 12 | 8 | 3.7 | 3.6 | 3.7 |
| 15 | 9 | 11 | 3.5 | 3.6 | 3.5 |
| 16 | 6 | 17 | 3.5 | 3.6 | 3.5 |
| 17 | 9 | 14 | 3.5 | 3.6 | 3.7 |
| 18 | 9 | 17 | 3.5 | 3.6 | 3.7 |
| 19 | 12 | 8 | 3.5 | 3.6 | 3.7 |
| 20 | 12 | 14 | 3.5 | 3.6 | 3.9 |
| 21 | 12 | 11 | 3.5 | 3.6 | 3.9 |
| 22 | 9 | 17 | 3.5 | 3.6 | 3.9 |
| 23 | 12 | 14 | 3.7 | 3.4 | 4.1 |
| 24 | 12 | 11 | 3.7 | 3.4 | 4.1 |
| 25 | 12 | 14 | 3.7 | 3.6 | 4.1 |
| 26 | 12 | 17 | 3.7 | 3.6 | 4.1 |
| 27 | 15 | 8 | 3.7 | 3.6 | 4.1 |
| 28 | 12 | 14 | 3.5 | 3.6 | 4.1 |
| 29 | 12 | 17 | 3.5 | 3.6 | 4.1 |

the thickness parameters again yield a crustal thickness in the range 36 km to 42 km. All the models are characterized by a thick pile (13-16 km) of low velocity material, in agreement with the results of reflection and refraction studies in this region. As in the first inversion, it is apparent that a structure with abrupt velocity variations, including velocity inversions within the lower crust is most consistent with the data.

The data inversion from event no.3 yielded 29 possible solutions (Table 5). The reason for this large number is that several of the models represent an almost homogeneous S-velocity structure with depth. When this happens, there is very little control over the layer thicknesses. As an example, models no 11,12,13,14 represent very nearly the same structure. It must be observed however that solutions such as 1,2,3,4, with very distinct velocity layering are also consistent with the data. The crustal thickness given by this inversion is again in the range 32 km - 42 km, the value of

32 km being derived from models 23 to 29, again on the basis of a crust-mantle transition region. For this event (no 3), it must be kept in mind that the wave path is more than 1000 km long and samples a very heterogeneous structure in as much the first half of the wave path represents a complex tectonic situation (subduction zone, e.g. Caputo et al., 1970). Therefore the models resulting from this inversion must at best be regarded as a very average representation of the real structure of the Ionian sea.

## DISCUSSION AND CONCLUSION

If we consider acceptable only those models with a sharp velocity discontinuity between crust and mantle, then the whole set of solutions from the 3 inversions yields a crustal thickness restricted in the range 36 km-46 km, which may be classified as a normal to thick continental crust. It should be pointed out here that even if one removes the maximum thickness, 16 km, of low-velocity material (which occurs in the Crete-Malta models), one is still left with 24 km of crust, much thicker than normal oceanic crust. This argues against the hypotheses of an oceanic crust overlain by a thick sedimentary sequence (Erickson et al.,1977). If on the other hand, we accept the idea a crust-mantle transition zone, we still find a crust of not less than 32 km.

This crust overlies a 12 km thick transition layer with an S-velocity of at least 4.1 $km.s^{-1}$.

The well-defined stratification observed in almost all the found models is consistent also with several recent models of stable continental crusts, notably that by Mueller (1977) in which he emphasizes the complexity of the velocity-depth function and postulates the presence within the crust of low-velocity layers and a high-velocity tooth ($V_P$ = 7.1 $km.s^{-1}$, which corresponds to $V_S$ = 4.1 $km.s^{-1}$ if we assume the relation $V_P = \sqrt{3}\, V_S$). We believe that the discrepancy which appears between results from surface wave dispersion and those from reflection and refraction seismology, concerning crustal thickness, may derive from a misinterpretation of such a high-velocity layer as the beginning of mantle material. The discrepancy would thus be more apparent than real.

It is interesting to compare the data obtained in the present investigation with other group velocity data for the Mediterranean. In Fig.2, group velocity dispersion curves are shown for two profiles in the Ionian Sea (events no.4 and 5) lying almost perpendicular to those in this study (D'Ingeo, 1978). The two sets of curves are very similar at periods longer than 20 s, indicating a continuity of continental structure at depth . The profile from event no.5 runs directly along the Mediterranean ridge. The fact that this event produces markedly lower group velocities at periods shorter than 20 s, may be considered an indication that the ridge is only a surface feature, probably a pile of low-velocity, sedimentary material.

Another interesting observation results from a comparison of our group velocity curves with those for the Balearic and Tyrrhenian bathyal plains in the Western Mediterranean (Mantovani et al., 1979) both of which have been shown to possess the characteristics of young ocean basins (Panza and Calcagnile,1979). These curves are also shown in Fig.2. The obvious difference between the two types of curve leaves little doubt about the non-oceanic character of the crust beneath the Ionian Sea.

Acknowledgements

We are grateful to Prof.St.Mueller and Prof.R.Schick for their initiative and help in making available the seismograph to the University of Malta, and to Dr. E.Wielandt and Mr.F.Marillier for their assistance with its installation.

Part of this work was carried out at the Physics Dept.,University of Malta. In this respect, one of us (P.F.) is grateful for the help of Dr.P.J.Vella and Dr.M.Attard. The same author is thankful for the scholarship support given by the Italian Ministry of Foreign Affairs for her stay at the Istituto di Geodesia e Geofisica of the University of Bari.

References

Caputo, M., Panza, G.F., and Postpischl, D., 1970, Deep structure of the Mediterranean basin, J.Geophys.Res., 75:4919

D'Ingeo, F.,1978, Regionalizzazione della crosta e del mantello superiore nell'area del Mediterraneo centro-orientale, Tesi in Scienze Geologiche, Università degli Studi di Bari.

Erickson, A.J., Simmons,G. and Ryan, W.B.F.,1977, Review of heatflow data from the Mediterranean and Aegean Seas, International Symposium on the Structural History of the Mediterranean Basins, Split(Yugoslavia), B.Biju-Duval and L. Montadert, eds., Editions Technip. Paris:263.

Levshin, A.L., Pisarenko, V.F. and Pogrebinsky,G.A.,1972, On a frequency-time analysis of oscillations., Ann.Geophysics, 28:211.

Mantovani, E., Nolet, G. and Panza G.F.,1979, Struttura crostale ed eterogeneità laterale nella regione italiana dallo studio della dispersione delle onde di Rayleigh di breve periodo (risultati preliminari).Seminario sull'integrazione dei dati di sismologia attiva e passiva. Milano Dec.12,1979. CNR-P.F.Geodinamica-S.P.Reti sismiche.Pubbl.n.315.

Mueller, S., 1977, A A new model of the continental crust, in "The Earth's Crust" J.G.Heacock, ed., Geophys.Monogr.Series, 20. A.G.U. Washington D.C.289.

Panza, G.F. and Calcagnile, G.,1979, The upper mantle structure in Balearic and Tyrrhenian bathyal plains and the Messinian salinity crisis., Paleogeogr., Palaeclimatol.,Palaecol.,29:3.

Prodehl, C., 1977, The structure of the crust-mantle boundary beneath North America and Europe as derived from explosion seismology, in "The Earth's Crust", J.G.Heacock, ed.,Geophys. Monogr.Series,20 A.G.U. Washington D.C. 349.

## ON THE "400-KILOMETERS" DISCONTINUITY IN THE MEDITERRANEAN AREA

G. Scalera,[1] G. Calcagnile,[2] and G.F. Panza[2]

(1) Istituto Nazionale di Geofisica - Roma
(2) Istituto di Geodesia e Geofisica - Università di Bari

### INTRODUCTION

It has been shown by Schwab et al.(1974) and Calcagnile and Panza (1974) that the study of Sa waves propagation is an extremely powerful tool for investigating the properties of the so called 400-km discontinuity. Nolet (1977) by using stacking techniques, has obtained the dispersion relations for the first seven Rayleigh modes in Western Europe in the period range typical for Sa waves. Over tectonically heterogeneous areas, like the European-Mediterranean area (Panza et al. 1980), it may be preferable the use of the standard two-station method (Panza, 1976), which allows the sampling of more limited areas, even if by this procedure the accuracy obtainable in phase velocity determinations is less than the accuracy proper of dispersion relations obtained by stacking.

In a previous paper Calcagnile et al.(1979) have proposed a set of models for the upper mantle beneath North-central Italy, based on fundamental Rayleigh mode dispersion. Two main questions are left open by their analysis, that is the existence or not of a clear lid to channel contrast and the S-wave velocity in the subchannel. In order to resolve better the upper mantle models we have made a simultaneous inversion of the dispersion relations available for the fundamental mode (Calcagnile et al. 1979) and for the first two higher modes (Panza and Scalera, 1978) of Rayleigh waves.

### INVERSION

The inversion was made in a 7-dimension parameters space: lid, channel and subchannel thicknesses and S-wave velocities, "spinel" layer S-wave velocity. The results of the inversion are listed in Table 1. A common feature to all solutions of the inversion problem is the rather low velocity in the sub-Moho material and this is an interesting result, since on the basis of higher modes dispersion data we can reject the class of models with a clear lid-to-channel contrast which can not be disregarded using only fundamental mode data (Calcagnile et al. 1979). Furthermore the channel velocity is fixed at 4.2 km/sec . Thus we may conclude that the litosphere is about 30-50 km thick with a crust of about 30 km . The lid overlies a low velocity layer extending to depth slightly exceeding 300 km . At this depth the S-wave velocity reaches values exceeding 5.0 km/sec . The result of the inversion thus indicates that the strong velocity gradient known as the "400-km" discontinuity, possibly related to the olivine-spinel phase transition, seems to take place at a depth of only about 300 km . In Fig.1 are summarized the informations at present available on the position of the discontinuity possibly related to the olivine-spinel phase transition. With the available data it is possible to bound the anomalous region to the south in corrispondence of the North African coasts (Mseddi, 1976) and to the north in corrispondence of the Baltic Shield (Calcagnile and Panza, 1978), while the eastern and western boundaries are still unresolved.

Fuchs et al.(1971) have proposed the existence of a correlation between the position of the "400-km and 600-km discontinuities" and the geoid ondulation. Were these discontinuities are at "normal depth" the geoid shows negative ondulation, while geoid elevations are found in corrispondence of areas where these discontinuities are detected at shallower depths. More recent and detailed geoid models

Table 1. Acceptable models for hedgehog inversion of fundamental and higher Rayleigh modes

| Solution n. | P1 km | P2 km | P3 km/sec | P4 km/sec | P5 km/sec | P6 km/sec | P7 km |
|---|---|---|---|---|---|---|---|
| 1 | 15 | 280 | 4.20 | 4.20 | 5.25 | 5.10 | 10 |
| 2 | 15 | 280 | 4.20 | 4.20 | 5.35 | 5.10 | 10 |
| 3 | 15 | 280 | 4.30 | 4.20 | 5.25 | 5.10 | 10 |
| 4 | 15 | 280 | 4.30 | 4.20 | 5.15 | 5.10 | 10 |
| 5 | 15 | 280 | 4.30 | 4.20 | 5.35 | 5.10 | 10 |
| 6 | 15 | 280 | 4.40 | 4.20 | 5.25 | 5.10 | 10 |
| 7 | 15 | 280 | 4.40 | 4.20 | 5.15 | 5.10 | 10 |
| 8 | 15 | 280 | 4.40 | 4.20 | 5.35 | 5.10 | 10 |
| 9 | 15 | 280 | 4.40 | 4.20 | 5.05 | 5.10 | 10 |

P1, P2, P7 indicate lid, channel and subchannel thicknesses respectively. P3, P4, P5, P6 indicate lid, channel, subchannel and "spinel" layer velocities respectively.

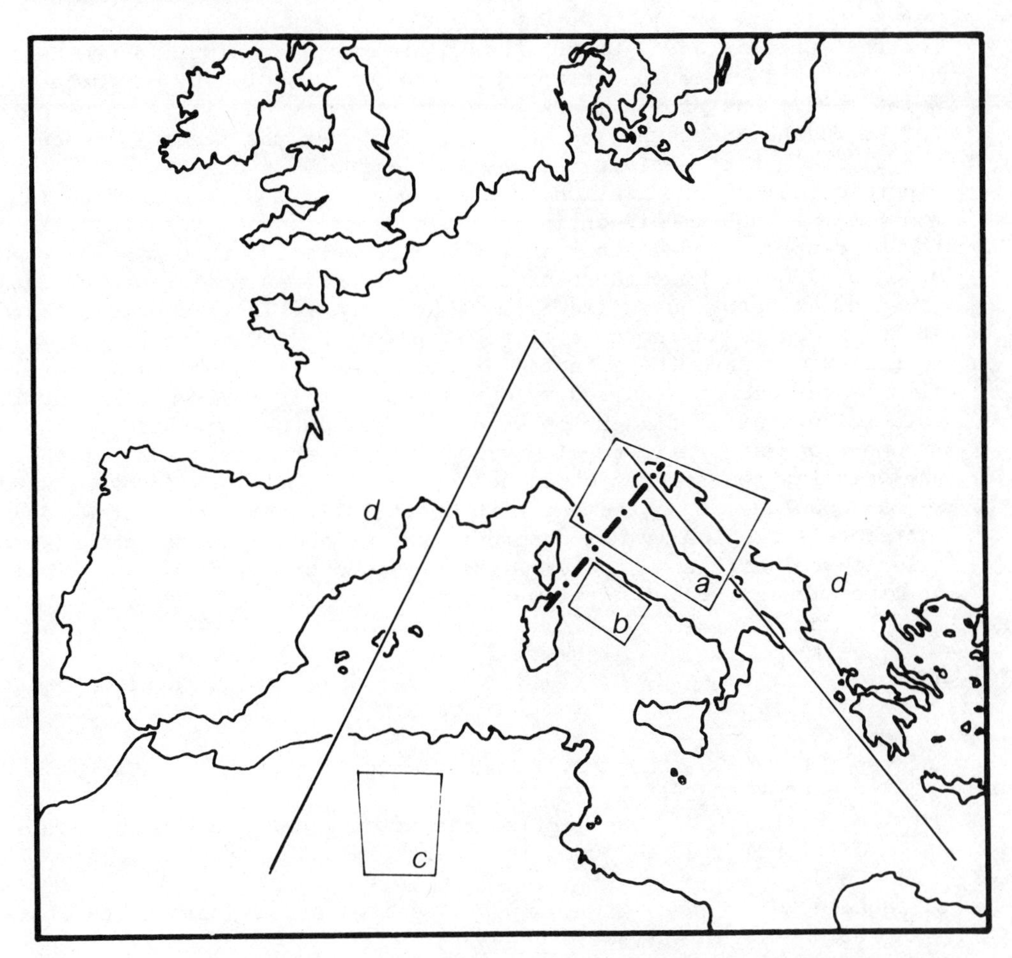


Fig. 1. a),b),c) represent areas with a rising 400-km discontinuity, a well developed L.V.Z., a normally located discontinuity, respectively (from Mseddi, 1976); d) represent large areas with a rising 400-km discontinuity from Mayer-Rosa et al. (1973); the dashed line represents our dispersion profile.

(e.g. NASA, 1977) and seismological data confirm Fuchs et al. observation. However it must be observed that a rise of about 80 km of the so called "400-km" discontinuity would generate long wavelength gravity anomalies exceeding by at least one order of magnitude the observed global gravity anomalies (Bott, 1971). Thus some compensating effect in the upper part of the upper mantle must be admitted. In other words a low density zone must be present within the first 200-250 km of the Earth Mantle in the areas where the discontinuity is at a shallower depth. The mass deficiency in the low density zone could be partly substained by termal expansion (Kaula, 1970), as also suggested by the global heat flow map, given by Chapman et al. (1979)

for the European area, which shows a remarkable maximum in the heat flow values in a region surrounding the path sampled by our data. Nolet (1977) has proposed the presence of a low density zone below 100 km in the European area and concluded that the thermal affect is not sufficient to explain the mass deficiency, invoking chemical or mineralogical stratification in the upper 300 km of the mantle. The rise of the "400-km discontinuity" can be explained by an increase of the percentage iron content in the mantle material at depths of the order of 300-400 km (Anderson, 1968). In fact extrapolating the data given by Akimoto and Fujisawa(1968) it is possible to conclude that at these depths an increase of about 10% of $Fe_2SiO_4$ with respect to $Mg_2SiO_4$ would have the effect of rising the "olivine-spinel" transition by the amount required by our data. If this is true than the low velocities that we find below about 200 km can be justified by the iron enrichment, while the low velocities in the upper part of the low velocity zone can be explained by partial melting. In other words we may conclude, in agreement with Nolet(1977), that even if we detect a seismologically homogeneous layer of about 300 km in thickness, this low velocity zone is very likely inhomogeneous from a chemical or mineralogical point of view.


Akimoto, S., Fujisawa, H., 1968, Olivine-spinel solid solution equilibria in the sistem $Mg_2SiO_4$-$Fe_2SiO_4$, J.Geophys.Res., 73, 1467-1479.

Anderson, L., 1968, Chemical inhomogeneity of the mantle, Earth and Planetary Sci.Lett., 5, 89-94.

Bott, M. H. P., 1971, The mantle transition zone as possible source of global gravity anomalies, Earth and Planetary Sci.Lett., 11, 28-34.

Calcagnile, G., Panza, G. F., 1974, Vertical and SV components of Sa, Geophys.J.R.Astr.Soc., 38, 317-325.

Calcagnile, G., Panza, G. F., 1978, Crust and upper mantle structure under the Baltic shield and Barent's sea from the dispersion of Rayleigh waves, Tectonophysics, 47, 59-71.

Calcagnile, G., Panza, G. F., Knopoff, L., 1979, Upper mantle structure of North-Central Italy from Rayleigh waves phase velocities, Tectonophysics, 56, 51-63.

Chapman, D. S., Pollack, H. N., Cermak, V., 1979, Global heat flow with special reference to the region of Europe, in: "Terrestrial heat flow in Europe," V. Cermak and L. R. Rybach, ed., Springer-Verlag, Berlin.

Fuchs, K., Mayer-Rosa, D., Liebau, F., 1971, Lateral inhomogeneities of the earth's mantle and their petrological interpretation, Zeits.Fur Geophysik, 37, 937-942.

Kaula, W. M., 1970, Earth's gravity field: relation to global tectonics, Science, 169, 982-985.

Mayer-Rosa, D., Mueller, S., 1973, The gross Velocity-Depth distribution of P- and S-waves in the upper mantle of Europe from earthquake observations, Zeits.Fur Geophysik, 39, 395-410.

Mseddi, R., 1976, Determination sismique de la structure du manteau superieur sous le Mediterranee, PhD Thesis, Université P. et M. Curie, Paris VI.

NASA, 1977, Global detailed gravimetric geoid based upon a combination of the GSFC GEM-10 Earth model and 1 X 1 surface gravity data, Goddard Space Flight Center Greenbelt, Maryland.

Nolet, G., 1977, The upper mantle under western Europe inferred from the dispersion of Rayleigh modes, J.Geophys. 43, 265-285.

Panza, G. F., 1976, Phase velocity determination of fundamental Love and Rayleigh waves, Pure and Applied Geophys., 114, 753-763.

Panza, G. F., Calcagnile, G., Scandone, P., Mueller, S., 1980, La struttura profonda dell'area mediterranea, Le Scienze, 24, n 141, 60-69.

Panza, G. F., Scalera, G., 1978, Higher mode dispersion measurement, Pure and Applied Geophys., 116, 1283-1294.

Schwab, F., Kausel, E., Knopoff, L., 1974, Interpretation of Sa for a shield structure, Geophys.J.R.Astr.Soc., 36, 737-742.

# THE THICKNESS OF PHASE TRANSFORMATION LAYERS IN THE EARTH'S MANTLE

Jacek Leliwa-Kopystyński

Institute of Geophysics
Polish Academy of Sciences
00-973 Warszawa, Pasteura 3, Poland

The most important phase transformations in the Earth's mantle are those related to silicates of magnesium and iron at depths of about 400 km (olivine → spinel transformation) and 650 km ($Mg_2SiO_4$ spinel → 2MgO periclase + $SiO_2$ stishovite transformation according to Sung and Burns (1975) and Navrotsky (1977), or transitions of $MgSiO_3$ leading to a perovskite structure according to Liu (1977)). In spherically symmetric models of the Earth (e.g. in model 1066B of Gilbert and Dziewoński, 1975) phase transformations correspond to the surface of discontinuity in density $\rho$ and seismic wave velocities $v_p$ and $v_s$. There are several causes which might change the position depth of phase equilibrium surfaces or which might transform those surfaces into layers with a defined thickness:

(i) Differences in regional temperature distributions in the mantle induce deviations of phase equilibrium surfaces from the spherically symmetrical shape.

(ii) A complex mineral composition of transformable materials (e.g. olivine$(Mg_xFe_{1-x})_2SiO_4$, $x \approx 0.9$) results in the formation on the p,T (pressure, temperature) plane, instead of the phase equilibrium line determined by the Clausius-Clapeyron equation, of an area of low- and high-pressure phase coexistence in the Earth's interior (Fig. 1, a).

(iii) In areas with vertical displacements (mid-oceanic ridges, subduction zones) phase transition kinetics leads to the transformation of the equilibrium surfaces into a layer even for minerals of simple chemical composition like $Mg_2SiO_4$ (Fig. 1, b).

Two independent approaches to the thickness determination of a phase transformation layer in the mantle were used by the author (Leliwa-Kopystyński et al., 1979; Leliwa-Kopystyński, 1980):

(i) The phase transformation occuring in homogeneous material along the Clausius-Clapeyron phase equilibrium line. Under some

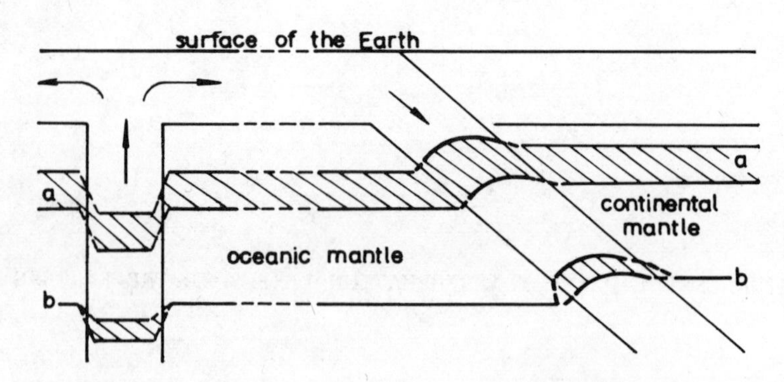


Fig. 1. Diagram of phase separation layers in the Earth's mantle.

simplifying assumptions the total thickness Z of the transformation layer measured in the direction of its motion and the temperature change $\Delta T$ in this layer, caused by the transformation, are

$$Z = b \frac{\rho_2 c_{tr}^2}{c_1 g T_{T,0} (\rho_2 - \rho_1) \sin\theta} \qquad T = b \frac{c_{tr}}{c_1} \tag{1}$$

where $c_{tr}$ denotes the latent heat of transition, $c_1$ is the specific heat of a low-pressure phase, $\rho_1$ and $\rho_2$ are low- and high-pressure phase densities, $T_{T,0}$ is the absolute preliminary temperature, g is the terrestial acceleration ($g \approx 10$ m s$^{-2}$ in the entire mantle), and $\theta$ denotes the inclination angle of the motion. The coefficient $b = 1$ for the adiabatic case, $b = 0$ for the isothermic one, and $b<1$ for the other cases. The isothermic value has a simple explanation: for very slow motion the heat produced during the transition is dispersed and the phase transition can be considered as an isothermic one, and the transition layer is, therefore, reduced to the surface, similarly as it is outside the areas in motion (Fig. 1, b). From formula (1) it follows that the thickness of the $Mg_2SiO_4$ olivine → spinel transition layer in the middle of a descending slab is $Z\sin\theta \approx 10$ km.

(ii) The transformation from the metastable state. A simple model of phase transformation kinetics, assuming thet the transformation probability in a given area is proportional to the difference between the actual pressure and the phase equilibrium pressure corresponding to the actual temperature, leads to a formula determining the fraction $\nu$ of the medium mass which has undergone the transformation at the depth $\zeta$ measured in the direction of motion in relation to the phase equilibrium level. Under an assumption that the downgoing area is divided into small domains of equal size which can be trabsformed independently and instantaneously in whole volume (Fig. 2), the following relation was formed

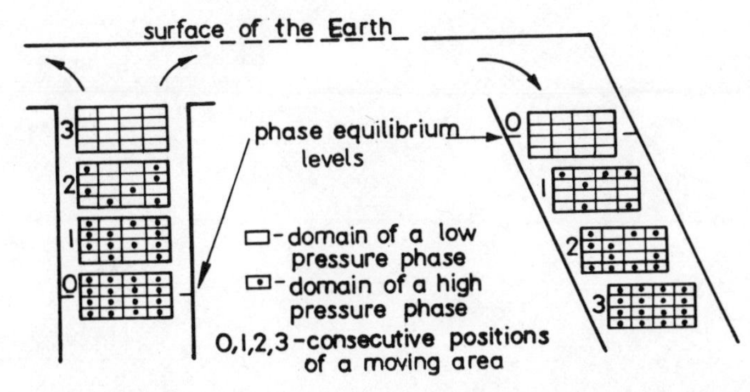


Fig. 2. Stepwise phase transitions in moving areas.

$$\nu = 1 - \exp[ - (A\rho g \sin\theta/2v)\zeta^2] \quad (2)$$

where v is the local velocity of a down-or upgoing area, and A is the kinetic parameter which is determined from seismic data.

References:

Gilbert, F., Dziewoński, A. M., 1975, An application of normal mode theory to the retrieval of structural parameters and source mechanism from seismic spectra, Phil. Trans. Roy. Soc. London, A, 278:187.

Leliwa-Kopystyński, J., Lacam, A., and Peyronneau, J., 1979, Polymorphic transformation in the Earth's mantle structure of the phase boundary, in: "High-Pressure Science and Technology," K. D. Timmerhaus and M. S. Barbes, ed., Plenum Press, New York and London.

Leliwa-Kopystyński, J.,1980, Probability of the earthquake initiated by the phase transformation from the metastable state, to be published in Proceedings of the 7th High Pressure AIRAPT Conference, Pergamon Press, Oxford.

Liu, L. G., 1976, The high pressure phases of $MgSiO_3$, Earth Planet. Sci.Lett., 31:200.

Navrotsky, A., 1977, Calculation of effect of cation disorder on silicate spinel phase boundaries, Earth Planet. Sci. Lett., 33:437.

Sung, C. M., Burns, R. G., 1975, Kinetics of high pressure phase transformations: implications to the evolution of the olivine → spinel transition in the downgoing lithosphere and its consequences on the dynamic of the mantle, Tectonophysics, 31:1.

# A CONCISE INVERSION SYSTEM FOR REFRACTION OBSERVATIONS OVER PLANE INTERFACES WITH ARBITRARY STRIKE AND DIP

K. Helbig

Vening Meinesz Laboratory, Rijksuniversiteit Utrecht
P.O. Box 80.021, 3508 TA Utrecht, The Netherlands

## INTRODUCTION: THE PROBLEM

Assume that in a refraction survey arrival times of refracted waves have been observed along at least two lines of different azimuth, and on at least one of the lines, the "main" line, observations in reversed directions have been taken. The set of observed arrival times is complete (i.e. there is a one-to-one relationship between interfaces in the ground and branches of the travel time curves and all segments of travel time curves are straight (thus the media between the interfaces are homogeneous). Moreover, we assume isotropy for all layers (this assumption can be replaced by that of independent information about shape and orientation of the slowness surfaces of the media).

This information is sufficient to obtain the spatial disposition of all interfaces and the velocities below all interfaces in closed form. However, such closed solutions are so complex that the convey but little insight into the problem. The method explained here is of the iterative type. The main advantage of such methods is didactic economy: it is only neccessary to show (a) the construction of the first interface and (b) the construction of the $(m+1)$th interface provided the interfaces with numbers 1 to $m$ are known (together with the corresponding velocities).

The derivation is nearly exclusively based on geometric arguments and geometric constructions. Geometric inversion often is accurate enough, at least if the scale is large and drawing is done carefully. Today a pocket calculator with angular functions costs less than a good drafting set, therefore, the graphical constructions should be viewed alternatively as a concise description of a numerical algorithm to calculate the parameters of the model.

The iterative nature of the processes described allows one to construct *any* number of layers. This statement, though correct in a mathematical sense, should not be taken literally: the method is applicable only to plane layers, and plane layers of different spatial attitude must intersect at a finite distance. Thus the assumption that there is a one-to-one relationship between interfaces and observed travel time branches can be satisfied only in a limited region. This in turn limits the depth from which critically refracted rays can be observed.

## THE BASIC TOOLS

### The Slowness Vector

Slowness is the inverse of the velocity. It is numerically equal to the time needed to cover the unit distance. The propagation of a plane wave can be described by the *slowness vector*, whose magnitude is proportional to the slowness and whose direction is perpendicular to the wave plane. The end points of all slowness vectors possible in a homogeneous medium lie on the *slowness surface*. In isotropic solids, for example, the slowness surface consists of two spherical sheets, an outer one corresponding to shear waves and an inner one corresponding to compressional waves.
The projection of a slowness vector on a reference is the *trace slowness* describing direction of the projection and "time to cover unit distance" of the trace (line of intersection of the reference plane and the wave plane).
The projection of a slowness vector on an arbitrary line (e.g. a geophone spread) is the *apparent slowness* (e.g. the time difference between two geophones separated by a unit distance along the line). The slope of a time-distance graph is equal to the apparent slowness. For plane parallel horizontal layer the apparent slowness is identical with ray parameter $p$.

### Snell's Law

A plane wave falling on an interface causes the generation of reflected and transmitted waves. To be causally connected these waves must have a common trace line. Therefore, they must have common trace slowness, i.e. common projection of the slowness vectors. Thus Snell's law can be stated as

"The endpoints of the slowness vectors of all waves participating in *one and the same* reflection/refraction process lie on a common normal to the interface and on the corresponding slowness surfaces".

Remark: Common trace slowness is necessary for causality, but not sufficient. There is always at least one non-causal incident wave with the same trace slowness, and there are situations of waves that are strictly not coupled, like P and SH waves.

Remark: the above derivation does not make any assumption about isotropy. Thus Snell's law in the form expressed here is generally valid.

## THE SOLUTION

### Construction of the first interface in the main construction plane

Figure 1 shows the reversed travel time graph along the line $AB$ (arbitrarily - without loss of generality - assumed to be of length 2). Slownesses can be read off directly at the auxiliary line (distance 1 from $A$ and $B$). $t_R$ is the "reciprocal time" that must be the same for both observation directions, $t_{i,A}$ and $t_{i,B}$ are the intercept times, i.e. the intersections of the backward extension of the high velocity branches of the travel time curve with the time axis in the source points.

The low velocity branches correspond to direct waves along the earth's surface, their slope is thus the slowness itself, not a projection. Thus the slowness surface is the sphere with the radius $n_1$. The two high velocity branches yield the apparent slownesses $p_r$ and $p_\ell$ for the observations towards right and left, respectively. They are projections of the slowness vectors corresponding to the waves arriving at the surface at $A$ and $B$, respectively (figure 2 and 3). The *construction* plane used for figures 2 to 5 is perpendicular to the refractor and contains the line $AB$. The construction plane is vertical only if $AB$ is at right angles to the strike of the refractor.

The two slowness vectors in figure 2 and 3 are the results of different experiments. However, the principle of reciprocity lets us conclude that the ray arriving at $B$ has started at $A$ with the slowness vector $-p_\ell$ ($OA''$ in figure 4). The ray starting at $A$, the ray along the interface, and the ray arriving at $B$ "participate in one refraction process", therefore the endpoints $B'$ and $A''$ lie on a common normal to the interface. Moreover, the slowness vector corresponding to the ray along the interface ($OC$) must be perpendicular to the common normal, which therefore is tangent to the slowness surface in the second medium.

We now know the attitude of the refractor in the construction plane and the slowness $n_2$. To obtain the position of the refractor - and thus the actual raypath - several possibilities exist. The simplest is based on the intercept times: The high-velocity branch

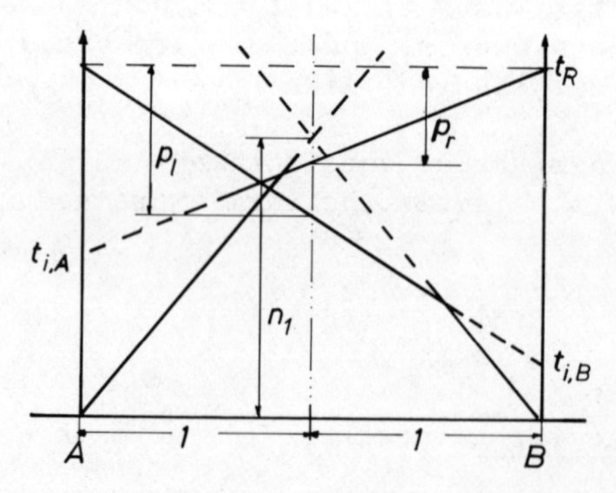


Fig. 1. Time - distance graph of reversed refraction observations over a dipping interface. Separation $AB$ is two distance units. $n_1$ is slowness in overburden, $p_\ell$ and $p_r$ apparent slownesses in observations towards left and right, respectively.

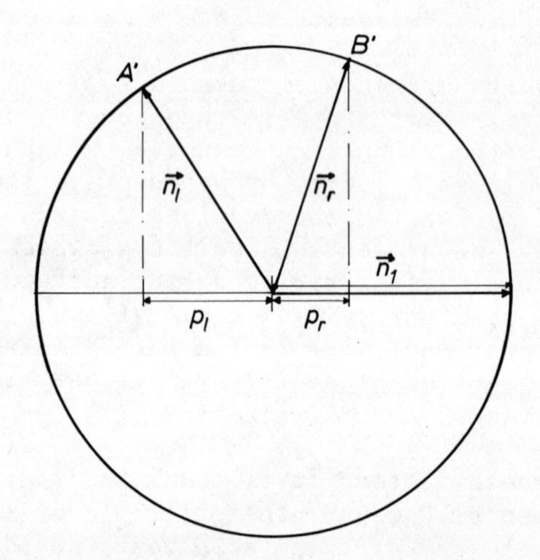


Fig. 2. Construction of direction of rays arriving at $A$ and $B$, respectively.

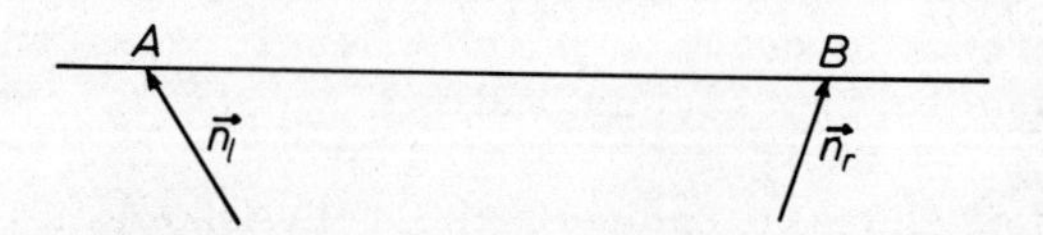


Fig. 3. First step in constructing ray paths.

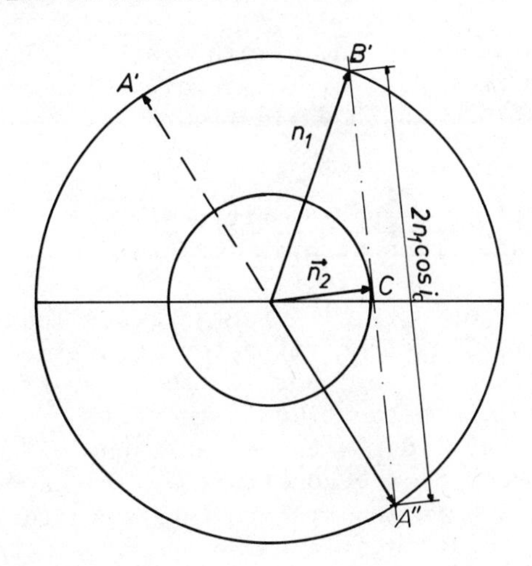


Fig. 4. The slowness vectors belonging to the ray from $A$ to $B$ have their endpoints on a common normal to the refracting interface. Length of $B'A''$ is equal to the conversion factor $2.n_1 \cdot \cos i_c$.

of the travel time curve belonging to the wave traveling from $A$ to $B$ can be expressed as

$$t(x) = t_{i,A} + p_r . x.$$

We divide the ray $A - C_A - C_B - B$ into three segments ($\bar{A}\bar{\bar{A}}$, $\bar{A}\bar{B}$, $\bar{B}C$) (figure 5). Since $A\bar{\bar{A}}$ and $\bar{B}\bar{\bar{B}}$ are wave fronts, the time needed for $\bar{A}\bar{B}$ is equal to that needed for $\bar{\bar{A}}\bar{\bar{B}}$, viz. $n_2.x.\cos\alpha$. The time needed for $A\bar{A}$ and $B\bar{B}$ is $h_A.\cos i_c.n_1$ and $h_B.\cos i_c.n_1$, respectively. Since $h_B = h_A - x.\sin\alpha$ we have

$$t(x) = 2.h_A.\cos i_c.n_1 + (n_2.\cos\alpha - n_1.\sin\alpha.\cos i_c).\ x,$$

thus $t_{i,A} = h_A.2.\cos i_c.n_1$. In other words: to obtain $h_A$ one has to divide $t_{i,A}$ by the "conversion factor" $2.n_1.\cos i_c$. But from figure 4 we see that the length of the "parameter line" $B'A''$ is

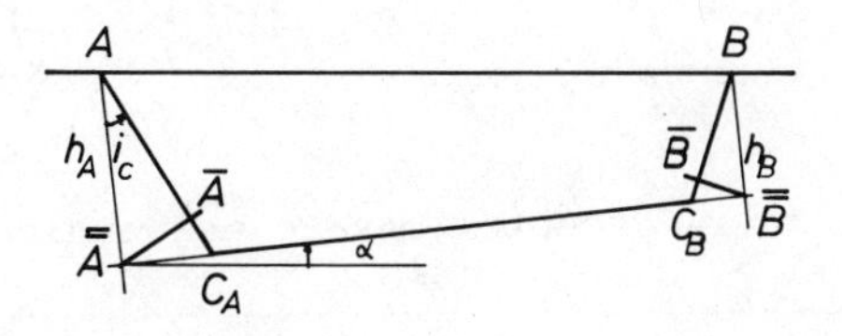


Fig. 5. Relations between perpendicular distance $h$, separation $x$, and dip $\alpha$.

$2.n_1.\cos i_c$. This leads to the simple graphical construction for $h_A$ (and $h_B$) shown in figure 6. The completed construction (within the construction plane) is shown in figure 7.

## Construction of the true spatial attitude of the interface with the help of an observation with different azimuth

A separate observation with different azimuth leads to a different construction plane. Since the refractor is assumed to be plane over the range of the survey, normals to the refractor have to be parallel. The simple construction is the following (figure 8): draw two intersecting lines with the azimuths of the two observation lines. In each of the construction planes draw a unit vector parallel to the (apparent) normal to the refractor. Plot the horizontal in-line distances of the endpoint of these unit vectors on the respective lines and draw perpendicular lines through the endpoints. Their intersection is the horizontal position of the endpoint of a unit vector normal to the refractor.

For this observation with different azimuth one need not reverse the observations since the true velocity in the second medium is already known from the observation in azimuth of the first observation (the "main line").

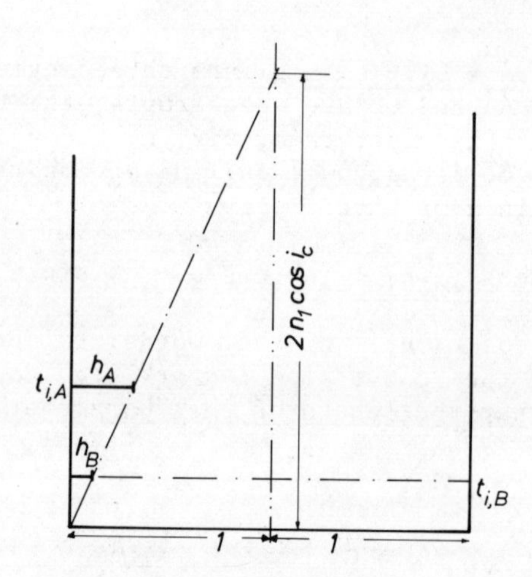


Fig. 6. Conversion of intercept times to perpendicular distances.

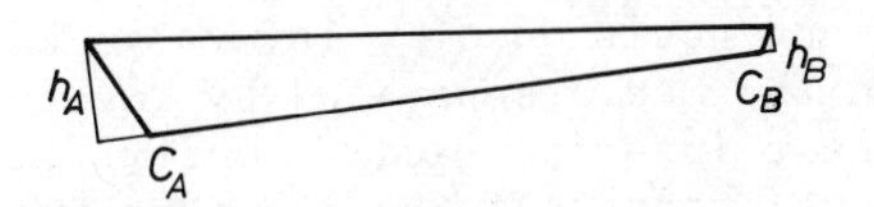


Fig. 7. Complete section in the construction plane.

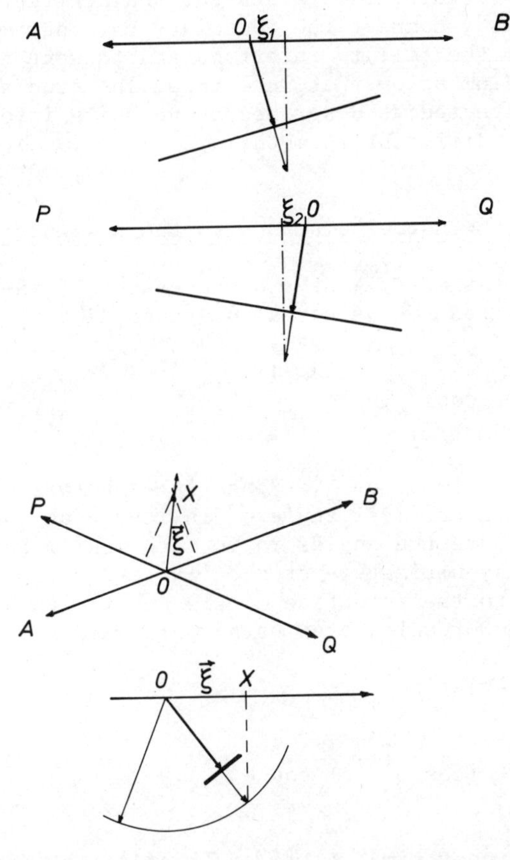


Fig. 8. Construction of true spatial attitude of refractor from observation along lines *AB* and *PQ* with different azimuth intersecting in *O*. *OX* is hypothetical line at right angles to strike. $\vec{\xi}$, $\xi_1$, and $\xi_2$ are the horizontal components of unit vectors parallel to the normal to the refractor in the different construction planes.

## Construction of the attitude of, and velocity below, the next interface

If attitude and position of $m$ layers is known together with the slownesses in the media separated by the interfaces, the attitude of the next interface and velocity below it can be constructed as follows: construct the slowness vector of the waves arriving at $A$ and $B$, respectively. Using Snell's law construct the slowness vectors in the next medium, continue until you obtain the slowness vectors arriving at interface $m$ from below. Invert one of the two slowness vectors to obtain a set that corresponds to "one refraction process", connect the endpoints by a parameter line that is normal to the $(m+1)$th interface and tangent to the slowness sphere of the medium below that interface. The true spatial attitude is constructed as described above using information from observations with different azimuth.

## Determination of the normal distance to the new interface

The intercept time $t_{i,A}^{(m)}$ of the $m$th branch of the travel time curve at $A$ is related to the normal distance $h_A^{(m)}$ by (see figure 9):

$$t_{i,A}^{(m)} = h_{A,m}^{(m)} \cdot 2 \cdot \cos i_m \cdot n_m + \sum_{\ell=1}^{m-1} h_{A,\ell}^{(m)} \cdot (\cos \beta_{A,\ell}^{(m)} + \cos \beta_{B,\ell}^{(m)}) \cdot n_\ell ,$$

where $h_{A,\ell}^{(m)}$ is the segment of $h_A^{(m)}$ that falls between interface $m$ and $m-1$, $i_m$ is the critical angle of incidence at interface $m$, and $\beta_{A,\ell}^{(m)}$ and $\beta_{B,\ell}^{(m)}$ are the angles in the $\ell$th medium between the segments of the ray ultimately critically refracted at interface $m$ and the normal to the interface $m$. All quantities are known (or can be constructed or calculated using information already at hand) except $h_{A,m}^{(m)}$.
We write for abbreviation

$$t_{i,A,\ell}^{(m)} = h_{A,\ell}^{(m)} (\cos \beta_{A,\ell}^{(m)} + \cos \beta_{B,\ell}^{(m)}) \cdot n_\ell .$$

The "partial intercept time" $t_{i,A,\ell}^{(m)}$ related to the segment of normal distance $h_{A,\ell}^{(m)}$ by the conversion factor

$$n_\ell (\cos \beta_{A,\ell}^{(m)} + \cos \beta_{B,\ell}^{(m)})$$

which is the projection of the two slowness vectors between interface $\ell-1$ and layer $\ell$ on the normal to interface $m$.
With this we can write

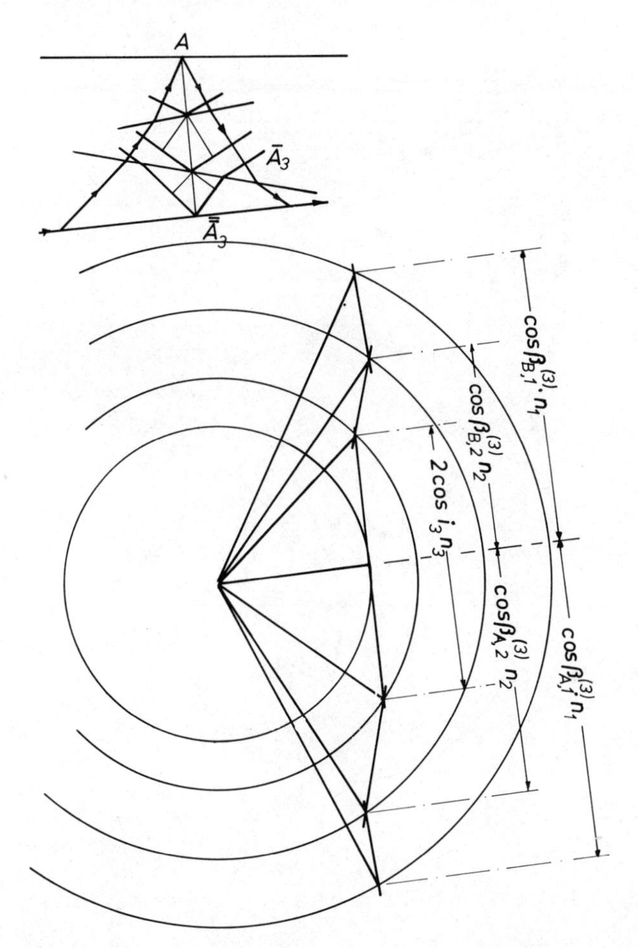


Fig. 9. Relation between intercept time and segments of perpendicular distance. The conversion factors for the different segments are equal to the projection of the corresponding slowness vectors on normal to the critically refracting interface.

$$h_A^{(m)} = \sum_{\ell=1}^{m-1} h_{A,\ell}^{(m)} + \frac{t_{i,A}^{(m)} - \Sigma\, t_{i,A,\ell}^{(m)}}{2 \cos i_m . n_m}$$

Example of a two-layer problem

Figs. 10 and 11 show the application of the method described in the previous section to a two layer problem. It has been assumed

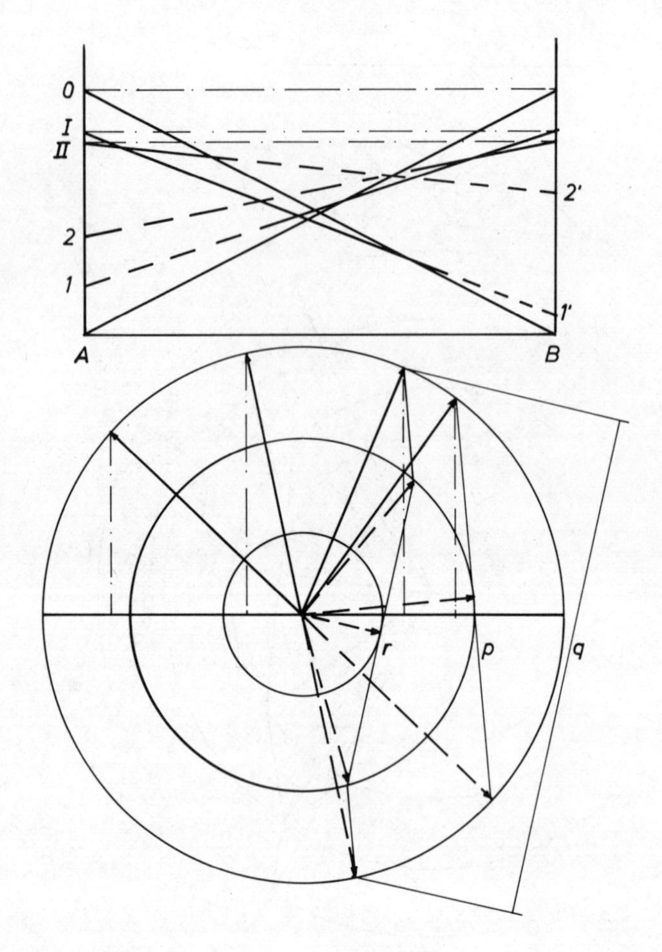


Fig. 10. Time-distance graph (above) and slowness surface construction for a plane (dipping) three-layer case. Roman numbers indicate reciprocal times, arabic ones intercept times.

that the observations were carried out at right angles to the (common) strike direction.

In figure 10 a pair of reversed travel time curves are shown (first arrivals heavily outlined). 0, I, II are reciprocal times for the three branches, 1, 2 and 1', 2' intercept times at $A$ and $B$, respectively. The distance $AB$ is arbitrarily chosen as unit. With this unit all apparent slownesses are numerically equivalent to the differences between the corresponding reciprocal times and intercept times.

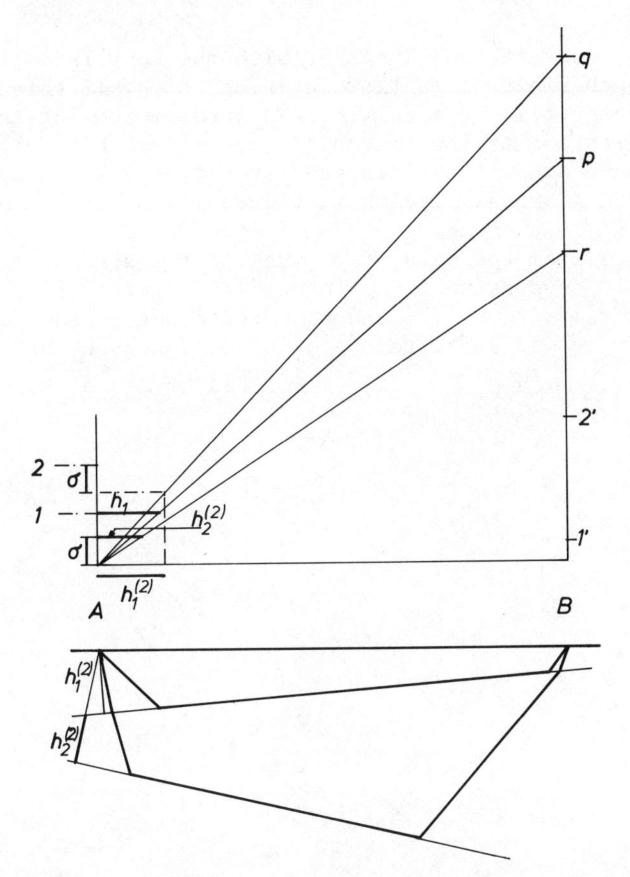


Fig. 11. Conversion of (partial) intercept times to (partial) perpendicular distances (above) and the complete section (below).

Figure 10b shows the graphical determination of the slowness vectors: slowness vectors arriving at the surface (deduced from apparent slownesses) are in heavy solid lines, and derived slowness vectors in broken heavy lines. The length of the parameter lines $p$ and $r$ and of the projection $q$ are the conversion factors needed to obtain depths from intercept times.

Figure 11a shows the construction of depths from intercept times and figure 11b the resulting structure: intercept time 1 and conversion factor $p$ yield $h_1$, thus the first interface can be constructed. The direction of $h_2$ is known from figure 10b, thus

$h_1^{(2)}$ can be constructed. Together with the conversion factor $q$ this partial depth yields a partial intercept time. If this is subtracted from intercept time 2, $\sigma$ remains. This remaining intercept time $\sigma$ combined with conversion factor $r$ gives $h_2^{(2)}$. This completes the construction, since the slownesses are already known as the radii of the three slowness circles in figure 10.

To move from the data (fig. 10a) to the structure (fig. 11b) one has only to construct one intermediate graph (fig. 10b). The construction of figure 11a can be carried out on the data graph (fig. 10a). It was separated here only for reasons of clarity.

# MAGNETOTELLURIC RESISTIVITY IN AN INHOMOGENEOUS TWO-LAYER MODEL

V. Iliceto and G. Santarato

Istituto di Mineralogia

Università di Ferrara, Italy

The propagation of natural electromagnetic (e.m.) waves in a homogeneous and isotropic tabular half-space was studied in detail by Cagniard (1953). For such a half-space he obtained the expression of the wave impedance and thus of the apparent resistivity at the surface as a function of the period of the e.m. wave.

The propagation of e.m. waves is instead radically modified if the medium is electrically inhomogeneous. This inhomogeneity may occur for various reasons. Of particular interest is the variation of the electrical conductivity $\sigma$ linked to a strong temperature gradient, which is the case in a geothermal area.

In order to evaluate the effects of these inhomogeneities, we studied a two-layer tabular model in which the overburden, of thickness h, is inhomogeneous with conductivity $\sigma$ linearly variable with the depth z:

$$\sigma = \sigma_0 \, (1 + \alpha z) \qquad (1)$$

while the substratum is homogeneous.

Let us consider a plane monochromatic e.m. wave, at frequency in the typical magnetotelluric (MT) range, normally incident on the ground surface. Assuming a time dependence $X = X_0 \exp(-i\omega t)$, the Maxwell equations turn out to be particularly simplified:

$$\begin{aligned} \vec{rot}\,\vec{E} &= i\mu\omega\vec{H} \qquad & \mathrm{div}\,\vec{H} &= 0 \\ \vec{rot}\,\vec{H} &= \sigma\vec{E} \qquad & \mathrm{div}\,\vec{E} &= 0 \end{aligned} \qquad (2)$$

where $\mu$ is the absolute magnetic permeability of the medium and $\omega$ the angular frequency.

Equations (2) provide 4 linear homogeneous differential equations for the 4 horizontal components $E_x$, $E_y$, $H_x$, $H_y$ of the e.m. field. As in the above conditions $E_z = H_z = 0$ we have (for further details see, for example, Porstendorfer, 1975):

$$\frac{\partial^2 E_x}{\partial z^2} + k^2 E_x = 0; \qquad \frac{\partial^2 H_x}{\partial z^2} - \frac{1}{\sigma}\frac{d\sigma}{dz}\frac{\partial H_x}{\partial z} + k^2 H_x = 0$$

$$\frac{\partial^2 E_y}{\partial z^2} + k^2 E_y = 0; \qquad \frac{\partial^2 H_y}{\partial z^2} - \frac{1}{\sigma}\frac{d\sigma}{dz}\frac{\partial H_y}{\partial z} + k^2 H_y = 0 \qquad (3)$$

In the first layer, since $k^2 = i\mu\omega\sigma_o(1 + \alpha z)$, equations (3) have variable coefficients. Having carried out the change of variable:

$$\xi = (\sqrt{i\mu\omega\sigma_o}/\alpha)^{2/3}\,(1 + \alpha z)$$

the solution is Airy's well-known integral (Watson, 1962). We obtain, for example, for $E_x$:

$$E_x(z) = \xi^{1/2}\,[A\,J_{-1/3}(2\,\xi^{3/2}/3) + B\,J_{1/3}(2\,\xi^{3/2}/3)] \qquad (4)$$

where $J_{\pm 1/3}$ are the Bessel functions of the first kind and of order $\pm$ 1/3. In the substratum $k_1^2 = i\mu\omega\sigma_1$ is constant and thus (3) may easily be integrated. For $E_x$ we have:

$$E_x(z) = C\,e^{-ik_1 z} + D\,e^{ik_1 z}.$$

$E_x(z)$ being known, we can determine $H_y$ from (1), since:

$$\frac{\partial E_x}{\partial z} = i\mu\omega H_y.$$

In particular, in the first layer, since $k_o^2 = i\mu\omega\sigma_o$, we have:

$$H_y(z) = \frac{\xi\alpha^{1/3}}{i\,\mu\,\omega}\,[B\,J_{-2/3}(2\,\xi^{3/2}/3) - A\,J_{2/3}(2\,\xi^{3/2}/3)].K^{2/3} \qquad (5)$$

The four integration constants A, B, C and D are determined using the boundary condition $\lim_{z\to\infty} E_x(z) = 0$, with continuity conditions of the function and its prime derivative at the discontinuity surface z = h, and remembering the homogeneity of the solution.

At every angular frequency $\omega$ the wave impedance $Z_{xy}(0)$ at the surface must coincide with the apparent impedance $\hat{Z}_a$ of an electrically equivalent homogeneous half-space. Since we know (Cagniard, ibid.) that:

$$|Z_a(0)| = (\mu\omega/\sigma_a)^{1/2} \tag{6}$$

from a comparison between (6) and the expression of $Z_{xy}(0) = E_x(0)/H_y(0)$, where $E_x$ and $H_y$ are given respectively by (4) and (5), we finally obtain the apparent resistivity $\varrho_a$ of the model:

$$\frac{\varrho_a}{\varrho_o} = \left| \frac{A\, J_{-1/3}(\beta_o) + B\, J_{1/3}(\beta_o)}{B\, J_{-2/3}(\beta_o) - A\, J_{2/3}(\beta_o)} \right|^2 \tag{7}$$

where $\beta_o = \frac{2}{3}[\xi(o)]^{3/2}$ .

Equations (4) and (5) also give formula of the phase, according to Cagniard's well-known definition (ibid).

The shape of the $\varrho_a$ curve versus frequency depends on $\alpha$ and $\sigma_1$. Having thus fixed a certain value for $\alpha$ , we obtain infinite curves for varying $\sigma_1$. The notable property of this ensemble of curves lies in the fact (see Fig. 1) that they remain coincident for small values of $\sqrt{T}$,$T = 2\pi/\omega$), chosen as abscissa according to Cagniard's formulation, until they move away one from each other towards the respective asymptotic values $\sigma_1 = 1/\varrho_1$ with gradually increasing values of $\sqrt{T}$. Before moving towards the asymptotes, they oscillate, all crossing in a very narrow band of frequencies which, for example on a bilogarithmic scale with one cycle in 7.5 cm, practically coincides with a point. To the right of this point the trend of the curves coincides with that of the Cagniard master curves.

The coordinates $T_{cp}$ and $\varrho_{cp}$ of this point have the same properties as the cross-point of Cagniard's two-layer master curves. Moreover, to that model our model is reduced when $\alpha$ tends towards zero, since $T_{cp}$ and $\varrho_{cp}$ give thickness h of the overburden:

$$h \approx (10\, T_{cp}\, \varrho_{cp})^{1/2}/8 \quad \text{(practical units)} \tag{8}$$

The interpretation of an experimental curve may thus be conducted with the usual method of matching curves using only 2 standard graphs. One contains all the families of curves obtained by varying $\alpha$ , limited to the right of the "cross-point", since the thickness of the overburden is obtained from the coordinates of this point. The other is the usual standard two-layer graph to determine the thicknesses and resistivities of the possible succession of homogeneous layers underlying the inhomogeneous overburden.

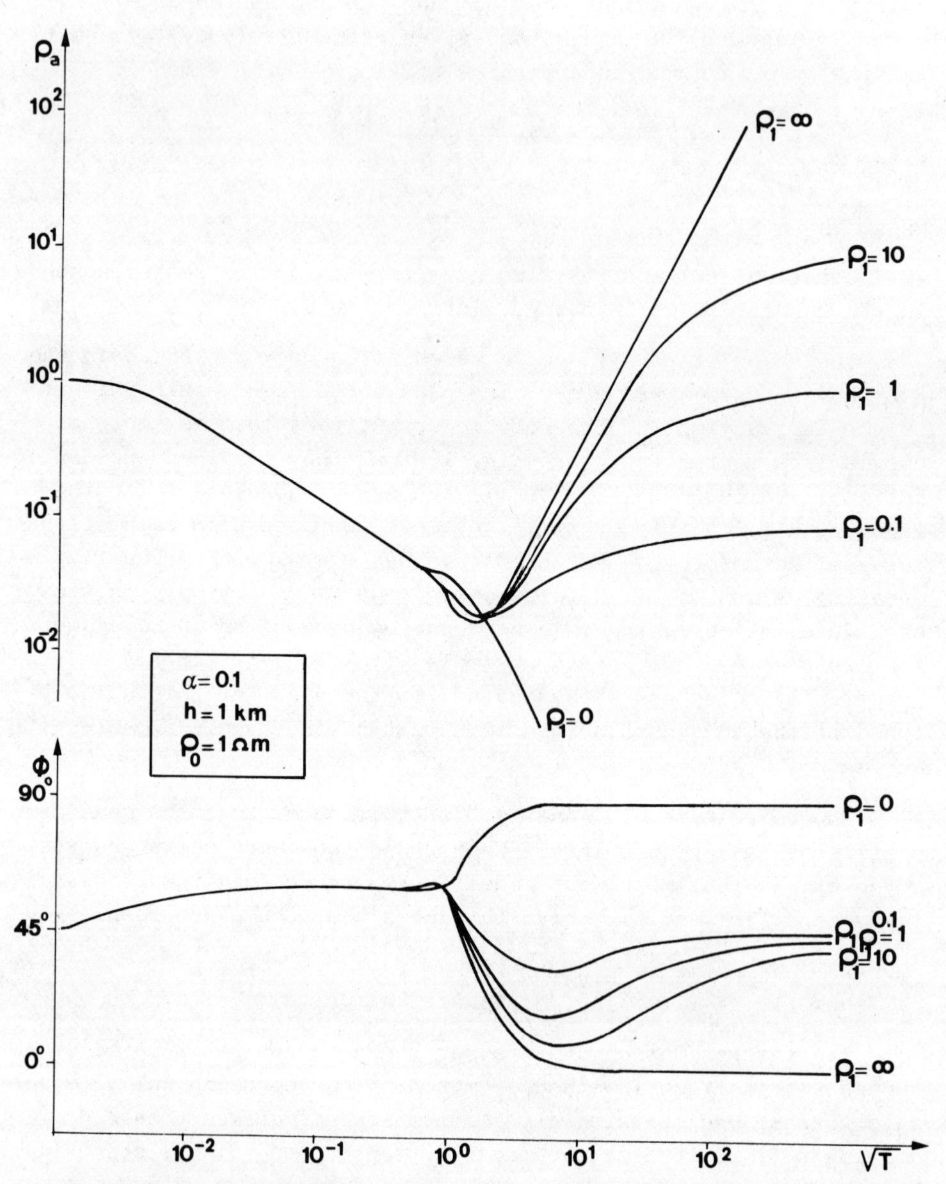


Fig. 1: Apparent resistivity and phase curves of a two-layer tabular half-space with an unhomogeneous overburden with respect to depth.

However, by using formula (7), it is possible to find directly the values of $\varrho_h$ (and thus of $\alpha$ ) and h, relative to the overburden. This can be done on a computer by means of non-linear regression with the least-squares method, applied to that branch of the experimental curve that the operator recognizes as belonging to the family of curves of which an example is given in Fig. 1. In this case, the only value which must be approximately known is $\varrho_0$, that is, the use in the field of sufficiently high frequencies is required, so as to have good information on the asymptotic trend of the apparent resistivity.

REFERENCES

Cagniard, L., 1953, Basic theory of the magneto-telluric method of geophysical prospecting, Geophysics, 18:605-635.

Porstendorfer, G., 1975, Principles of magnetotelluric prospecting. Gebrueder Borntraeger, West Berlin.

Watson, G.N., 1962, A treatise on the theory of Bessel functions. Cambridge University Press, 2nd edition.

ON THE BEHAVIOUR OF A LINEAR ELECTRICAL SYSTEM

EXCITED BY A PARTICULAR INPUT

Marcello Ciminale

c/o Istituto di Geodesia e Geofisica

Università degli Studi, 70100 Bari (Italy)

SUMMARY

We study the behaviour of a linear electrical system characterized by a frequency-dependent resistivity function when it is excited by a current-step input. The result we obtain gives indications about the treatment of discharge curves in the Induced Polarization method.

---

The model of the frequency-dependent resistivity $\rho(\omega)$ which at present appears the most suitable to fit experimental data is:

$$\rho(\omega) = \rho_0 \left[ 1 + \sum_{j=1}^{n} \frac{i b_j \omega}{a_j - i\omega} \right] \tag{1}$$

where $a_j$, $b_j$, are real positive constants, $\sum_{j=1}^{n} b_j$ is less than one and $\rho_0$ is the resistivity of the medium for a zero frequency solicitation. The model satisfies all the properties characterizing the transfer function of a linear and causal filter (Patella and Ciminale, 1979). The inverse Fourier transform of (1) is:

$$\overline{\rho}(t) = \begin{cases} 0 & t < 0 \\ \rho_\infty \delta(t) + \rho_0 \sum_{j=1}^{n} b_j a_j e^{-a_j t} & t > 0 \end{cases} \tag{2}$$

where $\rho_\infty$ is the limit of $\rho(\omega)$ for $\omega$ going to $\infty$.

From a physical point of view, formula (2) characterizes a relaxation-type electrical system. Indeed, the filter response to an impulsive input is the output (2) which decays (relaxes ) toward zero.

We wish to study now, in the time -domain, the behaviour of the above system when it is excited by an input signal of the type:

$$J(t) = \begin{cases} 0 & t<0 \\ J_o & t>0 \end{cases} \qquad (3)$$

where J is the current density.

This input is largely used both in field and in laboratory.

The corresponding electric field is given (Patella and Ciminale, I979) by:

$$E(t) = \int_o^{\infty} \rho^T(\tau)J(t-\tau)d\tau \qquad (4)$$

By means of (2) and (3) we have:

$$E(t) = J_o \rho_o \left[ 1 - \sum_{J=1}^{n} b_j e^{-a_j t} \right] \qquad (5)$$

The term $\rho_o$ times the quantity in brackets is connected to the current density by a simple algebraic product.

Moreover,it has the physical dimension of a resistivity. We are thus allowed to introduce the concept of a time dependent resistivity and consequently to see (5) as a generalization of Ohm's law in the time-domain.

The D.C. galvanic prospecting methods(in which (3) is the typical input) are based on the hypothesis of a stationary resistivity: (5) shows that this is not the case.

Nevertheless,the usual D.C. prospecting can enter into (5), provided that the electric field measurement is carried out after an infinitely long time from the instant of current injection; only in this case, the medium response is purely ohmic; in fact,

$$\lim_{t \to \infty} \rho_o \left[ I - \sum_{J=1}^{n} b_j e^{-a_j t} \right] = \rho_o \qquad (6)$$

Expression (3) is the typical solicitation also of the Induced Polarization prospecting method. In fact, the electric field (5) contains the information required in the IP method.

In conclusion,and at least in principle, we may consider the D.C. geoelectrical prospecting as the limit for t going to $\infty$ of the IP method.

In practice,however, the above limit is well approximated even after some milliseconds and moreover in general the polarization contribution to the electric field (5) has a small amplitude with respect to the ohmic response of the medium.

For this last reason in the IP prospecting it is most useful to study the discharge process provoked by an input of this type:

$$J(t) = \begin{cases} J_o & t<0 \\ 0 & t>0 \end{cases} \qquad (7)$$

To this input the system gives the following response:

$$E(t) = J_o \int_{-\infty}^{0} \varrho^{T}(t-\tau)d\tau \quad (8)$$

that is to say:

$$E(t) = J_o \varrho_o \sum_{j=1}^{n} b_j e^{-a_j t} \quad (9)$$

The physical parameter generally adopted in IP method is the well known chargeability. This parameter is simply connected to the electrical status of the system at some definite instants of charge and/or discharge processes without giving information on the evolution of the same processes. Owing to (5) and for practical reasons also to (9) we retain most useful in IP method to study the whole transient curve of the electric field in order to evaluate the parameters $a_j$ and $b_j$ which fully control such generalised electrical phenomenology.

## Reference

Patella, D., and Ciminale, M., 1979, The linear filter theory applied to the electrical response of earth materials, Boll. di Geof. Teor. ed Appl., XXI, 83.

# INVERSE PROBLEMS ON THE GLOBAL SCALE: GEOMAGNETIC POLARITY SEQUENCE AS AN INVERSE PROBLEM FOR GEOCHRONOLOGY

Giovanni Napoleone

Geology Department
University of Florence
Florence, I-50121

## INTRODUCTION

Earlier in this course, Gasparini et al. (1980) discussed the inversion approach to inference of crustal structures at great depths from magnetic field anomalies over large continental areas.

Other kinds of inverse problems are met with when considering the size of geomagnetic anomalies at even larger scale, i. e. at the global scale, and their distribution over the oceanic areas.

Here are emphasized some of them, as used to establish a new chonometer for the earth's history.

## INVERSION FOR THE GEOMAGNETIC ANOMALIES ON THE OCEAN

The anomalies surveyed on the oceans take the form of positive and negative stripes of varying wavelengths, symmetrically disposed relative to the mid-oceanic ridge, and of amplitudes in the order of 1,000 gammas (Vine, 1968).

Profiles made by oceanographic vessels along thousands of tracks have shown that this is the pattern for the whole oceanic region, with major offsets along the fracture zones (Vacquier, 1965). Therefore, the geophysical mechanism that has produced such a display must have been a general one, and has been acting on the planetary scale (Le Pichon, 1968; Heirtzler et al., 1968).

If this is the geophysical survey, the direct problem resides in the uppermost structures of the oceanic crust, the basalt layer forming the ocean floor and namely in the magnetic properties of these rocks. Magnetic properties of land-exposed basalts have been investigated since long ago: a considerable amount of their residual magnetization contributes to the distribution of the magnetic field on the earth's surface, as for instance on volcanoes (Yokoyama, 1962). Here, anomalies as high as 10% of the normal field are common, and successive flows in piles of basaltic lavas have recorded directional changes in time: historical lava flows of Mt. Aetna have imprinted secular variations (Chevallier, 1925) comparable to those recorded in the oldest magnetic observatories since 16th century. For older ages, radiometrically determined, the secular variation goes as far as a complete reversal of the magnetic field; and this is again an inverse problem, the magnetization of the rocks depending upon the dipolar inducing field at the time of their formation.

## DISTANCE SCALE INVERTED INTO TIME SCALE

Time can play a decisive role in geomagnetic phenomena. In the initial problem the inversion consists in modelling the ocean floor as a succession of magnetized basalt stripes, of appropriate dimensions, that fit the anomalies as measured on both sides of the oceanic ridge (Vine, 1968). In this feature, if time is taken into account, another basic result comes out for the inverse problems: the older stripes are progressively more distant from the mid-oceanic ridge. This means that a time scale can be plotted along side of the distance scale, along the magnetic profiles, if at least some points are dated (Vine, 1968; Heirtzler et al., 1968). This immediately provides the measure of the ocean-floor spreading along the profile and, if comparable profiles are averaged, a mean velocity can be deduced for the whole ocean, assuming the basalts have been produced at the mid-oceanic ridge and a driving mechanism has successively moved them aside (Tuzo Wilson, 1963). The whole inverse problem is represented in Figure 1.

As in a 3-D representation (like that on the box used by professor Helbig, in his earlier lectures in this course, for visualizing the process), the image on one plane leads to the depiction of the actual body, so in our case the distance from the reference point becomes inverted to the time occurrence of a geomagnetic reversal. If spreading rates are normalized and assumed, therefore, to have been constant for the whole earth, similar anomalies lay at a constant distance from the origin point, namely from the mid-oceanic ridge, and can simply be expressed by a progressive number such as "anomaly 34", or "anomaly 29", etc.).

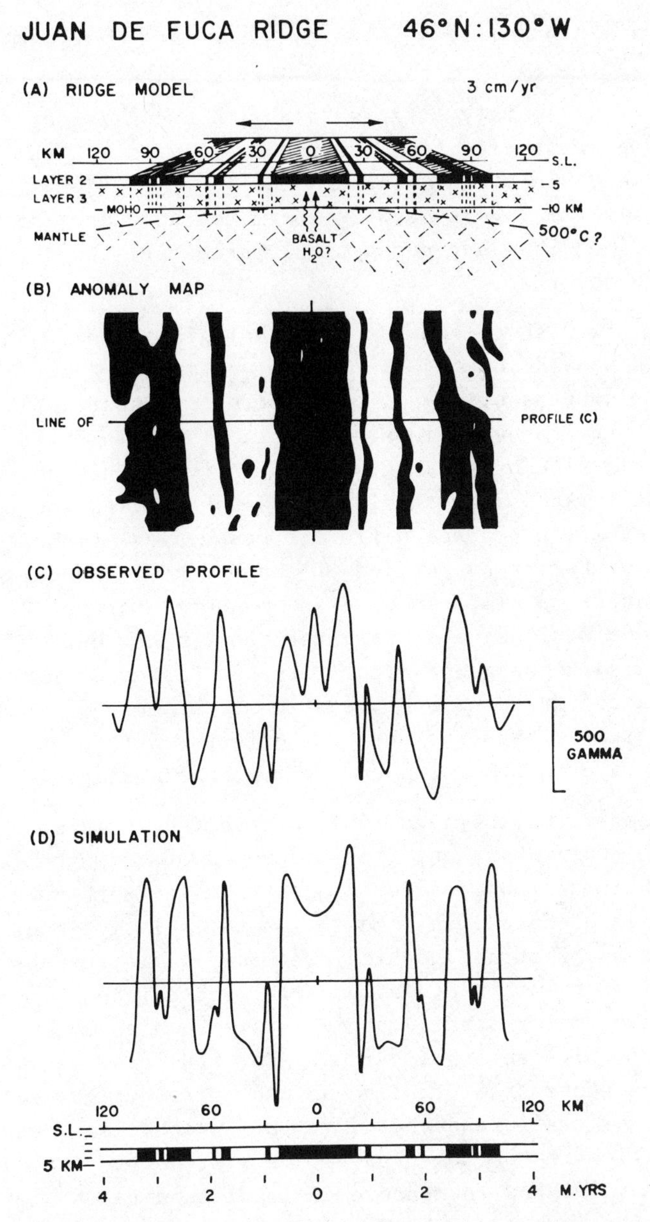


Fig. 1. Implications of time in the oceanic magnetic anomaly profiles. (a) crustal model applied to the Juan de Fuca Ridge, south-west of Vancouver Island, (b) as inverted from the anomaly map through (c) observed and (d) computed profiles over an oceanic basalt layer (at the bottom) where a symmetrical time scale is plotted against the distance scale. (After F. J. Vine, 1968)

## MAGNETIC STRATIGRAPHY ON SEDIMENTS: INVERSION FOR SEDIMENTATION RATES

Continuous anomaly stripes relate to a continuous time scale, which, therefore, marks the continuous evolution of the earth's magnetic field since beginning of sea-floor spreading (Larson and Hilde, 1975). The inverse problem permits the quantitative evaluation of a time-dependent geophysical parameter: the duration of a geomagnetic reversal.

If we now wish to consider the geological constraints to our problem, we should look at them as a function of time, in the sense that the parameters of a geological structure that imply time thereafter identify its evolutionary aspects. The simplest example is to consider a pile of sediments, originally deposited in the ocean and presently exposed on the land as a layered sequence. During this course we have learned how different geophysical techniques are used to measure different properties of such rocks by checking their densities, elastic moduli, electric and magnetic parameters, as well as their combinations, and interpret their contribution to the images detected in the geophysical surveys. Different methods, moreover, are used for such interpretations; the method we deal with, as already stated, applies the time effects on the deposition process of the layered sequence, in the mentioned example.

Considered the thickness of the pile and the span of time elapsed for its deposition, the succession of geomagnetic reversals recorded in it can be considered as its image. Both relative and absolute ages are used for such sediments. Relative ages are determined by means of stratigraphical methods, while the absolute ones are determined by means of radiometric methods, both yielding the recognition of major stratigraphic events such as the boundaries between successive geologic periods. Nevertheless, sedimentation processes, although semi-continuous and quite periodical, undergo abrupt changes that are detected (by all geophysical methods) as discontinuities in the measured parameters. How to evaluate such changes? That is, how to measure sedimentation rates?

Some more inversion problems ought to be solved. Here the final step, as brought out from the series of problems involved in a sedimentary sequence, is reported.

Figure 2 summarizes the cumulative picture worked out for the Gubbio standard.

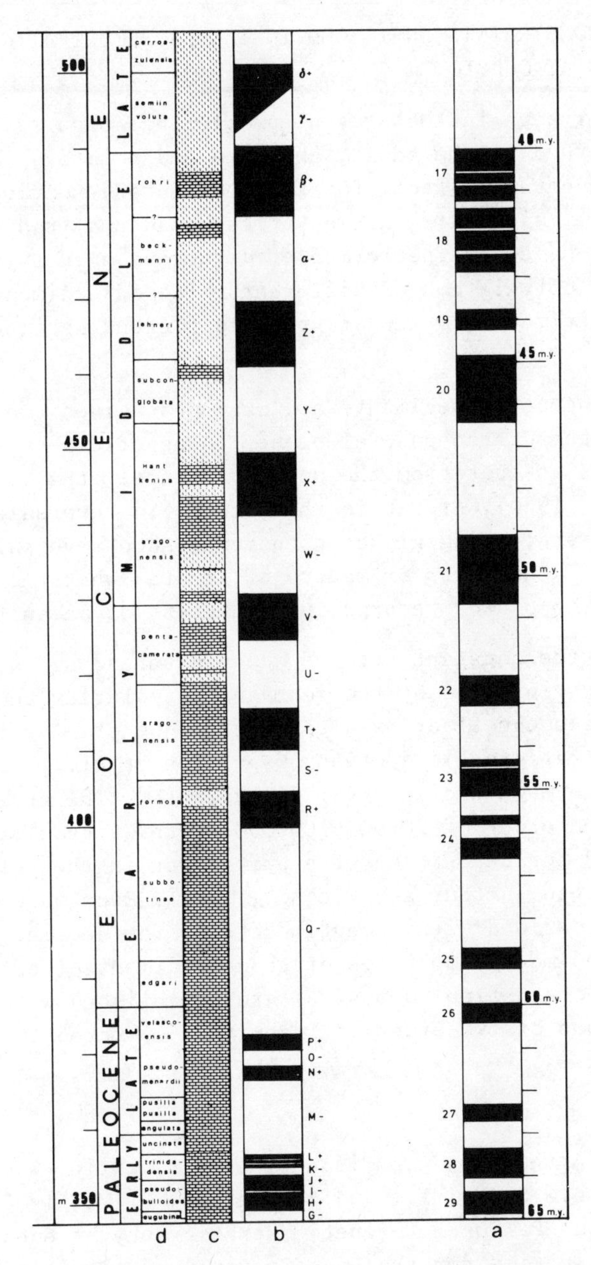


Fig. 2. Magnetostratigraphy at Gubbio for the Paleocene-Eocene. (a) time scale from the ocean-floor anomalies, starting with anomaly 29, is correlated with (b) the Gubbio reversals. (c) the pelagic limestones contain (d) biozones whose duration is computed from the span of magnetic reversals involved. (After G. Napoleone et al., 1980)

GEOLOGICAL CONSTRAINTS ENVISAGED IN THE MAGNETOSTRATIGRAPHICAL TYPE-SECTION AT GUBBIO

The challange in the present problem has been to find an alternative method to the basic, absolute and relative, dating of geological events. In fact, for the sedimentary sections so far investigated, major difficulties arise because a) radiometric ages are poor and have been performed at different places for different ages, and b) correlation of different piles of sediments is difficult and sometimes only tentative for different sections in the same pile.

The sequence of sediments at Gubbio has given the standard section for the Cretaceous-Paleogene geomagnetic reversal sequence, as continuous as that from the oceans (Alvarez et al., 1977; Napoleone et al., 1980). At Gubbio then, magnetic reversals define groups of layers; being global phenomena, such reversals define, therefore, in decimeters or meters of strata, what the accretion at the edges of plates records in kilometers of ocean floor.

Assumed the constant rate of oceanic spreading as the most viable process inferred by the geomagnetic polarity sequence recorded in the ocean-floor basalts (La Brecque et al., 1977) and in the Gubbio stratigraphic section (Napoleone et al., 1980), the duration of each magnetic interval is evaluated. The attainable accuracy is in the order of 20,000-100,000 years, approximately the time required for deposition of a unit layer in the pelagic limestones at Gubbio (Arthur and Fischer, 1977). Each magnetic interval at Gubbio is defined within centimeters to few decimeters and, finally, the sedimentation rates of those sediments (representing on the land the ocean-floor deposits) are established with an accuracy improved by one or two orders of magnitude compared to the available figures.

REFERENCES

Alvarez, W., Arthur, M. A., Fischer, A. G., Lowrie, W., Napoleone, G., Premoli Silva, I., and Roggenthen, W. M., 1977, Upper Cretaceous-Paleocene magnetic stratigraphy at Gubbio, Italy, V. Type section for the Late Cretaceous-Paleocene geomagnetic reversal time scale, Geol. Soc. Amer. Bull., 88:383.

Arthur, M. A., and Fischer, A. G., 1977, Upper Cretaceous-Paleocene magnetic stratigraphy at Gubbio, Italy, I. Lithostratigraphy and sedimentology, Geol. Soc. Amer. Bull., 88:367.

Chevallier, R., 1925, L'aimantation des laves de l'Etna et l'orientation du champ terrestre en Sicile du 12e au 17e siècle, Ann. de Physique, 10:5.

Gasparini, P., Mantovani, M. S. M., and Shukowski, W., 1980, Interpretation of long wavelength magnetic anomalies, This volume.

Heirtzler, J. R., Dickson, G. O., Herron, E. M., Pitman, W. C., III, and Le Pichon, X., 1968, Marine magnetic anomalies, geomagnetic field reversals and motions of the ocean floor and continents, Journal Geophys. Res., 73:2119.

La Brecque, J. L., Kent, D. V., and Cande, S. C., 1977, Revised magnetic polarity time scale for Late Cretaceous and Cenozoic time, Geology, 5:330.

Larson, R. L., and Hilde, T. W. C., 1975, A revised time scale of magnetic reversals for the Early Cretaceous and Late Jurassic, Journal Geophys. Res., 80:2586.

Le Pichon, X., 1968, Sea-floor spreading and continental drift, Journal Geophys. Res., 73:3661.

McElhinny, M. W., 1973, "Palaeomagnetism and plate tectonics", Cambridge Univ. Press, Cambridge.

Napoleone, G., Premoli Silva, I., Heller, F., Cheli, P., Corezzi, S., and Fischer, A. G., 1980, Eocene magnetic stratigraphy at Gubbio, Italy, Geol. Soc. Amer. Bull., in the press.

Tuzo Wilson, J., 1963, Evidence from islands on the spreading of ocean floors, Nature, 197:536.

Vacquier, V., 1965, Transcurrent faulting in the ocean floor, Phil. Trans. Roy. Soc. London A, 258:77.

Vine, F. J., 1968, Magnetic anomalies associated with mid-ocean ridges, in: "The history of the earth's crust", R. A. Phinney, ed., Princeton Univ. Press, Princeton.

Yokoyama, I., 1962, Geomagnetic anomalies on three Italian volcanoes, Ann. Oss. Vesuv., 5:173.

INDEX

Date Due

| | | | |
|---|---|---|---|
| NOV 2 1 1982 | | | |
| | | | |
| | | | |
| | | | |
| | | | |
| | | | |
| | | | |
| | | | |
| | | | |
| | | | |
| | | | |
| | | | |
| | | | |
| | | | |
| | | | |
| | | | |
| | | | |
| | | | UML 735 |